March 27–30, 2011
Santa Barbara, California, USA

I0050877

**Association for
Computing Machinery**

Advancing Computing as a Science & Profession

ISPD'11

Proceedings of the 2011 ACM/SIGDA
International Symposium on Physical Design

Sponsored by:
ACM SIGDA

Technical Co-Sponsored by:
CAS

Supported by:
**ATopTech, Cadence, IBM Research, Intel,
National Taiwan University, SpringSoft, Synopsys,
Taiwan Intelligent Electronics Consortium, & TSMC**

Association for
Computing Machinery

Advancing Computing as a Science & Profession

The Association for Computing Machinery
2 Penn Plaza, Suite 701
New York, New York 10121-0701

Copyright © 2011 by the Association for Computing Machinery, Inc. (ACM). Permission to make digital or hard copies of portions of this work for personal or classroom use is granted without fee provided that copies are not made or distributed for profit or commercial advantage and that copies bear this notice and the full citation on the first page. Copyright for components of this work owned by others than ACM must be honored. Abstracting with credit is permitted. To copy otherwise, to republish, to post on servers or to redistribute to lists, requires prior specific permission and/or a fee. Request permission to republish from: Publications Dept., ACM, Inc. Fax +1 (212) 869-0481 or <permissions@acm.org>.

For other copying of articles that carry a code at the bottom of the first or last page, copying is permitted provided that the per-copy fee indicated in the code is paid through the Copyright Clearance Center, 222 Rosewood Drive, Danvers, MA 01923.

Notice to Past Authors of ACM-Published Articles
ACM intends to create a complete electronic archive of all articles and/or other material previously published by ACM. If you have written a work that has been previously published by ACM in any journal or conference proceedings prior to 1978, or any SIG Newsletter at any time, and you do NOT want this work to appear in the ACM Digital Library, please inform permissions@acm.org, stating the title of the work, the author(s), and where and when published.

ISBN: 978-1-4503-1380-3

Additional copies may be ordered prepaid from:

ACM Order Department
PO Box 11405
New York, NY 10286-1405

Phone: 1-800-342-6626 (USA and Canada)
         +1-212-626-0500 (all other countries)
Fax: +1-212-944-1318
E-mail: acmhelp@acm.org

Printed in the USA

Foreword

It is our great pleasure to welcome you to the 2011 International Symposium on Physical Design (ISPD) held at Hotel Mar Monte, Santa Barbara, California. Evolving from a series of intermittent ACM/SIGDA Physical Design Workshops, ISPD has established itself as the premier forum for exchanging ideas and results on VLSI physical design. In the past, ISPD showcased seminal papers that advanced the state-of-the-art of physical design technology. In addition, the celebrated ISPD contests inspired numerous research studies that led to technical breakthroughs.

This year's symposium continues its dedication to high-quality despite its small scale. Papers submitted from all over the world are carefully reviewed by highly experienced technical program committee members and external reviewers. After a rigorous selection procedure that involves vigorous discussions, only a small number of papers are accepted. These papers cover various aspects of physical design, from traditional placement and routing to issues related to the most leading-edge technology, such as variability and manufacturability. These papers are complemented by invited talks by prestigious speakers.

The symposium begins with a keynote speech by Professor Massoud Pedram from University of Southern California. Professor Pedram is a highly accomplished and world-renowned scholar on electronic design automation as well as a former general chair of ISPD. His speech addresses power-efficiency of modern and future VLSI designs. A feature event in this year is the commemorative session for Professor Ernest Kuh, a groundbreaking researcher/educator on physical design and a Phil Kaufman Award recipient. We are delighted to have a talk by Professor Kuh and retrospections on his accomplishments by his former students and associate. Tuesday includes invited talks on 3D ICs, placement and DFM routing, given by speakers from Intel, Tezzaron, SpringSoft and Mentor Graphics. The physical design contest result is announced in the afternoon of Tuesday. This year's contest is a revisit to placement, a core physical design problem, with particular emphasis on routability, which is a growing challenge to sub-65nm technology chip designs. Along with the contest, a new set of industrial benchmarks are released and will facilitate scientific evaluations of related research results. Finally, Wednesday features invited talks on design for manufacturing by renowned academic scholar and industrial expert.

We would like to thank the authors and the keynote/invited speakers for the contributions to the high-quality program. We would like to express our gratitude to the technical program committee members and external reviewers, who spent substantial effort on the paper selection. We would also like to thank the steering committee, chaired by Prashant Saxena, who put together the wonderful invited talks and carefully selected the best paper, as well as the Publications Chair Cheng-Kok Koh, the Publicity Chair Bill Halpin, and the Contest Chair Natarajan Viswanathan, for their tremendous organizational services. We are also grateful to the sponsors for financial assistance. The symposium is sponsored by the ACM SIGDA (Special Interest Group on Design Automation) with technical co-sponsorship from the IEEE Circuits and Systems Society. Generous financial contributions have also been provided by (in alphabetical order): ATopTech, Cadence, IBM Research, Intel Corporation, National Taiwan University, SpringSoft, Synopsys, Taiwan Intelligent Electronics Consortium, and TSMC. Last but not least, we thank Lisa Tolles of Sheridan Printing Company for her expertise and enormous patience during the production of the proceedings.

On behalf the organizing committee, we sincerely hope that you enjoy ISPD 2011 and look forward to your participation in future editions of ISPD.

Yao-Wen Chang   **Jiang Hu**
ISPD 2011 General Chair   *Technical Program Chair*

Table of Contents

Keynote Address
Host: Yao-Wen Chang *(National Taiwan University)*

Session 1: Commemoration for Professor Ernest Kuh
Session Chair: Prashant Saxena *(Synopsys)*

Session 2: Clock Network Synthesis and Routing
Session Chair: Saumil Shah *(Magma)*

Session 3: Routing
Session Chair: Igor Markov *(University of Michigan at Ann Arbor)*

Session 4: Physical Design for 3D ICs
Session Chair: Azadeh Davoodi *(University of Wisconsin-Madison)*

ISPD 2011 Symposium Organization

General Chair: Yao-Wen Chang *(National Taiwan University)*

Program Chair: Jiang Hu *(Texas A&M University)*

Past Chair: Prashant Saxena *(Synopsys)*

Steering Committee Chair: Prashant Saxena *(Synopsys)*

Steering Committee: Charles Alpert *(IBM Research)*
Patrick Groeneveld *(Magma)*
Malgorzata Marek-Sadowska *(University of California at Santa Barbara)*
Shishpal Rawat *(Intel)*
Prashant Saxena *(Synopsys)*
PV Srinivas *(Mentor Graphics)*

Program Committee: Hongyu Chen *(Mentor Graphics)*
Hung-Ming Chen *(National Chiao Tung University)*
Tung-Chieh Chen *(SpringSoft)*
C.-K. Cheng *(University of California at San Diego)*
Salim Chowdhury *(Oracle)*
Azadeh Davoodi *(University of Wisconsin-Madison)*
Helmut Graeb *(Technische Universitaet Muenchen)*
Bill Halpin *(Google)*
Masanori Hashimoto *(Osaka University)*
Jiang Hu *(Texas A&M University)*
Shiyan Hu *(Michigan Technological University)*
Cheng-Kok Koh *(Purdue University)*
Igor Markov *(University of Michigan at Ann Arbor)*
Seungweon Paek *(Samsung)*
Saumil Shah *(Magma)*
Rupesh Shelar *(Intel)*
Yiyu Shi *(Missouri University of Science and Technology)*
Cliff Sze *(IBM Research)*
Chin-Chi Teng *(Cadence)*
Ting-Chi Wang *(National Tsing Hua University)*
Martin D. F. Wong *(University of Illinois at Urbana-Champaign)*
Evangeline F. Y. Young *(Chinese University of Hong Kong)*
Yaping Zhan *(AMD)*

Publications Chair: Cheng-Kok Koh *(Purdue University)*

Publicity Chair/Webmaster: Bill Halpin *(Google)*

Contest Chair: Natarajan Viswanathan *(IBM Corporation)*

Additional reviewers:

Saurabh Adya	Jingwei Lu
Sangmin Bae	Guojie Luo
Shashank Bujimalla	Qiang Ma
Tuck-Boon Chan	Malgorzata Marek-Sadowska
Kai-hui Chang	Tarun Mittal
Timmy Chang	Yu-Yen Mo
Samson Chen	Namita Negi
Tai-Chen Chen	Ricky Pan
Xiaodao Chen	Rajendran Panda
Jun Cheng Chi	Akshay Sharma
Wonjoon Choi	Hamid Shojaei
Yuelin Du	Dong Sup Song
Zhuo Feng	Jia Wang
Yuhong Fu	Renshen Wang
Tsung-Yi Ho	Reshen Wang
Jin Hu	Pei-Ci Wu
Kwangok Jeong	Qinghong Wu
Zhe-Wei Jiang	Bingjun Xiao
Da-Cheng Juan	Linfu Xiao
Pradip Kar	Xiaojian Yang
Myungchul Kim	Bo Yao
Johann Knechtel	Guo Yu
George Konstadinidis	Ting Yu
Dongjin Lee	Hongbo Zhang
Shuai Li	Tianpei Zhang
Zheng Li	

ISPD 2011 Sponsors & Supporters

Sponsor:

siG da
special interest group on
acm
design automation

Technical
Co-sponsor:

CAS **IEEE CIRCUITS AND SYSTEMS SOCIETY**

Supporters:

ATopTech

cādence®

IBM Research

(intel)

National Taiwan University

SpringSoft 思源科技

SYNOPSYS®

Taiwan Intelligent Electronics Consortium

tsmc

Robust Design of Power-Efficient VLSI Circuits

Massoud Pedram
University of Southern California
Department of Electrical Engineering
Los Angeles, CA 90089
pedram@usc.edu

Abstract

Digital information management is the key enabler for the unparalleled rise in productivity and efficiency gains experienced by the world economies. Computing and information processing systems are important elements of the world's digital infrastructure by providing ever-present and ever-increasing general purpose and data-driven processing and storage capabilities for both wired and mobile users. As such, they are also significant drivers of economic growth and social change. However, continued expansion of computing and information processing systems is now hindered by their unsustainable and rising power needs, with associated electrical energy costs and peak power draw requirements. Moreover governments, people, and corporations are becoming increasingly concerned about the environmental impact of these systems i.e., their carbon footprint. Separately from all this, with the increasing levels of variability in the characteristics of nanoscale CMOS devices and on-chip interconnects and continued uncertainty in the operating conditions of VLSI circuits, achieving power efficiency and high performance in computing and information processing systems under process, voltage, and temperature variations as well as interconnect wear-out and device aging has become a daunting, yet vital, task.

It is against this backdrop of rising power demands and energy costs as well as increased device- and circuit-level variability and aging effects that I present a number of best practices and methods for improving the power-performance efficiency of VLSI circuits and systems. The reviewed techniques range from dynamic power management to design of power-aware circuits, and from power/clock gating to leakage power minimization. A key issue to be addressed is how to deal with process and environment-induced variability of circuit parameters through statistical modeling and robust optimization and how to manage uncertainty about the workload and input data characteristics through observations and closed feedback loop control.

Categories & Subject Descriptors: B.8 Performance and Reliability

General Terms: Design

Copyright is held by the author/owner(s).
ISPD'11, March 27–30, 2011, Santa Barbara, California, USA.
ACM 978-1-4503-0550-1/11/03.

Professor Ernest Kuh's Talk

Ernest Kuh
University of California at Berkeley
kuh@eecs.berkeley.edu

Abstract

I very much appreciate the honor given to me at this conference and wish to thank the organizers for their kindness. According to the preliminary program the invited talks include placement, routing, clock distribution, etc by well-known authors. Their contributions to the field have made a great impact to the advancement of physical design. Some of them are my former students. I want to thank all of them for coming to the beautiful city of Santa Barbara for this meeting. In addition there are other related areas which could have been included at this meeting. For example, floor-planning, timing and performance driven layout, the power and ground problem are all important areas which have been studied. Also, interconnect is certainly of current and future interest. The multi-dimensional problem which makes placement, routing more interesting. As technology advances active layers will provide more possibility for efficient design. We are now in the deep sub-micron stage, there are limitations as to what one can do. Also, nano-devices and circuits are within our reach. In essence, the field of Physical Design has enormous opportunity for researchers in universities and industry to make contributions. Collaborative efforts and joint work are needed to speed up the advances in this field. I am certain that these topics will be covered in future ISPD meetings. I look forward to seeing you and many others at the next meeting. Finally, many thanks to all the sponsors and especially to Prof. Yao-Wen Chang for inviting me.

Categories & Subject Descriptors:

B.7.2 [**Integrated Circuits**]: Design Aids

General Terms:

Algorithm, Design, Performance

Keywords:

Clock distribution, floor-planning, physical design, placement, routing, timing

Bio

Ernest S. Kuh is the William S. Floyd, Jr. Professor Emeritus in Engineering and a Professor in the Graduate School of the Department of Electrical Engineering and Computer Sciences at the University of California, Berkeley. He joined the EECS Department faculty in 1956. From 1968 to 1972 he served as chair of the department; from 1973 to 1980 he served as Dean of the College of Engineering. From 1952 to 1956 he was a member of the Technical Staff at Bell Telephone Laboratories in Murray Hill, New Jersey.

Prof. Kuh attended Shanghai Jiao Tong University from 1945 to 1947; received the B.S. degree from the University of Michigan in 1949; the S.M. degree from the Massachusetts Institute of Technology in 1950; the Ph.D. degree from Stanford University in 1952; the Doctor of Engineering, Honoris Causa, Hong Kong University of Science and Technology in 1997; and the Doctor of Engineering degree from the National Chiao Tung University, Taiwan in 1999.

Prof. Kuh is a member of the National Academy of Engineering, the Academia Sinica, and a foreign member of the Chinese Academy of Sciences. He is a Fellow of IEEE and AAAS. He has received numerous awards and honors, including the ASEE Lamme Medal, the IEEE Centennial Medal, the IEEE Education Medal, the IEEE Circuits and Systems Society Award, the IEEE Millennium Medal, the 1996 C&C Prize (Japan Society for Promotion of Communication and Computers), and the 1998 EDAC Phil Kaufman Award.

Copyright is held by the author/owner(s)
ISPD'11, March 27–30, 2011, Santa Barbara, California, USA.
ACM 978-1-4503-0550-1/11/03.

Placement and Beyond in Honor of Ernest S. Kuh

Chung-Kuan Cheng

CSE Department, UC San Diego
La Jolla, CA 92093-0404
+1-858-534-6184

ckcheng@ucsd.edu

ABSTRACT

Professor Kuh is a pioneer and giant in physical layout. In this talk, we will describe his influence in placement. His pioneering work from interval graph for one dimensional gate assignment, BBL (Building-Block Layout System for Custom Chip IC Design) [2, 3, 6, 8], BEAR [7] layout system, BAGEL (Gate Array Layout) [17], RAMP (Resistive Analog Module Placement) [5], PROUD (Sea of Gates Placement) [28] to congestion, timing, and low power driven placement, Prof. Kuh always starts with innovative theoretical construction, software system building, and applications with impact on productivity.

Physical layout is an indispensable software system for VLSI Design with millions of modules. Placement is the key component of about 500 million dollars market for physical synthesis. In the layout design flow, the placement is the core of the system integrating with other synthesis and analysis tools.

For building block layout, Kuh's group tackled the nonslicing architecture. They devised the tile plane to represent the topology of the floorplan and the bottleneck of the routing. A routing order is derived to guarantee 100 percent routing completion [12]. The methodology and algorithms of building block placement [2] were adopted by companies such as Digital Equipment Corporation and ECAD, which was later renamed as Cadence.

For standard cell placement, RAMP placement is devised using the analogy of a resistive network [12]. The minimization of the circuit power corresponds to the quadratic wire length reduction. The approach provides a convergent solution in the era of interconnect dominance. As the technology scales, interconnect becomes dominating the system performance in terms on delay and power consumption. Performance driven logic synthesis, signal interconnect with repeater insertion, power ground and clock distribution strongly rely on the physical layout information. On the other hand the placement requires the synthesis result to perform the task. This mutual dependence has caused serious design convergence issues in the 1990s.

The work was performed at a time when simulated annealing method became a fashion in the EDA field. In 1983, Kirpatrick et al. [15] adopted the annealing method for placement and achieved excellent results. The annealing method takes the analogy of thermal annealing to perturb the partial solution with a probability according to the simulated temperature of the process. The strategy derives excellent results on some difficult problems at the expense of many random trials of the perturbations. Since most of EDA problem are known to be NP complete, annealing approach was considered as the right tool to solve the problem. Significant resources and efforts have thus been focused on annealing method in placement and other EDA subjects. The RAMP placement

method falls into the category of the analytical method. In 1979, Quinn and Breuer [23] introduced a force model to determine the state of equilibrium. They applied Hook's law to attract the modules connected by signal nets and repulsive forces to separate the modules with no connection. The repulsive forces cause a large set of nonlinear equations in the formulation, which complicates the calculation. An improvement has been proposed by Antreich, Johannes, and Kirsh [1] using the same force-directed method but with a more systematic formulation of equations. Kuh's approach removes the repulsive forces to simplify the nonlinear programming problem. His group demonstrated the feasibility of the global optimization approach via fast sparse matrix solvers [27].

The RAMP placement package was first installed at Hughes Aircraft Company to automate the placement in an in-house EDA project in 1983. The system had been used by the gate array design group since then. In 1988, in Kuh's group, Ren-Song Tsay developed PROUD to replace the matrix solver with successive over relaxation method. In 1992, at Cadence, CAD engineer, Louis Chao, implemented the algorithm and coined the placement tool as Qplacer to emphasize the usage of the quadratic cost function in the formulation. The Qplacer is the main placement tool of the ASIC design community for a long time. In the same year (1992), Ren-Song Tsay implemented the quadratic placement package at a start-up company which was later renamed as Avanti and acquired by Synopsys. The Qplacer became the main placement tool of the ASIC design community. Up to now, all major EDA companies, i.e. Cadence, Magma, Mentor Graphics, and Synopsys adopted the quadratic placement strategy and its extension by the group of Johannes. The placement is integrated with timing analysis and interconnect synthesis to provide convergent solutions for the designers.

For the usage of the quadratic placement, it took 104 seconds at 1MIPs machine to place 136 modules as the tool was installed at Hughes in 1983. In 1991, Kleinhans et al. [16] reported to take 2500 seconds at 15MIPs machine to place 6417 modules. In 1998, Eisenmann and Johannes [9] took 2031 seconds at Alphastation 250 4/266 (266MHz) to place 25K modules. As the quadratic placement method become popular, more features are integrated into the software package. In current designs, a state of art package can handle 6-7M components with 30-40 transistors. The limit is set by the memory capacity used for the analysis routine. Kuh's group has hosted a thriving set of Ph.D. students and visiting scholars to study placement. Margaret Sadowska has been a mentor to many in the group. The building block layout was contributed by Nang-Ping Chen, Chi-Ping Hsu, Chao-Chiang Chen, Wayne Dai, Bernhard Eschermann, Massoud Pedram, Yasushi Ogawa, and Margaret Sadowska. The channel ordering scheme for the layout [6] was published by Wayne Dai and Tetsuo Asano. The gate array layout [17] was constructed by Margaret Sadowska, Jeong-Tyng Li and C.K. Cheng. The standard placement has been studied by C.K. Cheng, Ren-Song Tsay; the low power placement by Massoud

Copyright is held by the author/owner(s).
ISPD'11, March 27–30, 2011, Santa Barbara, California, USA.
ACM 978-1-4503-0550-1/11/03.

Pedram; the timing driven placement [9, 20] by Shen Lin, Srinivasan Arvind, Michael Jaskson, Henrik Esbensen, and Margaret Sadowska; the IO assignment [22] by Massoud Pedram, Narasimha Bhat, Kamal Chaudhary, Deborah Wang, and Margaret Sadowska. Dong-Min Xu developed gate matrix layout [29, 30]; Minshine Shih studied partitioning [29, 30]; Pinhong Chen worked on floorplan sizing [4]; Hiroshi Murata extended a floorplan representation, sequence pair, to handle a mixture of soft and hard modules [18].

As a second wave, renowned visiting scholars went back and continued to contribute to placement. Prof. Hidoshi Onodera used branch and bound approach for building block placement at Kyoto University in 1991 [21]. Prof. Xianlong Hong devised an efficient floorplan representation, corner block list at Tsinghua University in 2000 [10]. Prof. John Lillis devised a new placement tool, Mongrel using hybrid techniques for standard cell placement at University of Illinois, Chicago in 2000 [11]. Prof. Andrew B. Kahng's group at UC San Diego developed an APlacer which won ACM International Symposium on Physical Design placement contest in 2005 [13, 14]. At Kuh's group, we had the privilege to meet many leaders in placement, e.g. Satoshi Goto, Tatsuo Ohtsuki, Brian Preas, David Liu, T.C. Hu, Jerry Lee, Scott Kirkpatrick, Melvin A. Breuer, Ulrich Lauther, and Kurt Antreich.

The placement with analogy of resistive network was inspired by Kuh's book entitled "Basic Circuit Theory". Kuh's group continues to investigate circuit analysis and synthesis. In 1991 [31], Shen Lin and Margaret Sadowska constructed an efficient timing simulator using a Step Wise Equivalent Conductance (SWEC) for nonlinear devices. In 1999 [32], Janet Wang and Qingjian Yu simplified transmission line analysis using model order reduction. In recent years (2005-2009), Kuh has advised analysis and synthesis of transmission lines using passive equalizer [33-38] and circuit simulation techniques [39-41].

VLSI placement is a challenge task with the increasing complexity of the circuit. The size of the modular placement cases scales with the advance of VLSI technologies. In 1961, Steinberg introduced a case of 34 modules. In 1972, Stevens in his thesis posted the ILLIAC IV boards with 67-151 modules [26]. In 1983, the test case from Hughes ranges from 300 to 500 modules. By 1991, the largest test case released by MCNC Center for Microelectronics is 15K modules. Moreover, subjects in placement such as mixed module placement, placement of heterogeneous circuits, placement integrating with behavior synthesis, 3-dimensional placement, and parallel placement are related to geometry handling, circuit performance, and advancement of the technologies and call for further study in the field.

Categories and Subject Descriptors
J.6 [**Computer Applications**]: Computer-aided design

General Terms
Algorithms, Design, Theory

Keywords
Placement, Physical Layout, Building Block Layout

REFERENCES

[1] K. J. Antreich, F. M. Johannes, and F. H. Kirsh, "A New Approach for Solving the Placement Problem using Force Models", *IEEE Intl. Symp. Circuits Systems*, 1982, pp. 481-486.

[2] C. C. Chen and E. S. Kuh, "Automatic Placement for Building Block Layout", *Proc. IEEE Intl. Conf. on Computer-Aided Design*, 1984, pp. 90-92.

[3] N. P., Chen, C. P. Hsu, E. S. Kuh, C. C. Chen and M. Takahashi, "BBL: A Building Block Layout System for Custom Chip Design", *Proc. IEEE Intl. Conf. on Computer-Aided Design*, 1983, pp. 40-41.

[4] P. Chen and E. S. Kuh, "Floorplan Sizing by Linear Programming Approximation", *Proc. ACM/IEEE Design Automation Conf.*, pp 468-472, June 2000.

[5] C. K. Cheng and E. S. Kuh, "Module Placement Based on Resistive Network Optimization", *IEEE Trans. on Computer-Aided Design* 3(3) (1984), pp. 218-225.

[6] W-M. Dai, T. Asano and E. S. Kuh, "Routing Region Definition and Ordering Scheme for Building-Block Layout", *IEEE Trans. on Computer-Aided Design* 4(3) (1985), pp. 189-197.

[7] W-M. Dai, H. Chen, R. Dutta, M. Jackson, E. S. Kuh, M. Marek-Sadowska, M. Sato, D. Wang and X-M. Xiong, "BEAR: A New Building-Block Layout System," *Proc. IEEE Intl. Conf. on Computer-Aided Design*, 1987, pp. 34-37.

[8] W-M. Dai and E. S. Kuh, "Hierarchical Floor Planning for Building Block Layout", *Proc. IEEE Intl. Conf. on Computer-Aided Design*, 1986, pp. 454-457.

[9] H. Esbensen and E.S. Kuh, "An MCM/IC Timing-Driven Placement Algorithm Featuring Explicit Design Space Exploration," *Proc. IEEE Multi-Chip Module Conference*, 1996, pp. 170-175.

[10] X. Hong, G. Huang, Y. Cai, J. Gu, S. Dong, C.K. Cheng, J. Gu, "Corner block list: an effective and efficient topological representation of non-slicing floorplan", *Proc. IEEE Intl. Conf. on Computer-Aided Design*, 2000, pp. 8-12.

[11] S.W. Hur and J. Lillis, "Mongrel: hybrid techniques for standard cell placement", *Proc. IEEE Intl. Conf. on Computer-Aided Design*, 2000, pp. 165-170..

[12] M. A. B. Jackson, E. S. Kuh, and M. Marek-Sadowska, "Timing-Driven Routing for Building Block Layout", *Proc. of IEEE Intl. Symp. on Circuits & Systems*, 1987, pp. 518-519.

[13] A. B. Kahng, S. Reda and Q. Wang, "APlace: A General Analytic Placement Framework", *Intl. Symp. on Physical Design*, 2005, pp. 233-235.

[14] A. B. Kahng, B. Liu and Q. Wang, "Supply Voltage Degradation Aware Analytical Placement", *IEEE Intl. Conf. on Computer Design*, 2005, pp. 437-443.

[15] S. Kirkpatrick, C. D. Gelatt, Jr. and M. P. Vecchi, "Optimization by Simulated Annealing", *Science* 220(4598) (1983), pp. 671-680.

[16] J. M. Kleinhans, G. Sigl, F. M. Johannes and K. J. Antreich, "GORDIAN: VLSI Placement by Quadratic Programming and Slicing Optimization", *IEEE Trans. on Computger-Aided Design*, 1991, pp. 356-365.

[17] J. T. Li, C. K. Cheng, M. Turner, E. S. Kuh and M. Marek-Sadowska, "Automatic Layout of Gate Arrays," *Proc. IEEE Custom Integrated Circuits Conf.*, 1984, pp. 518-521.

[18] H. Murata, E. S. Kuh, "Sequence-Pair Based Placement Method for Hard/Soft/Pre-placed Modules," *Intl. Symp. on Physical Design*, 1998, pp. 167-172.

[19] W. C. Naylor, R. Donelly and L. Sha, "Non-Linear Optimization System and Method for Wire Length and Delay Optimization for An Automatic Electronic Circuit Placer," *US Patent 6,671,859*, Dec. 30, 2003

[20] Y. Ogawa, M. Pedram, and E.S. Kuh, "Timing-Driven Placement for General Cell Layout," *Proc. Intl. Symp. on Circuits and Systems*, 1990, pp. 872-876.

[21] H. Onodera, Y. Taniguchi, and K. Tamaru, "Branch-and-bound placement for building block layout", *Proc. ACM/IEEE Design Automation Conf.* 1991, pp. 433-439.

[22] M. Pedram, K. Chaudhary, and E.S. Kuh, "I/O Pad Assignment Based on the Circuit Structure", *Proc. Intl. Conf. on Computer Design*, 1991.

[23] N. R. Quinn and M. A. Breuer, "A Force Directed Component Placement Procedure for Printed Circuit Boards", *IEEE Trans. on Circuits and Systems*, vol. CAS-26 (1979), pp. 377-388.

[24] M. Shih and E.S. Kuh, "Quadratic Boolean Programming for Performance-Driven System Partitioning", *Proc. ACM/IEEE Design Automation Conf.* 1993, pp. 761-765.

[25] M. Shih, E. S. Kuh, and R-S. Tsay, "Performance-Driven Partitioning on Multi-Chip Modules", *Proc. ACM/IEEE Design Automation Conf.*, 1992, pp. 53-56.

[26] J.E. Stevens, "Fast Heuristic Techniques for Placing and Wiring Printed Circuit Boards," Ph.D. Dissertation *Comp. Sci., Univ. of Illinois*, 1972.

[27] R-S. Tsay, E. S. Kuh and C-P. Hsu, "Module Placement for Large Chips Based on Sparse Linear Equations", *Intl. Journal of Circuit Theory and Applications*, vol. 16 (1988), pp. 411-423.

[28] R-S. Tsay, E. S. Kuh and C. P. Hsu, "Proud: A Fast Sea-of-Gates Placement Algorithm", *ACM/IEEE Conf. on Design Automation*, 1988, pp.318-323.

[29] D-M. Xu, Y.K. Chen, E.S. Kuh, and Z.J. Li, "A New Algorithm with Gate Matrix Layout", *Proc. IEEE Intl. Symp. on Circuits and Systems*, 1987, pp. 288-291.

[30] D-M. Xu, E. S. Kuh, and Y-K. Chen, "An Extended 1-D Assignment Problem: Net Assignment in Gate Matrix Layout," *Proc. of Intl. Symp. on Circuits and Systems*, 1990, pp. 1692-1696.

[31] S. Lin, M. Marek-Sadowska, and E.S. Kuh, "SWEC: A Step Wise Equivalent Conductance Timing Simulator for CMOS VLSI Circuits," Conf. on European Design Automation, pp. 142-148, 1991.

[32] J.M. Wang, E.S. Kuh, and Q. Yu, "The Chebyshev Expansion based Passive Model for Distributed Interconnect Networks," IEEE/ACM Int. Conf. on Computer-Aided Design, pp. 370-375, 1999.

[33] L. Zhang, W. Yu, H. Zhu, A. Deutsch, G. Katopis, D. Dreps, E.S. Kuh, and C.K. Cheng, "Low Power Passive Equalizer Optimization using Tritonic Step Response," IEEE/ACM Design Automation Conf., pp. 570-573, 2008.

[34] R. Shi, W. Yu, Y. Zhu, C.K. Cheng, and E.S. Kuh, "Efficient and Accurate Eye Diagram Prediction for High Speed Signaling," ACM/IEEE Int. Conf. on Computer-Aided Design, pp. 655-661, 2008.

[35] L. Zhang, W. Yu, Y. Zhang, R. Wang, A. Deutsch, G.A. Katopis, D.M. Dreps, J. Buckwalter, E.S. Kuh, and C.K. Cheng, "Low Power Passive Equalizer Design for Computer Memory Links," IEEE Hot Interconnects, Symp. on High Performance Interconnects, pp. 51-56, Aug. 2008.

[36] Y. Zhang, L. Zhang, A. Deutsch, G.A. Katopis, D.M. Dreps, J.F. Buckwalter, E.S. Kuh, C.K. Cheng, "On-Chip Bus Signaling Using Passive Compensation," IEEE Electrical Performance of Electronic Packaging, pp. 33-36, 2008.

[37] L. Zhang, Y. Zhang, A. A. Tsuchiya, M. Hashimoto, E.S. Kuh, and C.K. Cheng, "High Performance On-Chip Differential Signaling Using Passive Compensation for Global Communication," IEEE Asia and South Pacific Design Automation Conf., pp. 385-390, 2009.

[38] Y. Zhang, L. Zhang, A. Deutsch, G. Katopis, D. Dreps, E.S. Kuh, and C.K. Cheng, Design Methodology of High Performance On-Chip Global Interconnect Using Terminated Transmission-Line," IEEE Int. Symp. Quality of Electronic Design, pp. 451-458, 2009.

[39] Z. Zhu, K. Rouz, M. Borah, C.K. Cheng, and E.S. Kuh "Efficient Transient Simulation for Transistor-Level Analysis," Asia and South Pacific Design Automation Conf., pp. 240-243, 2005.

[40] Z. Zhu, R. Shi, and C.K. Cheng, E.S. Kuh, "An Unconditional Stable General Operator Splitting Method for Transistor Level Transient Analysis," Asia and South Pacific Design Automation Conf., pp. 428-433, 2006.

[41] Z. Zhu, H. Peng, K. Rouz, M. Borah, C.K. Cheng and E.S. Kuh, "Two-Stage Newton-Raphson Method for Transistor Level Simulation," IEEE Trans. on Computer Aided Design, pp. 881-895, May 2007.

From Academic Ideas to Practical Physical Design Tools

Ren-Song Tsay
National Tsing-Hua University, Taiwan
rstsay@gmail.com

ABSTRACT
In this paper, the author discusses how ideas from academic research are adapted into making physical design tools that are both successful and practical. Excluding recent developments, the author uses his past experiences to review the thinking process of selecting appropriate algorithms and the progressive optimization idea for creating effective design tools. The review mainly focuses on routability and timing optimization issues, though the ideas presented in the paper can be applied or extended to new tool development.

Categories and Subject Descriptors
C.5.4 [**Computer System Implimentation**]: VLSI Systems, B.7.2 [**Integrated Circuits**]: Design Aids; *Placement and routing, Layout*

General Terms
Design

Keywords
Physical design, placement, routing, clock tree synthesis, timing optimization

1. INTRODUCTION
As the original architect of the Astro physical compiler, one of the most successful commercial physical design tools, the author constantly faces challenging design styles and issues and needs to adapt research ideas into practical use. In general, academic ideas mostly address *point* issues and are effective under certain conditions while a production solution is required to address all concerns, including corner situations. Therefore, the main trial is to have appropriate algorithms at the proper steps in the design flow. To create an effective physical design solution, the author applies the concept of progressive optimization and successfully demonstrates the final design tool in practice.

A typical physical design flow starts from a gate-level netlist and goes through partitioning/floorplanning, power network planning, placement, clock-tree synthesis and routing steps. Each design step also considers performance (timing and power) analysis and optimization. In this paper, the author uses his past experiences and reviews mainly classical routability and timing optimization issues, but not recent development. It is assumed that readers are fairly familiar with basic physical design issues and algorithms.

Since almost all physical design problems are NP hard, good heuristic solutions are essential in practice. A good heuristic normally has insights into the kernel issue and is different from a hacking approach, which often scratches only the surface of an issue.

Some heuristics can be amazingly simple and effective, if they are executed at the proper steps.

In reality, it requires a combination of multiple heuristics to solve the sophisticated physical design problem. To select and integrate appropriate heuristics into an effective complete solution, the author applies the concept of progressive optimization that gradually refines the solution from global optimization to detailed optimization while using a consistent cost metric that bridges consecutive optimization steps.

At the global phase, usually an approximated cost function with strong optimization properties is used to search for a globally *good* solution. Design constraints, such as legal module location and non-overlapping constraints, which can easily trap a solution into local minimum, are usually ignored first and gradually considered along the optimization phase.

Then, at the detailed optimization phase, design constraints are carefully considered and greedy optimization approaches are usually applied to local smaller scale optimization to control computation complexity.

The author also finds that certain pre-processing steps are critical. Otherwise, certain issues are just impossible if they occur too late in the design phase. For instance, pin accessibility and cell porosity should be analyzed at cell library preparation phase. If a pin has only one access point and if it is blocked by P/G or other pre-wires, then no matter how powerful a placer and router tool is, it is almost impossible for a design to be done successfully.

One particular challenge is that the classical physical design is a net-oriented problem by nature, while timing optimization is a path-oriented issue. This requires an ingenious approach that handles the two concerns of different natures smoothly at the same time.

In the following, the author first discusses experiences in routability optimization. Next, practical timing optimization techniques are presented and followed by a brief conclusion.

2. Routability
Unlike undesirable timing constraint violations, if any net is not fully connected or if it contains physical design rule violations, then a design will not function at all. Therefore, the ultimate goal for a physical design tool is to place and route a design in a smallest possible area with no physical constraint violations. In the following, the author discusses quadratic placement and progressive routing approaches along with effective pre-processing and netlist restructuring optimization approaches.

2.1 Quadratic Placement
The first step of physical design is to place modules onto legal placement locations with no overlaps in a given chip design area. Since the routability objective cannot be formulated mathematically, the highly correlated total net length measure is used as an alternative objective.

Permission to make digital or hard copies of all or part of this work for personal or classroom use is granted without fee provided that copies are not made or distributed for profit or commercial advantage and that copies bear this notice and the full citation on the first page. To copy otherwise, or republish, to post on servers or to redistribute to lists, requires prior specific permission and/or a fee.
ISPD'11, March 27–30, 2011, Santa Barbara, CA, USA.
Copyright 2011 ACM 978-1-4503-0550-1/11/03…$10.00.

Usually, either bounding box length or quadratic length is used for net length measurement. Although bounding box length is easier to evaluate, the quadratic length measure presents a perfect mathematical global optimization structure, which can be converted to an equivalent force attraction balance problem in physics. If ignoring legal location constraints or considering only the first-order constraint, a linear system solver can efficiently find an optimal solution [10].

With legal location constraints, the solution space is a set of discrete points, not a continuous space. In the perspective of optimization, the cost value on the illegal spots is infinity. Therefore, most optimization approaches can easily get trapped at local *minimum* once legality constraints are considered.

The quadratic cost function is a convex approximation of the contour of the discrete cost function and can direct quickly to the global optimum area. Then, by recursive partitioning, legal location constraints can be gradually enforced [10].

In practice, mincut optimization is used to further optimize the partitioned result at each recursive partitioning step. Usually, 10 to 15% quality improvement in terms of total bounding-box length is observed.

Finally, a detailed placer is required to handle complicated legal location constraints and possibly congestion optimization. Since the design has been *globally* optimized, a detailed placer only has to focus on a local area optimization. Usually combinations of pairwise interchange and single module move are sufficient at this phase and another 15% quality improvement in average can be achieved.

In practice, wire length over-optimization may degrade routability as highly connected modules will naturally clump together and become a congested area that is hard to route. A solution is to specify a lower density on the congested areas and force highly connected modules to spread out. Usually, a fairly accurate congestion estimate [7] based on the two L-routes of each pin pair of the net spanning trees is used to guide placement density area setting. Actually, the deterministic nature of the quadratic placement algorithm is a key to make this iterative optimization approach converge.

In fact, the spanning-tree congestion estimation approach used in placer is exactly the first step of global router which is to be discussed next. Using a consistent cost measure between two consecutive tools is an important rule for progressive optimization.

2.2 Progressive routing approach

Traditionally, whole chip routing is divided into global route and detailed route phases. With the two-phase routing approach in the past, we often observed many zigzag and detoured wires. The net ordering issue is identified the main cause. Therefore the author introduces a net-order independent approach and additionally adds an intermediate router, which is similar to a channel router, for track assignment between the global routing and detailed routing steps.

2.2.1 Net ordering issue

Net ordering is a key concern for routing algorithm. Most algorithms route nets sequentially and routing tasks often become more and more difficult at the later phase. Although the linear assignment approach [11] is a good concurrent net routing attempt, it cannot see routing blockages behind the cut line.

To resolve the issue, the author suggests an effective progressive approach. First, we construct minimum length spanning trees for all nets. Then, we estimate congestion using the two L-routes of each pin pair of the spanning trees. Based on the congestion numbers, minimum congestion spanning trees are built to balance congestion. Afterward, we choose one of the two L-routes according to the updated congestion information and finally construct rectilinear

Steiner trees. Note that the above steps are net-order insensitive and can globally optimize and evenly distribute all nets. Then, it is ready to use a maze router to perform ripup-and-reroute to further optimize the result.

In practice, one challenge is how to determine proper global routing cell size and how to accurately calculate routing capacity. Due to limited space, details are skipped in this paper.

2.2.2 An intermediate router

Since we use spanning tree and Steiner tree construction in global router, the planned wire segments usually have a good chance to be connected in straight lines. Nevertheless, in the past if a detailed router, usually a maze router, was applied directly after the global routing phase, the sequential routing issue would appear. Usually, early runners took all the best routing spots and later participants got squeezed out and formed undesirable zigzags again.

To resolve this issue, the channel routing concept is borrowed to perform track assignment on pseudo-channels, which are usually defined by strips of global routing cells. Using vertical constraints and dog-leg analysis while allowing minimum violations, track assignment serves as an effective one-dimensional global optimization tool. In contrast, the global router is a two-dimensional global optimization tool.

With physical wires created by the intermediate router, a detailed router with ripup-and-reroute feature can now effectively complete the routing with minimum net ordering impacts.

2.3 Pre-processing

In practice, cell library preparation, routing track and routing space planning are equally important as the placement and routing optimization.

We observe that an *overly optimized* cell library can be detrimental to placement and routing tasks. For instance, if there are too many pins densely packed in one cell, these pins often cannot be easily connected. Also, if a pin has only one or very few access points, pre-wires may accidentally block the access points and cause an un-routable design.

A simple pre-processing procedure can effectively resolve these issues. For example, we can use a porosity analysis algorithm to check the internal pin and blockage distribution and improve

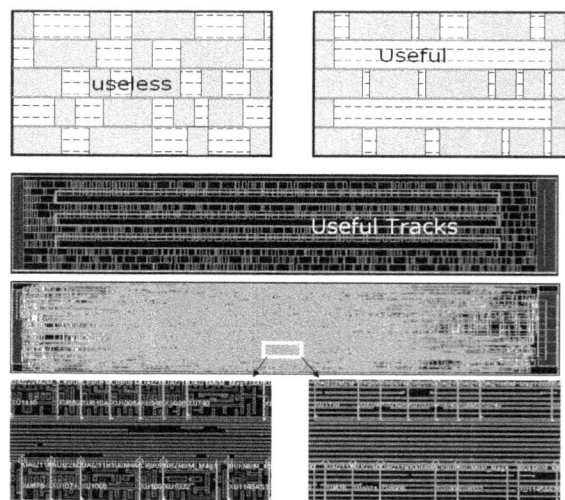

Figure 1. An illustration of useful track alignment and a sample real routing result.

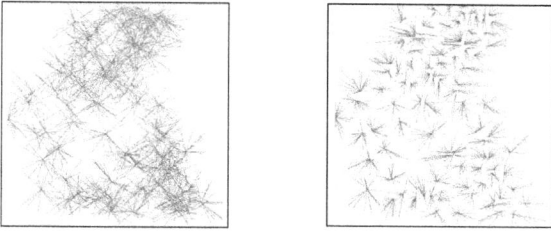

Figure 2. The bottom-level clock tree clusters before and after optimization.

routability by padding in extra white cell area. Similarly, for the pin access issue, we can pull those difficult pins out into more accessible areas at the cell library preparation phase.

Additionally, we find that routing track planning is extremely important for routing success. We can pre-check whether cell pins will be on track after placement. Experience tells us that off-track pins are extremely costly. In fact, only with track information, we can estimate congestion accurately and perform the progressive concurrent routing procedures previously discussed.

Some may be misled and think that tools such as shape-based routers can ignore tracks. Experience shows that without global track and wire planning, any conceivable routing algorithm will be subject to net ordering issues and fail to deliver satisfactory results.

For routing area planning, we take the *useful track alignment* approach [2] as an example to demonstrate how one simple planning action can greatly affect final routing result.

We observe that in practice, routers often prefer long free tracks. If modules are allowed to be freely placed at any location, the routing space for lower metal layers, particularly the M1 layer, is often fragmented and difficult to use effectively. If cell utilization rate is sufficient, we can intelligently preserve aligned routing tracks using placement blockages as demonstrated by the example shown in Figure 1. Details can be found in [2].

2.4 Equivalent netlist restructuring
Certain groups of nets, such as scan chain and clock tree, are constructed in the logical design phase and given in the input netlist. They should be re-structured based on the final physical information for best routability and performance. The new netlist is either logically equivalent or functionally equivalent to the old one and certain actions have to be done to ensure correctness. For example, after scan chain optimization, the chain sequence has to be updated for correct scan operation.

Usually these types of nets, along with high-fanout nets, do not need to be considered in placement optimization. If they are included in placement optimization, most high-fanout nets have little effect on placement results but will increase unnecessary computation complexity. In contrast, nets such as those in clock trees may mislead the placer since the original clustering does not consider physical information at the logical design phase.

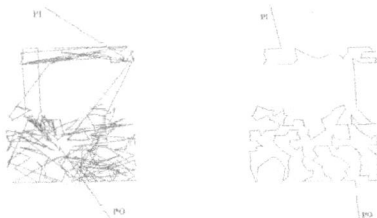

Figure 3. A scan chain example before and after

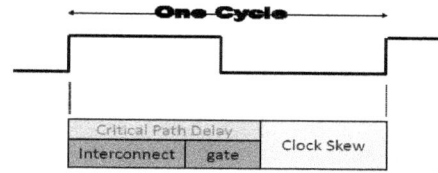

Figure 4. The clock period is determined by clock skew and critical path delay.

In Figure 2, we show the bottom-level of a clock tree before and after optimization using the physical information [3]. Clearly, the optimized clock tree has better routability. Similarly, shown in Figure 3 is a scan chain before and after optimization [4]. The total wire length is reduced almost four times for this particular example.

Additionally, logically equivalent ports can also be leveraged for better routability or timing performance. Details are not discussed in this paper.

3. Timing Optimization
The performance of a design determines its value while a complete connection is just a basic requirement. Timing and power are two most critical performance measures. In this paper, mainly timing optimization is addressed.

As for timing optimization, Figure 4 abstractly shows that the clock period is composed of clock skew and critical path delay, which includes interconnect and gate delays. In the following, the author discusses first the clock network optimization approaches and a unique slack graph timing optimization approach. Finally, some simple but effective logically-equivalent netlist restructuring techniques are discussed.

3.1 Clock network optimization
Besides the clock tree synthesis or optimization that balance wire load and phase delays discussed in previous section, a challenging task is to perform minimum clock skew clock routing. Before the invention of the zero-skew clock routing approach [6, 9], all clock routing approaches were geometry based approaches that could not accurately capture the nonlinear delay.

The key to generating exact zero-skew clock routing results is due to an interesting property of the Elmore delay model: the total delay to any end terminal of a tree can be calculated by summing up the delay of each resistor segment along the path. The resistor segment delay is calculated by multiplying the resistance value and the total downstream capacitance of the resistor. By perfectly combing the bottom-up clock routing construction and the bottom-up Elmore delay calculation, an exact zero-skew result can easily be constructed [6, 9].

Although there were many more sophisticated zero-skew clock routing algorithms developed afterward, in reality the routability constraint prefers a simpler and flexible routing approach that can effectively considers the effects of process, voltage and temperature variations.

3.2 Slack graph optimization approach
The main challenge of timing optimization in place-and-route is that timing is a path-oriented issue while place-and-route is a net-oriented issue by nature. In fact, timing optimization is an all-path problem. Optimizing just a few paths or nets cannot truly solve the problem.

To resolve the issue, the author observes that the timing slack graph is a perfect bridge between timing and physical optimization [1, 5]. The slack graph actually contains both timing and connectivity information. Essentially, it is a snapshot of the timing verification

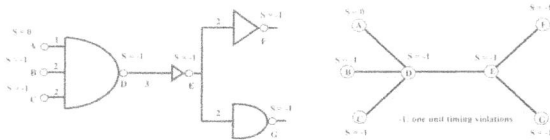

Figure 5. A sample circuit and its corresponding timing slack graph.

Figure 6. A buffer and repeater insertion example.

result in terms of a slack number on each pin and edge. The slack number is the difference of the required signal arrival time and the actual arrival time. Note that the required arrival time is usually derived from the late mode timing constraints.

As an illustration, the design shown in the left side of Figure 5 has a required arrival time of 6 units delay and the corresponding slack graph is shown in the right side. The slack number on each pin node or edge implies the timing information, and the graph itself is essentially a variation of the netlist structure that determines the connectivity information.

To optimize path-oriented timing objective in a net-oriented physical design tool, an intuitive approach is to convert timing constraints into net constraints and fit in the physical design environment. A timing budgeting idea based on a zero-slack was then developed. Heuristics are applied to compute maximum allowable timing budget on each net.

The question is that most budgeting algorithms do not consider connectivity and timing sensitivity (how per unit wire length affects delay). Therefore, the iterative timing optimization often may not converge in practice.

In addition to considering connectivity and timing sensitivity, the author applies the idea of minimum perturbation on the existing placement result and computes optimum net constraints. Essentially, if assuming small placement perturbation, the relationship of delta length change Δl_{ij} and delay change Δd_{ij} can be equated as the following

$$\Delta l_{ij} = \Delta d_{ij} / (R_i c + r C_j),$$

where r and c are the per unit wire resistance and capacitance. R_i and C_j are the driving resistance at node i and the loading capacitance at node j. By letting $D_{ij} = (R_i c + r C_j)$ and without going into detail, we have the following equation

$$min \sum_{i,j} \left(\Delta l_{ij} \right)^2 = min \sum_{i,j} \frac{1}{D_{ij}^2} \left[(x_j + s_j) - (x_i + S_{ij}) \right]^2,$$

where x_j is the arrival time improvement of node j, s_j is timing slack number of node j, and S_{ij} is the slack of edge connecting node i to node j. This equation is similar to the objective function of the quadratic placement optimization and can be efficiently solved. Then, $l_{ij} + \Delta l_{ij}$ becomes the wire length upper bound to drive the pacer and router for optimal timing results. For example, with the upper bound constraint, an effective analytical net weighting approach can be adopted in placement optimization [8].Most importantly, this optimized timing slack assignment will cause minimum perturbation to the optimal placement result and hence the optimization process can converge effectively.

3.3 Logically-equivalent netlist restructuring

Similar to the routability issue, certain timing violations can be resolved more effectively by restructuring netlist. Usually, this is done after the detailed placement phase and hence very local changes are expected. Therefore, instead of doing timing optimization on the whole slack graph, greedy algorithms focusing on critical paths can be applied. Usually, the most effective edge considering timing sensitivity and connectivity is selected for optimization. For instance, in Figure 5, the most effective edge is DE. One unit delay improvement on this edge can resolve all timing violations. Otherwise, it will require a total of two units delay improvement on both edges EF and EG to resolve the violations.

Combined with incremental slack update, cell sizing can be an effective timing optimization tool based on the slack graph. Buffer insertion is one effective operation to block off unnecessary capacitance load from the critical sink as shown in Figure 6. Also, proper repeater insertion on long wire to critical sink can reduce the quadratic increase effect of wire delay to wire length.

4. CONCLUSIONS

This paper presents a few practical approaches used in a popular physical design tool. A challenge is to select appropriate algorithms and execute them at the proper steps. More advanced ideas have been developed since the development of these approaches. Nevertheless, the progressive optimization concept has continued to be proven an effective idea.

5. ACKNOWLEDGMENTS

The author would like to thank Prof. Ernest Kuh for his teaching of research skills and colleagues at UC Berkeley, IBM research, ArcSys, Avant! and Axis Systems for numerable inspiring discussions.

6. REFERENCES

[1] Chwen-Cher Chang, J. Lee, M. Stabenfeldt and Ren-Song Tsay, "A Practical All-Path Timing-Driven Place and Route Design System", *Proc. Asia-Pacific Conf. on Circuits and Systems*, 1994, pp. 560-563.

[2] Fong-Yuan Chang, Ren-Song Tsay, Wai-Kei Mak and Sheng-Hsiung Chen, "Cut-Demand Based Routing Resource Allocation and Consolidation for Routability Enhancement," ASPDAC 2011

[3] J.-M. Ho and Ren-Song Tsay, "Clock Tree Regeneration, " *Prof. IEEE GLS-VLSI'92*, Feb. 1992.

[4] Kenneth D. Boese, Andrew B. Kahng and Ren-Song Tsay, "Scan Chain Optimization: Heuristic and Optimal Solutions", Research Report UCLA,1994

[5] Ren-Song Tsay, Chwen-Cher Chang, "Electronic Design Automation Tool for the Design of a Semiconductor Integrated Circuit Chip," US Patent 5461576, Oct. 24, 1995

[6] Ren-Song Tsay, "An exact zero-skew clock routing algorithm," IEEE Trans. on Computer-Aided Design of Integrated Circuits and Systems, 12(2):2.42-249, 1993.

[7] Ren-Song Tsay and S. C. Chang, "Early Wirability Checking and 2-D Congestion-Driven Circuit Placement." In *International Conference on ASIC*, pp. 50–53, 1992.

[8] Ren-Song Tsay and J Koehl, "An analytic net weighting approach for performance optimization in circuit placement", In Proc. ACM/IEEE Design Automation Conf., 1991, pp. 620-625.

[9] Ren-Song Tsay, "Exact Zero Skew", Proc. of International Conference on Computer Aided Design, pp. 336--339(1991).

[10] Ren-Song Tsay, E.S. Kuh, and C-P. Hsu, "PROUD: A Fast Sea-of-Gates Placement Algorithm," *Proceedings of 25th Design Automation Conference*, pp. 318-323, June 1988.

[11] T. Parng and Ren-Song Tsay, "A New Approach to Sea-of-gates Global Routing," *Proc. IEEE International Conference on Computer-Aided Design*, Nov. 1989, pp. 52-5

On Old and New Routing Problems

Malgorzata Marek-Sadowska
University of California
Santa Barbara, CA 93106
(805) 893 2721

mms@ece.ucsb.edu

ABSTRACT

The objective of this paper is to commemorate Professor Ernest S. Kuh's contributions to EDA and to bring attention to his research group's many accomplishments in physical design area. The focus is on routing and our goal is to trace the effects that the ideas which originated in Kuh's group have had on researchers then and now.

Categories and Subject Descriptors

B.7.2 [Integrated Circuits] Design Aids - Placement and Routing

General Terms

Algorithms, Performance.

Keywords

Routing, wire coupling, global routing, VeSFET, power grid.

1. INTRODUCTION

Physical design is complex because a huge number of transistors on a chip must be connected together. All intermediate steps of a design process optimize cost functions whose objective is to make routing simpler or to mitigate the effects that routing may have on circuit performance. Over the years, with technology advances, the effects of interconnect on design properties have been steadily increasing. Initially, interconnects were competing with transistors for area. This was particularly true when only a few metal layers were available. Then, the loading introduced by interconnects could no longer be ignored and timing driven routing was introduced. Further shrinking of transistor and wire dimensions brought some electrical problems into forefront, including crosstalk and power grid noise. These phenomena can be mitigated or amplified by routing. Nowadays, routing should not only satisfy all the constraints introduced by the earlier technologies, but should also produce manufacturable and reliable interconnect structures

In this paper we will revisit some of the problems that were studied in Kuh's group. These include single row routing, channel routing, global routing, post global routing crosstalk elimination, routing compaction and incremental routing optimization. The papers published by the Kuh's group were widely followed and most of the studied problems still retain their practical significance. We will conclude the paper by discussing some open problems in routing.

Permission to make digital or hard copies of all or part of this work for personal or classroom use is granted without fee provided that copies are not made or distributed for profit or commercial advantage and that copies bear this notice and the full citation on the first page. To copy otherwise, or republish, to post on servers or to redistribute to lists, requires prior specific permission and/or a fee.
ISPD'11, March 27–30, 2011, Santa Barbara, California, USA.
Copyright 2011 ACM 978-1-4503-0550-1/11/03...$10.00.

2. SINGLE ROW ROUTING PROBLEM

Historically, automated routing was first developed in the context of printed circuit boards (PCBs). One of the first systematically studied problems was that of single row routing formulated by H.C. So in [1]. The model assumes a board with fixed geometry and many layers. Each layer has fixed plated through holes, uniformly spaced on a rectangular grid. Every other column of holes consists either of conductor pins reaching all layers, or vias which are plated-through holes used for connections between layers. Circuit modules are mounted on top of the multilayer board. The terminals of the modules must be connected according to the specifications using wires, pins and vias. The problem is difficult because of the geometric constraints which must be satisfied by the wiring. These include the size of the multilayer board, the feasible number of layers, the minimum width of a wire and the minimum spacing between two adjacent parallel wires. The wiring can be completed by routing sequentially connections between rows of pins and vias [1].

The single row routing problem can be formulated is as follows: A set of nodes representing pins or vias are evenly spaced on a row R. We are given a list L of nets {Ni}, where each Ni is defined as a set of nodes to be connected on a plane. L contains only disjoint nets. The common objective is to make the connections such that the wiring congestions above and below R (referred also to as the upper and lower streets) are minimized. For example, let L={N1, N2, N3, N4}, where N1={1,4,7}, N2={2,9}, N3 = {3,5} and N4 = {6,8}. Figure 1 shows a possible routing realization. Here, the congestion above the line is 1 and below is 2. Because of its simple formulation, the single row routing problem had attracted attention of numerous researchers. First, it was studied in [2] and [3].

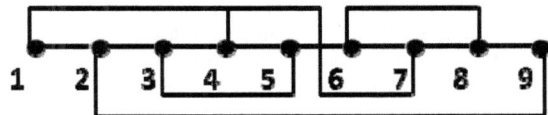

Figure 1. An example of a single row routing.

In [4] an elegant topological interpretation of the problem was given which allowed understanding the nature of the problem. Consider again the routing problem from Fig.1. We represent each net by an interval spanning between its leftmost and rightmost pins and draw the intervals in an arbitrary order. In Fig. 2(a), the intervals are ordered as N2, N1, N3 and N4. The nodes of the nets are connected by a reference line in the order they appear on the row. In Fig. 2(a) the reference line is shown with a broken line. Stretching the reference line and aligning it with the row of pins determines the topology of the routing solution.

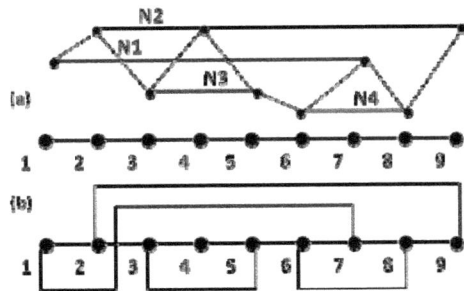

Figure 2. Interval graphical representation of the routing problem in Fig. 1. (b) The realization corresponding to the interval graph of (a).

Note, that this representation allows associating the realizations with interval ordering, and routing density with the number of nets at pin positions above and below the line. This is the first representation of a routing problem that allows quantifying and modifying a solution for the entire problem without determining the exact geometry of interconnects. It is not necessary to route connections sequentially to find the solution. Instead, one can manipulate the sequence of the intervals and measure its topological properties.

The interval graph representation was used to formulate the necessary and sufficient conditions for optimum routing [4],[5]; to study the computational complexity of the problem [6][8][10][12] and served as a backbone of numerous algorithms [5][7][9][11][13]. Most recently, new results on single row routing problem have been reported in [14] and in PhD thesis of D. Szeszler [15].

In [16], the authors show that single-row routing can model the channel assignment problem in the wireless cellular network system and the theoretical single-row multiprocessor network system. In [17], the single-row-routing-based technique is proposed for real-time simulation of channel assignments in a wireless cellular network.

3. CHANNEL ROUTING

Channel routing problem has been first formulated by Hashimoto and Stevens in 1971 [18], and owes its popularity to [19] where the famous "Deutsch difficult example" was published. The problem has a simple formulation stated as follows. The terminals of nets are situated on two opposite horizontal lines. We wish to connect all nets using two metal layers such that the number of horizontal tracks, or the channel height, is minimized. Typically, it is assumed that horizontal and vertical wires are placed on different metal layers.

The problem has been studied extensively for the past 40 years, with the most recent results stated in [15]. But undoubtedly, a classic in channel routing is the paper by Yoshimura and Kuh [21]. This paper departs from the left-edge algorithm and instead applies graph theoretical techniques. It proposes to use a weighted compatibility graph to model packing of the horizontal wire segments and selects the segments to be placed on the currently routed track by finding a min-cost path in that graph.

In [22], a different layering model has been studied. The idea was to allow both horizontal and vertical wires on the same layer in a way that cycles in vertical constraints graph were eliminated.

Also, in this paper a compacted channel routing was introduced for the first time.

In mid-80's, computational geometry techniques have been introduced into CAD tools. One of their applications was gridless routing. Gridless routing can take advantage of different technology and design rules. Terminals can be placed at arbitrary positions, not necessarily at grid points. In [22], a gridless channel routing was described. It departed from columns or tracks. Only wire widths, spacing, and via sizes were considered. Each net could have several wire widths to satisfy special design needs and improve performance of the circuits. The geometric routing technique developed there has been later applied to many other detailed routing problems.

4. ROUTING FOR BUILDING BLOCK LAYOUT STYLE

Channel routing is an essential step in connecting building blocks. In the mid-1970, the building-block design style was proposed at NEC and the early papers have been published by researchers from Japan [23], [24], [25]. Building block layout was inspired by the efficiency of manual designs. But its automation proved to be very complex and no commercial software was ever successful in producing layouts comparable to manual designs.

Building block layout automation has been pursued by Kuh's group in the 1980s and significant innovation and progress have been made in solving several of its sub-problems. The methodology developed at Berkeley was based on dividing the routing space determined by the placement and global routing into appropriate regions. Interconnects were sequentially completed within the regions without the necessity of redoing routing in the regions previously routed. Crucial to the quality of building block routing is the ability of proper assessment of routing resources accompanied by incremental placement and global routing adjustments as detailed routing progresses. The latter is a particularly hard problem formally addressed for the first time in [26].

The global routing adjustment problem is stated as follows. We are given placed rectilinear blocks with net terminals at their boundaries. Connections between the blocks' terminals are to be routed following the net topologies decided by the global router. Routing area is divided into regions whose dimensions may be adjusted. For example, if those regions are channels, their heights may be modified to accommodate wires. These region adjustments cause modifications of placement which in turn require modifications of global routes. To solve the problem, a proper geometric data structure to maintain information about the neighborhood of each block is needed. At the same time the representation of interconnect should be easily modifiable such that the effects of block movement could be accounted for. The method proposed in [26] satisfies these requirements.

For a placement of rectilinear-shaped building blocks, two tile planes are defined. In the horizontal tile plane all space tiles are maximal horizontal strips, and in the vertical tile plane, all space tiles are maximal vertical strips. Figure 3 illustrates an example of a horizontal tile plane.

A space tile is a bottleneck tile if both its sides are covered by the sides of its adjacent space tiles. A space tile is a dominant tile if none of its sides is covered by the side of its adjacent space tile. The shaded tiles in Figure 4 are horizontal bottleneck tiles corresponding to the building block placement in Figure 3. The set of dominant and bottleneck tiles are mutually disjoint.

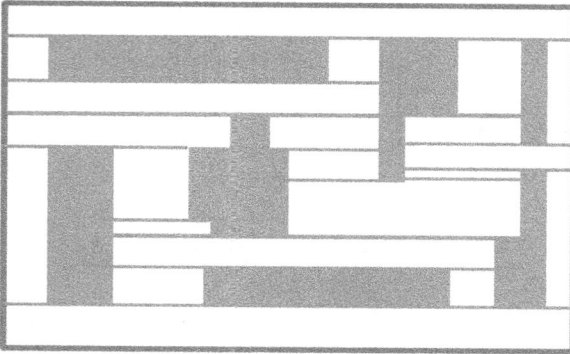

Figure 3. Building block placement and its horizontal tile plane.

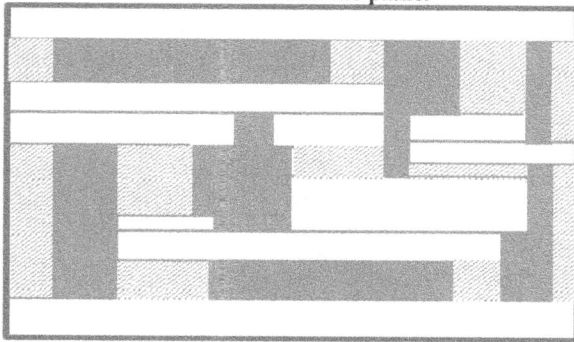

Figure 4. Bottleneck tiles in the horizontal plane.

A floor plan graph can be derived from the horizontal and vertical dominant tiles which define edges, referred to as walls, in the graph. Intersections of horizontal and vertical dominant tiles correspond to wall junctions connecting the tiles' walls. The region bounded by walls but containing no walls forms a room. A room either hosts a block or is empty.

The one-to one correspondence between a dominant tile in the tile plane and a wall in the corresponding floor plan graph is a basis for dynamic update of the floorplan graph which defines the topological structure of the given block placement. The global routing information is stored in the bottleneck tiles. Complete routing information can be stored in horizontal or vertical bottleneck tiles. When a block moves horizontally, its horizontal bottleneck tiles may change dimensions, but their topological structure remains intact. On the other hand, horizontal movement of a block may cause that a new vertical bottleneck tile appears or an existing vertical bottleneck disappears. Thus, when horizontal block movement occurs, routing information is kept only in the horizontal bottleneck tiles. After a block has been moved, the vertical tile plane is restored and routing information is recreated from the horizontal bottleneck tiles back to the vertical tiles. Similarly, when a block moves vertically, the routing information is kept in vertical bottleneck tiles and it is later recreated for the horizontal tiles.

Although building block layout style may now be less practical for chip design because modern IC technologies use many metal layers for connections, and wires often can be routed over predesigned blocks, many of the techniques and ideas originally developed for building blocks find applications with other layout

styles or technologies. In particular, the dynamic routing modification techniques have been further developed in the context of multichip modules (MCMs) [27][28][29].

5. WIRE COUPLING MINIMIZATION AND CROSSTALK CONTROL

5.1 Wire spacing

Compaction is the process of repeatedly removing spurious space from layout. It can be applied at a block level but it can also be used to adjust routing. The latter application is also called wire spacing. In earlier technologies, compactors were applied as a final step for cutting the chip area and thus improving the manufacturing yield. Early papers formulate the problem in the context of symbolic layout compaction [30], [31]. In [32], an algorithm which uses a graph-theoretic approach to solve efficiently the compaction problem with mixed constraints has been proposed. Later, the problem has been studied for channels [33] and for layout editors [34].

Kuh's group pioneering contribution to the wire spacing problem was to use for the first time routing compaction/decompation as means for controlling the capacitive delay and crosstalk effects. In [35], they proposed an algorithm that modifies a given routing. It determines first the critical paths by performing a static timing analysis on the layout. The timing constraints on the critical paths were modeled as repulsive and attractive constraints which were solved using Lagrangian relaxation technique. Both the coupled capacitance and the ground capacitance were minimized. The non-linear objective function was solved as an approximate linear programming problem for a fixed set of Lagrange multipliers λ and then updated λ based on the current solution. The problem was formulated such that the linear term (ground capacitance delay) was the objective function and the non-linear term (coupled capacitance delay) was modeled as a repulsive constraint. The repulsive constraints were determined based on the concept of the degree of proximity.

5.2 Post-global routing optimization

Often, the area cost of eliminating undesirable couplings for given wire adjacencies may be prohibitive. But in typical situations such adjacencies are quite random and can be modified with no, or with minimal adverse effects on the layout quality. Also, various nets may have various tolerances to crosstalk in particular routing regions they pass through. This was recognized in the works of Xue, Kuh and Wang [37] [38], where the authors proposed a post global routing crosstalk optimization approach. The idea was to depart from the net-level crosstalk estimation and reduction within detailed routed regions and instead to estimate the crosstalk risk for each routing region on the chip as a whole and reduce the risks by adjusting nets' routes globally among routing regions on the chip.

At a global routing step, the routing region is partitioned into a union of disjoint sub-regions and the topologies of all nets are determined such that the numbers of nets crossing boundaries of sub-regions do not exceed given capacities. Usually, global router does not determine the net ordering at the boundaries thus crosstalk and/or coupling control is not straightforward at this level. On the other hand, because the net adjacencies are not decided yet, it is possible to determine them appropriately to reduce coupling.

Let us assume that the routing region is divided by a grid. It implies a global routing graph $G(V, E)$ in which nodes V represent routing sub-regions and edges E correspond to horizontal or vertical boundaries on a routing layer. Denote as N the set of nets routed on the chip. A route of a net constitutes a connected subgraph of G. In this formulation, each net can to be routed in only one direction, either horizontally or vertically within each boundary region, and it occupies an entire track spanning a length equal to the distance between the centers of the neighboring regions. Nets N_s are labeled as sensitive, if they are sensitive to coupling to some other net. A sensitivity of a net i to a net j is measured by parameter S_{ij} and noise caused by net j on net i is captured by $noise(i,j) = S_{ij} \ len(i,j)$ where $len(i,j)$ is the length of wires of these nets placed on neighboring tracks. If $bound(i)$ denotes the amount of noise that net i can tolerate, then a net is safe from crosstalk violations if summation of noises acting on it is less than its bound.

The problem of determining if a feasible solution exists can be casted on finding feasible net orderings on each region boundary. To do this, for each net, its $bound(i)$ is divided into edge bounds $bound(i,e)$ which capture bounds that a net segment passing through an edge e can tolerate.

For a boundary e, its crosstalk dependencies can be modeled as a weighted *crosstalk risk graph* $CRG(N_s(e), E_s(e))$ where each node i in $N_s(e)$ corresponds to a sensitive net routed through e. The node weight represents the partitioned risk tolerance bound $B(i,e)$ of net i in e. The edge weight of an *edge (i,j,e)* reflects the potential crosstalk noise between nets i and j if these nets are routed on adjacent tracks in e. An *edge (i,j,e)* exists if and only if *noise* is less than the partitioned risk tolerance bounds of both nets in region. Now, if all nodes of $CRG(N_s(e), E_s(e))$ can be visited in such an order that for every node, the sum of weights of incoming and outgoing edges is less than the node weight, than a net ordering on e exists such that crosstalk noises on sensitive nets are less than the bounds they can tolerate. It is possible that such an ordering can be found only with shields and or empty tracks are inserted.

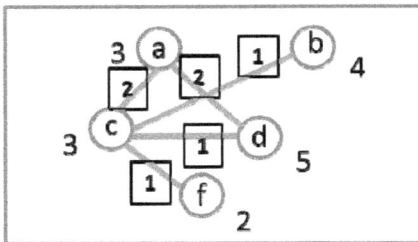

Figure 5. A crosstalk risk graph.

Consider an example of a crosstalk risk graph in Figure 5. The numbers next to nodes are node weights. The numbers in frames next to edges are edge weights. The objective is to find a simple path that visits all nodes and such that for each node the incoming and outgoing edges satisfy the noise constraints. Because some nodes in $CRG(N_s(e), E_s(e))$ might be isolated or a path that satisfies the constraints might not exist, a solution may need to be constructed with dummy nodes that represent shields of empty tracks. Figure 6(a) shows a simple paths subgraph of the graph in Figure 5 that satisfy the noise constraints. Figure 6(b) depicts the Hamiltonian path corresponding to the noise feasible order with an added dummy node g representing a shield.

Hamiltonian paths can be found for each boundary region. This graph-based model allows for reasoning about each region individually as well as for considering all regions jointly by properly deciding their bounds. It can also be used to decide which global routes to rip-up and reroute.

6. POST-ROUTING TOPOLOGY OPTIMIZATION OF CLOCK AND POWER NETWORKS
6.1 Clock link insertion
Routing of a clock network is a crucial step in circuit design and many papers have been devoted to this topic. The early papers considered only tree topologies for clocks and design automation efforts followed the path of first finding the tree topology, adjusting wire length, buffering and wire sizing. However, because it is often difficult to fit the tree branches into the given routing region and maintain balanced skew, Xue and Kuh in [39][40] proposed a method of incrementally changing the routed clock tree topology.

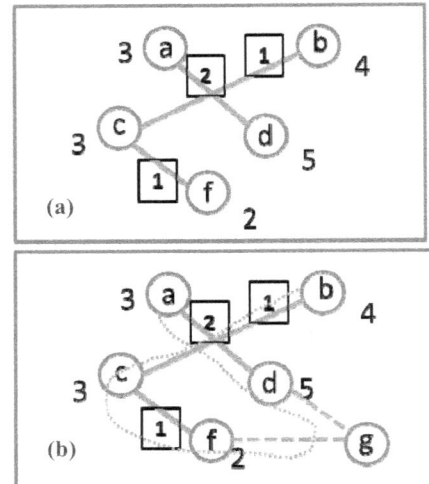

Figure 6. (a) A simple path constrained subgraph. (b) A Hamiltonian path with an added shield g.

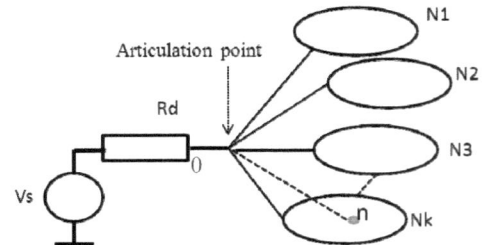

Figure 7. Clock topology modification for delay reduction.

The idea was to reduce both maximum delay and skew of an existing routing topology by tapered link insertion and non-uniform wire sizing. These were the first works that departed from the established methodology of restricting clock nets to tree or other predetermined topologies that would not be modified or changed after routing. Papers [39][40] were first works considering clock topologies with loops. They laid out theoretical foundations for skew variability reduction algorithms that were developed some 10 years later [41][42].

To understand the key results of [39] [40], consider Figure 7 depicting a graph N that is not bi-connected with an articulation point labeled 0. Removing node 0 splits N into k+1 disjoint parts. To study the effect of topology changes on maximum source-leaf delay D_{max} and ΔD_{ij}, we consider a new routing topology N' obtained by adding a link e with parameters R_e (resistance) and C_e (capacitance to ground) between node 0 and a chosen node n. It can be shown that with appropriate selection of R_e and C_e the delays to all nodes in N_k will decrease in comparison to their values before the modification and when R_e decreases, the delays to the leaf nodes of N_k decrease too. This result holds true also when 0 is not an articulation point, i.e. when k=1. It implies that it is always possible to reduce the maximum delay by properly adding new links into the existing topology.

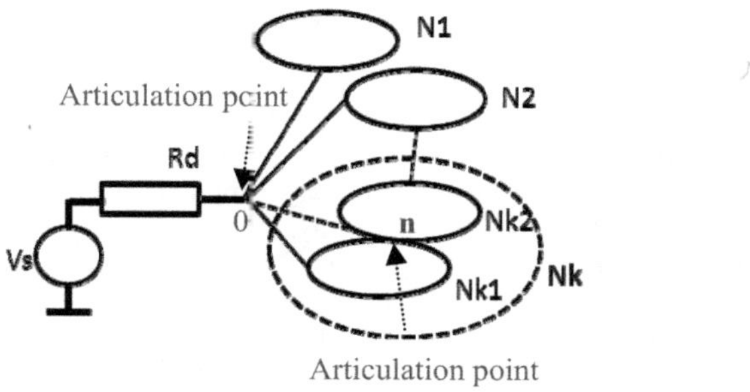

Figure 8. Clock topology modification for skew reduction

We will now consider the effects of topology changes on delay skew. Referring to Figure 8, let N_k be decomposed into two sub-components N_{k1} and N_{k2} connected at an articulation point n and the clock topology is modified by adding a link between n and 0. It is marked with broken line in Figure 8. Let the value of R_{ij} in Ohms, be equal to the potential in Volts at node i when a 1A current were injected into node j and all other nodes in N were open circuited. To determine how $\Delta R(i, j, m) = R_{im} - R_{jm}$ changes in the modified circuit, various cases must be considered. The final result is that it is always possible to reduce the maximum delay skew of an existing routing topology by adding a new link properly between the reference node 0 and the node with maximum delay (skew). Note that the node with maximum delay is also a node with maximum skew. This result is very important as it demonstrates that it is possible to reduce both maximum delay and maximum delay skew by decreasing delays. All clock routing algorithms were increasing delays to minimize delay skew.

6.2 Power Grid Topology Optimization

In [43], Mitsuhashi and Kuh formulated a problem of adding power enhancement buses to an existing power grid. The problem is particularly relevant in the context of standard cells which internally contain portions of power nets that can be connected by cell abutment. Usually, the intra-cell portion of a power net is not

sufficient, as it may suffer from electromigration (EM) effects or may lead to unacceptable voltage drops.

The problem studied in [43] can be stated as follows. Given a cell placement with an intra-cell power network, cell current demands, power pads, and possible positions of the enhancement power buses, the objective is to select a subset of enhancement buses and their conductance such that the area of added wires is minimal and both voltage drop and EM constraints are satisfied. This is a non-linear combinatorial problem. The solution method proposed in [43] relaxes some discrete variable constraints, obtains a feasible solution that meets the constraints and then constructs a sequence of improved solutions.

The authors of [43] made an interesting observation. They showed a simple example where indiscriminate power bus enhancement may cause current crowding leading to EM violation which can be resolved by removing some of the enhancement buses. To the author's knowledge, the example shown in [43] has not been

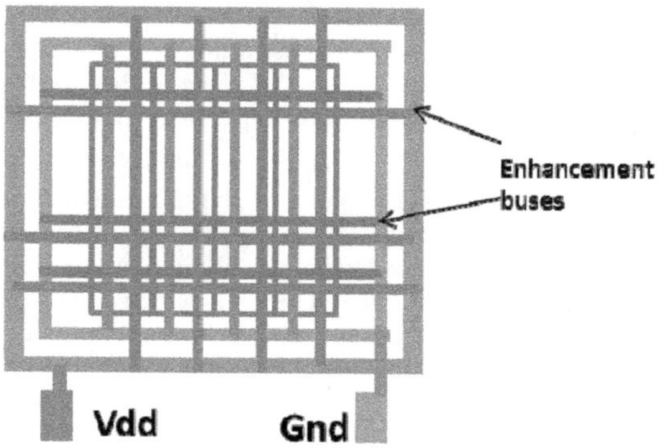

Figure 9. Vdd and Gnd networks with enhancement buses.

studied and/or explained in any follow up papers. The problem has some similarities to that studied in [44], but it is not the same. Electromigration caused wire degradation is now a serious reliability concern. Perhaps the problem pointed out in [43] deserves a fresh look?

7. A SUBJECTIVE LIST OF INTERESTING ROUTING PROBLEMS

Routing has been studied for the past forty years and although the area is considered mature, there are still many interesting routing related problems worth studying. They can be classified into the following categories: (1) known problems that require constant research and development effort because they are affected by the increasing design complexity and non-uniform geometry scaling; (2) routing for manufacturability; (3) routing for new devices; (4) routing for ECO and debugging; (5) power grid topology design.

7.1 Routing complexity affected problems

Global router plans net topologies on a chip. The problem is modeled as a routing problem on a grid graph $G(V,E)$. In G, a vertex v_i corresponds to a global routing cell (G-cell) – usually a rectangular region of the chip, and an edge e_{ij} represents the boundary between v_i and v_j with a given maximum routing capacity c_{ij}. The objective of global router is to determine routing topologies that satisfy the edge capacity constraints, control wire

17

lengths and congestion. A survey of global routing techniques can be found in [45].

Global routing problem has been studied for decades but its basic model has remained unchanged over the years. The existing global routers, including the newer ones [46] [47] do not provide a methodology of how to determine the edge capacity such that the global routing that satisfies the capacity constraints can indeed be mapped to tracks for the given number of layers and positions of obstacles. Terminal positions within the G-cells are not modeled and it is possible that a routing solution that can be distributed to layers may not be able to complete connections to the terminals within the G-cells. Additionally, the existing global routers do not model manufacturability related constraints, such as requiring wires to be strictly parallel on each metal layer or satisfy adjacency constraints to prevent crosstalk. With shrinking design rules, the existing global routing methods may require some major re-thinking.

7.2 Routing for Manufacturability

Potentially, many routing problems in this category may emerge. It is also possible that the lifetime of some of them will be short, as with changing technology, their importance may diminish.

At present, double patterning photo lithography (DPL) is one of candidate technologies for patterning in 22 nm and for more advanced technology nodes. DPL requires layout to be properly decomposed. But if the initial layout has native conflicts, proper decomposition does not exist. The first objective of DPL aware physical design is to ensure no such conflicts are present. Additionally, stitches affect yield, thus it is desirable to minimize their number and pattern density for both decomposed layout planes should be balanced. Most of the existing tools for double patterning begin with a given layout [49]. A double-patterning friendly router has been proposed in [48]. Routing for DPL still requires more effort – first a complete characterization and analysis of the problem is needed, then routing strategies for "well decomposable" layout could be developed. Once the nature of the problem is fully understood, routing techniques starting at the global routing level could be proposed.

7.3 Routing for New Technologies

As new devices and technologies emerge, new routing problems appear. For example, routing for VeSFET arrays [50] falls into such a category. A VeSFET canvas contains devices of identical geometric shapes and sizes. The entire chip area is occupied by double gate transistors. In the isolated transistor canvas shown in Figure 10, transistors don't share terminals, thus extra wires for drain-to-source connections are needed. All wires are above the active area and transistor pins. Any connected wire could possibly block other transistors pins' access. Vias are needed for bending connections. If two pins cannot be connected linearly, wires on at least two layers must be used. A layout generator for blocks composed of about 100 transistors embedded into such a canvas for transistors with tied gated has been proposed in [51].

Many configuration topologies of VeSFET canvases exist and routing for them has not been studied yet. Routing for VeSFET arrays opens a very rich research area. Although on the surface, the layout problem for VeSFET arrays appears to be similar to that for gate arrays; in fact it is very different and much harder due to extreme layout density unheard of in the gate array context. None of the existing methodologies developed for gate

arrays can be easily adapted to work with VeSFETs. VeSFET arrays can be routed on both sides. This option is particularly promising for designs with individually controlled gates for which dense layout with connections on one side of an array is often impossible. There is also a strong dependency between the circuit layout and performance for VeSFET-based designs. This problem for one-sided routing, tied gate designs implemented with isolated transistor arrays has been discussed in [52]. Presently, no routing methodologies exist for individually controlled gate designs on isolated or shared pillar transistor canvases.

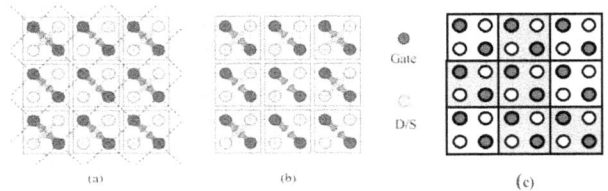

Figure 10. A uniform, isolated VeSFET transistor array with (a) diagonal and (b) horizontal-vertical grids. (c) Symbolic representation. Light circles are gate and dark are drain-source terminals. *P*-type transistors are yellow, *n*-type are blue.

Many configuration topologies of VeSFET canvases exist and routing for them has not been studied yet. Routing for VeSFET arrays opens a very rich research area. Although on the surface, the layout problem for VeSFET arrays appears to be similar to that for gate arrays; in fact it is very different and much harder due to extreme layout density unheard of in the gate array context. None of the existing methodologies developed for gate arrays can be easily adapted to work with VeSFETs. VeSFET arrays can be routed on both sides. This option is particularly promising for designs with individually controlled gates for which dense layout with connections on one side of an array is often impossible. There is also a strong dependency between the circuit layout and performance for VeSFET-based designs. This problem for one-sided routing, tied gate designs implemented with isolated transistor arrays has been discussed in [52]. Presently, no routing methodologies exist for individually controlled gate designs on isolated or shared pillar transistor canvases.

7.4 Routing for ECO and Debugging

Focused Ion Beam (FIB) technology has become widely used in production settings. It can be used for making small modifications in a manufactured circuit by cutting wires and/or depositing conductive materials to make connections. This technique allows for fixing bugs or adding features before remanufacturing the chip. It can reduce the re-spin risk and cost [54]. Presently, FIB-implemented layout corrections are done manually. The cost of FIB depends on the number of steps that must be executed. Figure 11 shows a sequence of steps that must be executed to remove a connection and add another connection in a different place. If several changes are required, minimizing the number of correction steps may be non-trivial. Automation of a FIB Engineering Change Order (ECO) process is an open problem.

7.5 Power Grid Topology Design

Power and ground networks are commonly designed for an assumed topology. The wires are properly sized such that the voltage drop and electromigration constraints are met. Such an approach worked well for single mode of operation. However, when large blocks of a chip are periodically disconnected, current distribution on the grid changes and care must be taken to avoid EM violations [44]. It is likely, that the problem could be solved efficiently by selecting a proper topology of the grid [43].

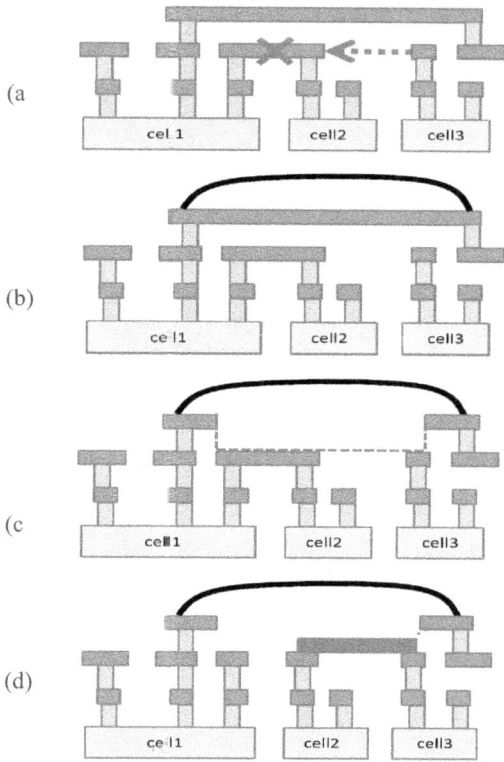

(a

(b)

(c

(d)

Figure 11. FIB-based correction steps. (a) Specification, (b) Bypass connection on the top surface, (c) Top metal cut, a wire to be cut at a lower layer exposed, (d) Desired connection made at the bottom of the drilled hole.

8. REFERENCES

[1] H. C. So, "Some theoretical results on the routing of multilayer printed-wiring boards," in *Proc. 1974 IEEE Int. Symp. on Circuits and Systems*, pp. 296-303.

[2] B. S. Ting, E. S. Kuh, and I. Shirakawa, "The multilayer routing problem: Algorithms and necessary and sufficient conditions for the single-row, single-layer case," *IEEE Trans. Circuits Syst.*, vol. CAS 23, pp. 768-778, Dec. 1976.

[3] B. S. Ting and E. S. Kuh, "An approach to the routing of multilayer printed circuit boards," in *Proc. 1978 IEEE Int. Symp. on Circuits and Systems*.

[4] E. S. Kuh, T. Kashiwabara, and T. Fujisawa, "On Optimum Single-Row Routing," *IEEE Transactions on Circuits and Systems*, vol. 26, no. 6, June 1979.

[5] S. Tsukyama, E. S. Kuh, and I. Shirakawa," An Algorithm for Single-Row Routing with Prescribed Street Congestions," *IEEE Transactions on Circuits and Systems,* vol. CAS-21, no. 9, September 1980, pp. 765-772.

[6] R. Raghavan and S. Sahni, "Single Row Routing," in *IEEE Transactions on Computers*, vol. C-32, no. 3, March 1983, pp. 209-220.

[7] B. B. Bhattacharya, J. S. Deogun, N. A. Sherwani, "A Graph Theoretic Approach to Single Row Routing Problems," *Proc. Int. Symp. On Circuits and Systems*, 1983, pp. 1437-1440.

[8] N. A. Sherwani and J. S. Deogun, "New Lower Bounds for Single Row Routing Problems," *Proc. of the 32nd Midwest Symposium on Circuits and Systems*, 1989, pp. 565-568.

[9] M. Hossain, N. A. Sherwani and J. S. Deogun, "Optimal algorithms for restricted single row routing problems," *Proc. of the 34th Midwest Symposium on Circuits and Systems*, 1991, pp. 823-826.

[10] D.H.C. Du, O.H. Ibarra, and J.F. Naveda, "Single-Row Routing with Crossover Bound," *IEEE Transactions on Computer-Aided Design of Circuits and Systems*, vol. 6, no 2, 1987, pp. 190 – 201.

[11] N.A. Sherwani, B. Wu, and M. Sarrafzadeh, "Algorithms for Minimum-bend Single Row Routing Problem," *IEEE Transactions on Circuits and Systems I: Fundamental Theory and Applications*, vol. 39, no 5, 1992, pp. 412 – 415.

[12] R. Raghavan and S. Sahni, "The Complexity of Single Row Routing," *IEEE Transactions on Circuits and Systems*, vol. 31. no.5, 1984, pp. 462 – 472.

[13] T. T.K. Tarng, M. Marek-Sadowska, and E.S. Kuh, "An Efficient Single-Row Routing Algorithm," *IEEE Transactions on Computer-Aided Design of Circuits and Systems, vol. 3, no. 3, 1984, pp. 178 – 183.*

[14] A. Y. Zomaya, R. Karpin and S. Olariu, "The Single Row Routing Problem Revisited: A Solution Based on Genetic Algorithms," *VLSI Design*, 2002 vol. 14 (2), pp. 123–141.

[15] D. Szeszler, Combinatorial Algorithms in VLSI Routing," PhD Dissertation, Muegyetem University, Budapest, Hungary, 2005.

[16] S. Salleh, S. Olariu, A. Y. Zomaya, K. L. Yieng, N. A. Aziz, "Single-row mapping and transformation of connected graphs," *The Journal of Supercomputing*, 2007, vol 39, no 1, pp. 73-89.

[17] S. Salleh and N. H. Sarmin, "Dynamic Single-Row Routing Technique for Channel Assignments," 2009 *Sixth International Conference on Information Technology: New Generations*, pp. 41-46.

[18] A. Hashimoto and J. Stevens, "Wire Routing by Optimizing Channel Assignment," *Proc. 8th Design Automation Conf.* (1971), 214-224.

[19] D. Deutsch, "A Dogleg Channel Router," *Proc. 13rd Design Automation Conf.*, 1976, pp. 425-433.

[20] A. Hashimoto and J. Stevens, "Wiring Routing by Optimizing Channel Assignment within Large Apparatus," *Proc. Design Automation Conference*, 1977, pp. 155-169.

[21] M. Marek-Sadowska and E.S. Kuh, "General Channel-Routing Algorithm," *IEE Proceedings,* Vol. 130, Ft. G, No. 3, June 1983, pp. 83-88.

[22] H.H. Chen and E.S. Kuh, "Glitter: A Gridless Variable-Width Channel Router," *IEEE Transactions on Computer-Aided Design*, vol. CAD-5, no. 4, October 1986, pp. 459-465.

[23] H. Kato, H. Kawanishi, S. Goto, T. Oyamada, and K. Kani, "On automated wire routing for building-block MOS LSI," *Proc. International Conf. on Circuits and Systems,* 1974, pp. 309-312.

[24] S.Kimura, N.Kubo,T.Chiba,and I.Nishioka," An automatic routing scheme for general cell LSI," *IEEE Trans. Computer Aided Design,* vol. CAD-2, no. 4, pp. 285-292, Oct. 1983.

[25] T. Kozawa, C. Miura, and H. Terai, "Combine and Top Down Block Placement Algorithm for Hierarchical Logic VLSI Layout," *Proc. 21st DAC,* 1984, pp. 667-669.

[26] W.-M. Dai, M. Sato, and E. S. Kuh, "A Dynamic and Efficient Representation of Building-Block Layout," *Proc. 24th ACM/IEEE Design Automation Conference,* 1987, pp. 376-384.

[27] W. W.-M. Dai, R. Kong, J. Jue and M. Sato, "Rubber Band Routing and Dynamic Data Representation," *Proc. Int. Conf. on Computer Aided Design,* 1990, pp. 52-55.

[28] D. Staepelare, J. Jue and W. W.-M. Dai, "Surf: A Rubber-Band Routing System for Multichip Modules," *IEEE Design and Test of Computers,* December 1993, pp. 18-26.

[29] J. Z. Su and W. W. Dai, "Post-Route Optimization for Improved Yield Using a Rubber-Band Wiring Model," *Proc. Int. Conf. on Computer Aided Design,* 1997, pp.700-706.

[30] S.B. Akers, J.M. Geyer, D.L. Roberts, "IC Mask Layout with a Single Conductor Layer," Proc. *7th Annual Design Automation Workshop,* 1970, pp. 7-16.

[31] A. E. Dunlop, "SLIM – The Translation of Symbolic Layouts into Mask Data," Proc. *17th Conference on Design Automation,* 1980, pp. 595 – 602.

[32] Y.Z. Liao and C.K. Wong, "An Algorithm to Compact a VLSI Symbolic Layout with Mixed Constraints," *IEEE Transactions on Computer-Aided Design,* vol. CAD-2, no 2, April 1983, pp. 62-69.

[33] D. N. Deutsch, "Compacted Channel Routing," *Proc. Int. Conf. on Computer Aided Design,* November 1985, pp. 223-225.

[34] J. K. Ousterhout, G. T. Hamachi, R. N. Mayo, W. S. Scott, G. S. Taylor, "Magic: A VLSI Layout System," *IEEE Design & Test ,* vol. 2, issue 1, Jan. 1985. pp. 19-30.

[35] K. Chaudhary, A. Onozawa and E.S. Kuh, "A Spacing Algorithm for Performance Enhancement and Cross-talk Reduction," *Proc. International Conference on Computer Aided Design,* 1993, pp 697 – 702.

[36] A. Onozawa, K. Chaudhary, and E.S. Kuh, "Performance Driven Spacing Algorithms Using Attractive and Repulsive Constraints for Submicron LSI's," IEEE Transactions on Computer Aided Design, vol. 14, no. 6, June 1995, pp. 707-719.

[37] T. Xue, E. S. Kuh and D. Wang, "Post Global Routing Crosstalk Risk Estimation and Reduction," *Proc. International Conference on Computer Aided Design,* 1996, pp. 302-309.

[38] T. Xue, E. S. Kuh and D. Wang, "Post Global Routing Crosstalk Synthesis," IEEE Transactions on Computer Aided Design, vol. 16, no. 12, 1997 , Page(s): 1418 – 1430.

[39] T. Xue and E.S. Kuh, "Post Routing Performance Optimization via Tapered Link Insertion," *Proc. International Conference on Computer Aided Design,* 1995, pp. 74-79.

[40] T. Xue and E.S. Kuh, "Post Routing Performance Optimization via Multi-Link Insertion and Non-Uniform Wiresizing," *Proc. of the Conference on European Design Automation,* 1995, pp. 575-580.

[41] A. Rajaram, J. Hu and R. Mahapatra, "Reducing Clock Skew Variability via Cross Links," *Proc. 43th Design Automation Conference,* 2004, pp. 18-23.

[42] A. Rajaram and D.Z. Pan, "Fast Incremental Link Insertion in Clock Networks for Skew Variability Reduction," Proc. Int. Symp. on Quality Electronic Design, 2006, pp. 84 – 89.

[43] T. Mitsuhashi and E. S. Kuh, "Power and Ground Network Topology Optimization," Proc. 29th *Design Automation Conference,* 1992, pp. 524-529.

[44] A. Todri and M. Marek-Sadowska, "Reliability Analysis and Optimization of Power-Gated ICs," *IEEE Transactions on Very Large Scale Integration (VLSI) Systems,* accepted for publication

[45] J. Hu and S. Sapatnekar. A Survey on Multi-net Global Routing for Integrated Circuits. *Integration, the VLSI Journal,* vol. 31, no. 1, pp. 1-49, 2002.

[46] M.Cho and D.Z. Pan, "Box Router: A New New Global Router Based on Box Expansion and Progressive ILP," *IEEE Trans. Computer Aided Design,* vol.26, no. 12, December 2007, pp. 2130-2143.

[47] M. Cho, K. Lu, K. Yuan, and D.Z. Pan, "BoxRouter 2.0: A Hybrid and Robust Global Router with Layer Assignment for Routability," *ACM Trans. Des. Autom. Elect. Syst.,* 14, 2, Article 32 (March 2009), pp. 32(1) – 32(21).

[48] M. Cho, Y.Ban, and D.Z. Pan, "Double Patterning Technology Friendly Detailed Routing", *Proc. International Conference on Computer Aided Design,* 2008, 506 – 511.

[49] J.-S. Yang, K. Lu, M. Cho, K. Yuan, and D. Z. Pan, "A New Graph Theoretic, Multi-Objective Layout Decomposition Framework for Double Patterning Lithography," *Proc. Asia and South Pacific Design Automation Conf.,* 2010, pp. 637 – 644.

[50] W. Maly, "Integrated Circuit Fabrication and Associated Methods, Devices and Systems," U.S. Non-Provisional Patent Application Serial Number CMU Docket 06-091; DMC Docket 06-001PCTCMU.

[51] Y.W. Lin, M. Marek-Sadowska, W. Maly, "Layout Generator for Transistor-Level High-Density Regular Circuits," *IEEE Transactions on CAD for VLSI,* vol. 29, issue 2, 2010, pp. 197 – 210.

[52] Y.W. Lin, M. Marek-Sadowska, W. Maly, "On Cell Layout-Performance Relationships in VeSFET-Based, High-Density Regular Circuits," *IEEE Transactions on CAD for VLSI,* vol. 30, no 2, 2011, pp. 229 – 241.

[53] Y.-R. Wu, S.-Y. Kao, and S.-A. Hwang, "Minimizing ECO Routing for FIB," *Proc. Int. Symposium on Design and Test,* 2010, pp. 351- 354.

Grid-to-Ports Clock Routing for High Performance Microprocessor Designs

Haitong Tian, Wai-Chung Tang,
Evangeline F. Y. Young
Department of Computer Science and
Engineering
The Chinese University of Hong Kong
{httian,wctang,fyyoung}@cse.cuhk.edu.hk

C. N. Sze
IBM Austin Research Laboratory
csze@us.ibm.com

ABSTRACT

Clock distribution in VLSI designs is of crucial importance and it is also a major source of power dissipation of a system. For today's high performance microprocessors, clock signals are usually distributed by a global clock grid covering the whole chip, followed by post-grid routing that connects clock loads to the clock grid. Early study [7] shows that about 18.1% of the total clock capacitance dissipation was due to this post-grid clock routing (i.e., lower mesh wires plus clock twig wires). This post-grid clock routing problem is thus an important one but not many previous works have addressed it. In this paper, we try to solve this problem of connecting clock ports to the clock grid through reserved tracks on multiple metal layers, with delay and slew constraints. Note that a set of routing tracks are reserved for this grid-to-ports clock wires in practice because of the conventional modular design style of high-performance microprocessors. We propose a new expansion algorithm based on the heap data structure to solve the problem effectively. Experimental results on industrial test cases show that our algorithm can improve over the latest work on this problem [10] significantly by reducing the capacitance by 24.6% and the wire length by 23.6%. We also validate our results using hspice simulation. Finally, our approach is very efficient and for larger test cases with about 2000 ports, the runtime is in seconds.

Categories and Subject Descriptors

B.7.2 [**Design Aids** : Placement and Routing

General Terms

Algorithm, Design

Keywords

Clock Routing, Grid, Non-tree, Microprocessor Designs

Permission to make digital or hard copies of all or part of this work for personal or classroom use is granted without fee provided that copies are not made or distributed for profit or commercial advantage and that copies bear this notice and the full citation on the first page. To copy otherwise, to republish, to post on servers or to redistribute to lists, requires prior specific permission and/or a fee.

ISPD'11, March 27–30, 2011, Santa Barbara, California, USA.
Copyright 2011 ACM 978-1-4503-0550-1/11/03...$10.00.

1. INTRODUCTION

In today's high performance systems, clock signals are distributed through a global clock grid [6–10], followed by post-grid routing that connects clock loads to the grid. Early studies showed that most of the clock power dissipation was due to three major categories of capacitances – clock load, clock twig and clock mesh wires, and clock grid buffers. The post-grid clock routing wires (i.e., lower mesh wires and clock twig wires) comprises 18.1% of the total capacitance dissipation [7]. This post-grid clock routing problem is thus a very important one, although not many previous works have addressed it.

Due to the high complexity of microprocessor design, the clock distribution network is usually synthesized and tuned at the same time when different design teams are working on their logic modules. In this case, the clock distribution between the clock grid and the block-level clock ports is subject to conflict of routing resources for data signals. To resolve this conflict and to facilitate simultaneous work between different design teams, a subset of routing tracks have to be reserved for this post-grid clock routing. As a result, this post-grid clock routing problem assumes a given set of reserved tracks, forming a virtual grid structure. The quality of this routing step is of significant importance as it will affect directly the total power consumption, the clock skews and slews at the input of the ports and finally the quality of the chip. These provide motivations to solve this multi-source multi-port post-grid clock routing problem with an objective to minimize the interconnect capacitance while meeting given delay and slew constraints. Traditionally, this step is done manually and iteratively to satisfy the constraints, resulting in a long time to market, especially when the problem size has increased to thousands of clock ports in the layout region. This also motivates the research of a fast algorithm to resolve this clock routing problem effectively.

This post-grid clock routing problem bears a fundamental difference with those previous works on clock tree construction, since the available routing tracks on different metal layers are given and can be very scarce. Besides, in our problem, there are multiple ports and multiple sources in the layout region. There is one very recent work addressing the same problem by Shelar [9, 10] and he proposed a tree growing algorithm to solve the problem with delay and slew constraints. In his algorithm, the clock network is generated by expanding from the sources step by step, and the frontier edge (detailed in Section 5.1) with the smallest wire capaci-

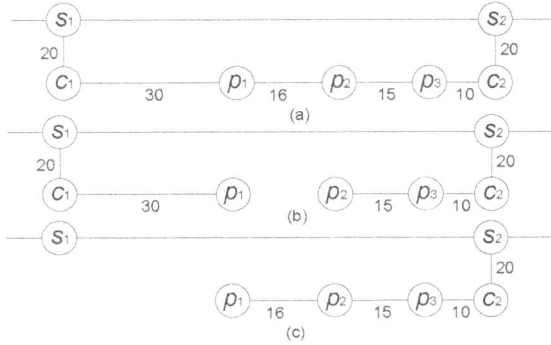

Figure 1: (a) Routing graph with sources s_1 and s_2, ports p_1, p_2 and p_3, and via nodes c_1 and c_2. The numbers near the edges denote the wire capacitances. (b) Routing solution of [10]. (c) Routing solution of our approach, with a 36% reduction in wire capacitance.

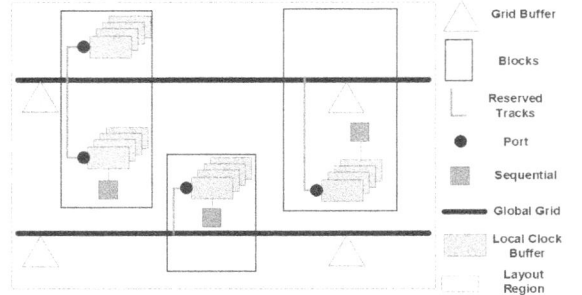

Figure 2: Post-grid Clock Network Distribution

tance is chosen every time. Checking against delay and slew constraints is done when a port is reached. However, the routing method in [10] has a couple of intrinsic problems. It uses a top-down tree growing heuristic in which the downstream capacitance information is not available when the trees are being constructed, and it thus can hardly optimize the delay value. In addition, its slew calculation is based on the lumped-RC model instead of the distributed RC model, and this may lead to accuracy and fidelity problems. In Fig. 1, we show a simple example to illustrate the differences between Shelar's approach and ours. In this example, three ports p_1, p_2 and p_3 are to be connected to the two sources s_1 and s_2. Fig. 1(b) and Fig. 1(c) show the routing topology obtained by the tree growing approach in [10] and by our approach respectively. In this example, our approach can achieve a 36% reduction in wire capacitance compared with the tree growing approach.

In this paper, we devised an efficient algorithm for this post-grid clock routing problem that can satisfy user given delay and slew bounds while minimizing the total wire capacitance. Note that similar to the formulation in [10], clock skew is optimized in the context of minimizing the maximum delay[1]. We compared our approach with the previous work [10] and can show that with the same delay and slew constraints, our approach can improve over [10] by 24.6% in wire capacitance and by 23.6% in wire length. To further verify the quality of our results, we simulate the constructed clock network using hspice and the simulation results confirm the effectiveness of our approach.

In the following, we will first give a preliminary overview in Section 2 of this hybrid clock network, which motivates this multi-source multi-port post-grid clock routing problem. Problem definition will be given in Section 3 while our approach will be presented in Section 4. Finally, experimental results, comparisons and discussions will be shown in Section 5, followed by a conclusion in Section 6.

2. POST-GRID CLOCK ROUTING

Clock signals are generated by a phase locked loop (PLL) and reach the global grid through grid buffers. The grid, typically lying on the topmost metal layer, is usually implemented using spines and will distribute clock signals to different regions of the chip. The grid and PLL are usually designed manually. The signals will further be routed through a set of reserved tracks on the lower metal layers to the clock ports of different blocks, and this step is called post-grid clock routing. The block-level ports will be created in such a way to align with the reserved tracks and the clock signals will be further sent to different sequentials inside the blocks. There can be thousands of ports in each layout region in a real post-grid clock routing problem. As shown in Fig. 2, the global grid wires are driven by multiple grid buffers and deliver the clock signals to block-level ports by routing along the reserved tracks on the lower metal layers.

A simple example of this post-grid clock routing problem is shown in Fig. 3(a). In this example, there are five metal layers (from layer 3 to layer 7) with six ports lying on metal layer 3 and the source grid is on metal layer 7. Routing can only be done on those reserved tracks (dashed lines). A sample routing solution is shown in Fig. 3(b). The target is to connect all the ports to the sources without exceeding a very stringent delay bound (which is also an upper bound of the skew), a slew bound and to minimize the total wire capacitance. Note that for this particular instance, our algorithm gets the *optimal* solution as shown in Fig. 3(b).

3. PROBLEM DEFINITION

In this post-grid clock routing problem, we are given (1) a set of reserved tracks (including the source grid which is always on the topmost metal layer) on different metal layers which have alternate routing directions, (2) the locations and capacitances of n ports $P = \{P_1, P_2, ..., P_n\}$ on some lower metal layers, and (3) the types of wires (with different capacitance/resistance tradeoffs) available on each metal layer. We assume that the clock grid on the topmost layer provides zero-skew clock signals. The objective of this post-grid clock routing problem is to connect all the ports to the sources[2] by making use of the reserved tracks and different wire types so as to satisfy the constraints on maximum delay bound D and maximum slew S, and to minimize the total wire capacitance. The delay here is computed accord-

---

[1]Notice that clock skew is upper bounded by the maximum delay. In our problem, this delay bound is set to be very stringent, e.g, within 5ps, which is in reality the limit for the clock skew.

[2]These sources are vias to the source grid on the topmost matal layer.

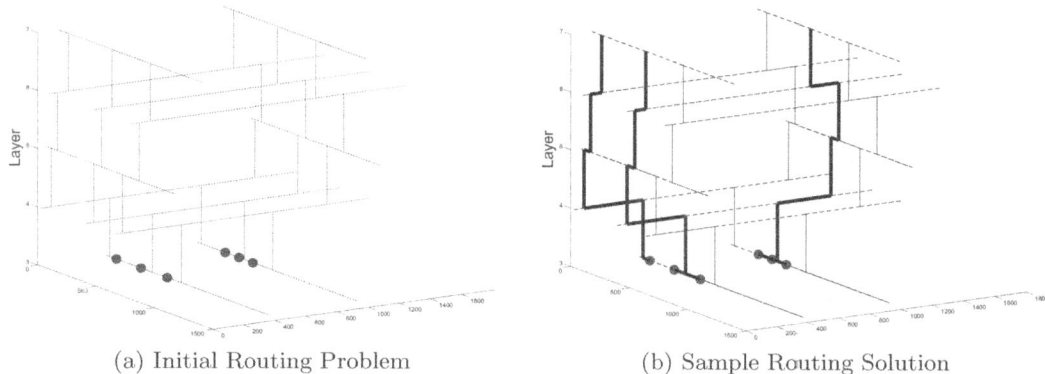

(a) Initial Routing Problem   (b) Sample Routing Solution

Figure 3: Post-grid Clock Routing Problem

ing to the Elmore delay model due to its simplicity and high fidelity, and the slew is estimated by $\sqrt{(2.2RC)^2 + (S_i)^2}$ according to [5], where R and C denote the resistance and capacitance of the wire segment respectively, and S_i denotes the input slew.

Similar to the previous work [10], we do not optimize the skew directly. This is because the grid-to-ports delay bound (also upper bound the skew) is very stringent and is set to be within 5ps for all the data sets, which is very small compared with the overall circuit skew budget. Therefore, it is not necessary to put the skew as another optimizing objective specifically. In addition, similar to [10], we do not consider buffer insertion in this post-grid clock routing. A very detailed explanation is provided in [10]. In fact, the well-defined grid and reserved tracks make buffer insertion unnecessary for this post-grid clock routing problem.

4. OUR APPROACH

This post-grid clock routing problem can be seen as a multi-source multi-sink[3] tree construction problem with a delay bound, a slew bound, and an objective to minimize the total wire capacitance. We first model the virtual grid of reserved routing tracks by a graph G. The set of vertices contains (1) the block-level clock ports (i.e., the sinks), (2) the possible via positions between reserved tracks on adjacent metal layers, and (3) the clock sources (which are the vias connecting to the source grid). The edges in G represent the wire segments on the reserved tracks connecting ports, vias or sources. Our approach includes a pre-processing step that performs segment merging, finding segment intersections and construction of the graph G and uses some techniques in [3,4] and it will not be detailed here.

To solve this clock routing problem, we devise a delay-driven *path expansion* algorithm that will propagate from each port in selected directions. A path is a routing between an intermediate node (a via node or a source node) and a block-port along the reserved tracks. In the expansion process, we will always select the path with the smallest Elmore delay (note that it is the total delay of the path) in the current path pool to be further processed. A path p will be *taken* when it reaches a source. Then, all the paths

---

[3]These "sinks" are block-level clock ports in our problem and are different from the "sinks", which are flip-flops or latches, in traditional clock routing problems.

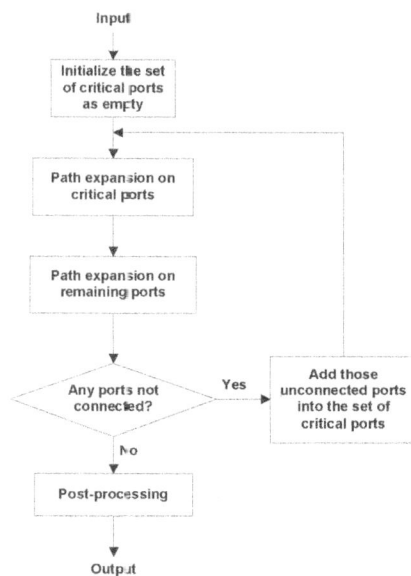

Figure 4: An Overall Flow of Our Approach

that intersects with p will also be considered and *taken* if no delay nor slew violation occurs. This path expansion step will be repeated until all the ports are connected, or no more ports can be connected without violating the delay and slew constraints. These are the basic steps of our delay-driven path expansion algorithm. It will be invoked repeatedly with a pre-processing step that will connect up some critical ports first. Finally, some post-processing techniques are performed to further reduce the total wire capacitance. A flow of our approach is illustrated in Fig. 4.

4.1 Delay-driven Path Expansion Algorithm

In this delay-driven path expansion algorithm, we will propagate from all the ports simultaneously along the reserved tracks to reach a source. A heap data structure H is used to store all the currently expanding paths sorted according to their Elmore delays. At the beginning, the heap H is initialized with all the ports, which can be regarded as zero length paths with zero delay.

In each step, we will pick a path p from the top of the heap, which has the smallest Elmore delay among all the paths in H. We will then check whether p has reached a source. If not yet, we will expand p vertically up if a via [4] exists at the endpoint $last(p)$ of p or will otherwise expand sideways (horizontally or vertically, depending on the track direction of the metal layer the last node of p is lying on) along the reserved tracks. We will first compute the Elmore delays of these new paths. Those new paths with Elmore delay smaller than the delay limit D will be inserted into the heap H. The path p will then be removed from H.

However, if the path p has reached a source, we will first check against the delay and slew constraints. If no violation occurs, we will take this path p into our routing solution. Suppose that the path p is expanded from a port $port(p)$, all the paths originating from $port(p)$ will be removed from H. Furthermore, we will process every path q where q intersects with p. All these paths will be considered in a non-decreasing order of their Elmore delays. For each of these paths q, we will check whether connecting q to p in the routing solution will violate any constraint at $port(q)$ as well as at any port in the current clock tree under construction. If any violation occurs, we will just neglect q and consider the next candidate. Otherwise, we will take q into the routing solution and connect it to p. We call these paths which do not come to the top of the heap but are processed *chain paths*. Note that once a path is taken into the routing solution, all the nodes on it will be regarded as "sources" for later expansions, and all the paths originating from its port will be removed from H.

Wire length reduction is not directly addressed in our algorithm. But as we always choose a path with the minimum delay to expand and delay is closely related to wire length, paths with shorter wire lengths will have a higher chance to be selected and processed. Therefore, we can expect a reduction in wire length using our approach. A pseudo-code of this path expansion algorithm is shown in Algorithm 1.

4.1.1 Processing of Chain Paths

In the above path expansion algorithm, after a path p is taken into the routing solution, we will process all the paths that intersect with p in the algorithm. First of all, we will initialize a current routing tree T_p as the single path p and initialize a set $chain(p)$ with all the paths in H that intersect with p. The paths in $chain(p)$ are sorted according to their Elmore delays in a non-decreasing order. We will then do the following recursively until the set $chain(p)$ becomes empty. First, we will pick and remove a path p_1 from $chain(p)$ that has the smallest Elmore delay. We will then check if connecting p_1 to T_p will violate the delay or slew constraints for $port(p_1)$ as well as for all the existing ports in T_p. If yes, p_1 will be neglected and the next path in $chain(p)$ will be considered. Otherwise, p_1 will be added into T_p and all the paths originating from $port(p_1)$ will be removed from H. Furthermore, all the paths in H that intersect with p_1 will be added into $chain(p)$ recursively.

----

[4] Note that the capacitance and resistance of the vias are neglected here for simplicity. The same assumption was made in the previous work [10]. However, the via capacitance and resistance can be easily incorporated into our framework by considering them when computing the delay of a path.

Algorithm 1: Path Expansion Algorithm

```
1  begin
2      while H is not empty do
3          p = delete_min(H);
4          if p connects to source and d(p) ≤ D and
             s(p) ≤ S then
5              T_p ← p;
6              clean up H;
7              //remove all paths in H that originate
                //from port p
8              foreach p' intersects with p do
9                  chain(p) ← p';
10             end
11             while chain(p) is not empty do
12                 q = delete_min(chain(p));
13                 if adding q to T_p does not violate D and
                    S constraints then
14                     connect q to T_p;
15                     foreach p' intersects with q do
16                         chain(p) ← p';
17                     end
18                     clean up H;
19                     // remove all paths in H that
20                     // originate from port q
21                 end
22             end
23             Store T_p as one clock tree in the solution;
24         else
25             H ← expansion of p in selected directions;
26         end
27     end
28  end
```

4.2 Pre-processing to Connect Critical ports

The path expansion algorithm does not guarantee connecting all the ports to the sources successfully especially when the user specified constraints are too stringent. If there are critical ports (far away from sources or with very large port capacitance) which are harder to satisfy the requirements, it will be better to generate smaller trees for them first before handling others. Therefore, our post-grid clock routing algorithm involves iterations of the path expansion algorithm and will identify critical ports that fail to be connected to a source in a previous iteration. Those critical ports will be given higher priority to be processed in the next path expansion iteration such that smaller clock trees are more likely to be generated to connect them.

The pseudo-code in Algorithm 2 summarizes the overall flow of our approach. We create a set of critical ports P_c which is initialized as ϕ. We then enter the path expansion iterations in which we first execute the path expansion algorithm on the set of ports in P_c. This gives the critical ports a higher priority to be routed to the sources. We will then execute the path expansion algorithm on the remaining ports $P - P_c$. Notice that these remaining ports may also be connected to the trees constructed for the critical ports. After that, all the ports that cannot be routed to a source in this round will be added to P_c. Priorities also exist in P_c in which a higher priority is given to those most recently added

Algorithm 2: Main Program

```
1  begin
2      P ← all ports;
3      P_c ← φ; //critical ports
4      k=0;
5      repeat
6          Initialize H as P_c;
7          path expansion() with H initialized as P_c;
8          Initialize H as P - P_c;
9          path expansion() with H initialized as P - P_c;
10         P_c ← P_c + ports that fail to be connected to a
           source;
11         k ← k + 1;
12     until all sinks are connected or k > K;
13     if all sinks are connected then
14         Post-process;
15         //wire replacement and topology refinement
16     else
17         No solutions under current constraints;
18     end
19 end
```

Algorithm 3: Wire Replacement

```
1  begin
2      T_r ← all trees;
3      while T_r is not empty do
4          T_i ← select one tree in T_r;
5          P_l ← port with the largest Elmore delay in P_x;
6          P_x ← all terminal ports in T_i except P_l;
7          while P_x is not empty do
8              P_i ← port node in P_x with the smallest
               Elmore delay;
9              P_a ← lowest common ancestor of P_l and P_i;
10             repeat
11                 Replace e(P_i) using the second type of
                   wire if no violation occurs;
12                 P_i ← parent(P_i);
13             until ∃P_k ∈ T_i where d(P_k) > D or P_i = P_a;
14             P_x ← P_x - P_i;
15         end
16         P_i ← P_l, P_a ← tree root of T_i;
17         repeat steps 10-13;
18         T_r ← T_r - T_i;
19     end
20 end
```

ports. We repeat these steps until all the ports are connected or the number of iterations exceeds a user defined limit K.[5]

4.3 Post-processing to Reduce Capacitance

For all the data sets, there are two types of wires on each layer with capacitance and resistance tradeoffs[6]. The first type has higher capacitance but lower resistance per unit length, while the second type has lower capacitance but higher resistance per unit length. The per unit length delay of type-one wire is less than that of type-two wire on all the layers. In our path expansion algorithm, we will first just use type-one wire on all layers to optimize delay as much as possible. A post-processing step is then performed to reduce the total wire capacitance as long as the delay and slew constraints are maintained by changing the wire types. Two techniques, *wire replacement* and *topology refinement*, are invoked in this post-processing step.

4.3.1 Wire Replacement

This refinement process is done for the trees in the clock network one after another with the following steps. First, all the terminal ports in the current tree are stored in a port pool P_x in which they are sorted in a non-decreasing order of their Elmore delays, and the port P_l with the largest delay in the tree will be recorded. We will then sequentially explore all the ports in P_x. Without loss of generality, lets assume that the currently processing port is P_i, and node P_j is the parent node of P_i in the tree. We use $e(P_i)$ to denote the edge connecting P_i and P_j. We will then check whether any violation occurs if $e(P_i)$ is replaced by the second type of wire. If not, we will replace it with the second type of wire and set $P_i = P_j$. This step is repeated until the delay or slew constraint is violated at any port in the current tree, or when P_i becomes an ancestor of the node P_l (since we do not want to increase the largest delay in this tree). Port P_l will be finally explored after all other ports

---

[5] In this case, the algorithm fails to converge to a feasible solution. Note that this may happen when the delay or slew constraints are too stringent.

[6] Our algorithm can also handle the case that multiple types of wire are available on each layer.

Algorithm 4: Topology Refinement

```
1  begin
2      P_y ← all terminal ports;
3      sort(P_y) in a non-increasing order of their Elmore
       delays;
4      while P_y is not empty do
5          P_i ← a port in P_y;
6          modified path expansion() on P_i;
7          // Paths expand toward all directions, and the
8          //path with smallest wire capacitance will be
9          //expanded first
10         P_y ← P_y - P_i;
11     end
12 end
```

in the tree have been processed. In our implementation, the above process is repeated three times, as we find that for most test cases, running more iterations of this wire replacement process brings little or no capacitance reduction. The pseudo-code in Algorithm 3 details the flow of this wire replacement process.

4.3.2 Topology Refinemen

In the path expansion algorithm, we will expand a path p upwards as long as the end node of p is at a via connecting to the upper layer. Besides, chain paths are greedily processed as long as the delay and slew bounds are maintained. Thus, there are still chances to bring down the capacitance by changing the topology of the initially constructed trees. To achieve this, we will employ a topology refinement step on all the terminal ports as follows. First, we will sort all the ports that are terminal nodes in the trees in a non-increasing order of their Elmore delays in a port pool P_y. These ports will be processed sequentially in the algorithm. For any port P_i being processed, we will first disconnect P_i from the tree it is currently connecting to, and record the total wire capacitance C_b of the removed path p_i. A new path expansion algorithm will then be invoked at P_i which

Table 1: Comparisons with TG

Test Cases	No. Sinks	Capacitance (pf)				Wire Length (mm)				Delay (ps)	Runtime (s)		
		TG x_1	Ours1 x_2	Improvement $\frac{x_1-x_2}{x_1}$ %	Ours	TG y_1	Ours1 y_2	Improvement $\frac{y_1-y_2}{y_1}$ %	Ours		TG	Ours1	Ours
test1	300	3.3	2.6 (2.8)	20.9 (16.0)	2.3	12.6	10.0 (10.6)	20.1 (15.6)	10.6	0.45	0.02	0.23	0.20
test2	1846	13.7	9.7 (10.6)	29.2 (22.4)	5.0	42.9	32.3 (34.9)	24.8 (18.6)	34.2	1.15	0.10	2.53	2.68
test3	836	8.1	5.2 (5.8)	36.3 (28.2)	4.2	32.2	20.5 (23.1)	36.3 (28.5)	22.6	0.80	1.35	2.37	2.20
test4	502	5.3	4.0 (4.5)	23.7 (14.6)	1.7	12.4	9.5 (11.0)	23.0 (11.0)	10.6	1.35	0.03	2.81	2.91
test5	137	1.4	1.1 (1.2)	21.1 (15.7)	0.5	3.4	2.7 (3.1)	19.6 (10.5)	3.0	1.10	0.01	0.07	0.09
test6	724	7.9	5.7 (6.2)	27.1 (21.7)	2.5	18.8	14.2 (15.5)	24.7 (17.4)	15.3	1.25	0.05	0.57	0.67
test7	981	9.9	7.5 (8.2)	23.8 (17.2)	3.1	23.2	17.9 (19.9)	23.0 (14.1)	19.5	1.45	0.05	0.87	1.02
test8	538	5.9	4.5 (4.8)	24.5 (18.0)	1.9	14.1	10.8 (12.2)	23.7 (13.3)	11.9	1.80	0.04	0.41	0.50
test9	1915	19.9	14.3 (15.6)	28.3 (21.5)	5.5	46.1	33.2 (37.0)	28.1 (19.7)	35.6	2.75	0.13	2.98	3.38
test10	1134	10.7	8.6 (9.4)	19.4 (12.4)	3.4	25.8	20.2 (22.0)	21.9 (14.8)	21.4	1.90	0.09	6.72	6.88
test11	724	6.6	4.9 (5.3)	24.8 (18.9)	1.9	13.5	10.4 (11.3)	23.1 (16.5)	11.2	1.05	0.04	2.84	3.00
test12	225	2.5	2.0 (2.1)	20.2 (13.8)	0.9	6.3	4.9 (5.4)	22.0 (13.7)	5.4	1.30	0.01	0.13	0.17
test13	859	9.5	7.2 (7.6)	24.0 (19.3)	3.3	24.1	18.8 (20.4)	22.0 (15.4)	20.1	1.10	0.06	0.81	0.95
test14	366	3.9	3.1 (3.3)	20.7 (15.9)	1.4	9.5	7.8 (8.5)	18.4 (10.8)	8.4	0.95	0.04	0.25	0.29
Ave.	792	7.7	5.7 (6.2)	24.6 (18.3)	2.7	20.4	15.2 (16.8)	23.6 (15.7)	16.4		0.14	1.69	1.79

Note 1: Both TG and Ours1 use just type one wire on every layer.
Note 2: "Ours" represents our regular approach of using both types of wire on each layer
Note 3: The figures inside brackets denote the results before the post-processing techniques.

is different from the previous path expansion algorithm that (1) only the second type of wire will be used during the path expansion process, (2) paths will be expanded in all possible directions and (3) the path with the minimum wire capacitance (instead of the minimum wire delay) will be selected and processed first in the expansion process. New paths with wire capacitance less than C_b will be inserted into the heap. Once a path reaches a source or a tree (note that all trees are connected to sources now), we will check whether any violation occurs if the new path is taken. This new path will be taken if no violation occurs. Otherwise, we will continue the modified path expansion algorithm until another path reaches a source or a tree, or when all the paths are exhausted. If all the paths are explored but no path is successfully connected, we will simply restore the original path p_i. The above steps are repeated twice in our implementation. Algorithm 4 shows the flow of this topology refinement process.

4.4 Extension to Handle Large Load Capacitances

In practice, there are cases in which a small number of ports have exceptionally large capacitances that even its shortest direct connection to the nearest source will have a delay exceeding the limit D. To handle these special cases, we have extended our algorithm to first connect those *problematic* ports by a non-tree structure to several sources to bring down the delay to within the limit D. The non-tree structure is constructed by connecting the problematic port to more than one sources by several paths and by adding cross links between those paths.

Consider a particular problematic port P_e. After a path p_1 is taken into the routing solution, we will do the following steps to create a non-tree structure. First, we will expand from node n_l in the opposite direction of the path p_1 to find a nearest source. Let p_2 be the new path. p_2 will be taken into the routing solution if it helps in reducing the delay of P_e. Then, all the crosslinks between p_1 and p_2

(note that crosslinks can only exist at locations with reserved tracks) will be recorded and examined. The computational model in [2] is used to calculate the delays at the ports when crosslinks exist. All the crosslinks that can reduce the delay of P_e will be taken into the routing solution one by one until the delay and slew constraints are met, or when all the crosslinks are exhausted. If the delay and slew constraints are still not met with p_2 and all the crosslinks added, we will set $n_l = parent(n_l)$ and repeat the above steps recursively with one edge up the original path p_1 to find more sources and crosslinks.

After handling all the problematic ports, other ports will be handled as usual according to Algorithm 2. Note that we also allow other ports to connect to the non-tree structures, as long as the delay and slew constraints are not violated.

5. EXPERIMENT RESULTS

The path expansion algorithm proposed in this paper is implemented in C++ and all the experiments are carried out on a Linux machine with 4GB RAM and a Pentium 4 microprocessor running at 3.2GHz. We have also implemented the tree growing approach (TG) in [10] using C++ for comparisons. In the experiments, we assume that the slew of the source signals is 10ps, and the slew bound of the output signals is set to be 15ps. The first three test cases (test1-3) are provided by industry. The remaining eleven test cases are obtained from the circuits used in the ISPD 2010 Clock Network Synthesis Contest [1]. For the ISPD test cases which have no layer information given, five layers of reserved tracks are added according to the track conventions used in test 1-3.

5.1 Comparisons with the Tree Grow (TG) Approach

In the paper [10], Shelar proposed a tree growing algorithm to construct a clock network on reserved tracks in the context of post-grid clock routing . A pool F of frontier nodes, which is initialized to be all source nodes, is used to store all the current nodes to be expanded. The following steps are performed recursively until all the ports are connected to the sources. First, all unexplored edges adjacent to a frontier node in F are stored in an edge pool E_f and sorted in an ascending order of their edge capacitances. Then, the clock network is built by a greedy edge expansion process, in which all the edges in E_f are sequentially added into the clock network. Furthermore, the frontier node pool F will be updated with the end nodes of the newly added edges. After all the ports are connected to the sources, the final clock network is obtained by deleting redundant edges in the trees. Delay and slew constraints are considered in the algorithm.

Since the approach in [10] considers only one type of wire on each layer, for fair comparison, we compare the result of our approach using just the first type of wire on every layer (i.e., without the wire replacement step and use only type one wire in all the other steps) with the result of [10] using the first type of wire on every layer. In these experiments, we first get the lowest achievable delays obtained by TG empirically on all the test cases and use these delays as our delay bounds. The same slew limit is applied to both methods. The results are shown in Table 1. Column 3 and 7 show the total wire capacitance and the total wire length generated by TG. The results of our approach are shown in column 4 and 8. On average, our approach provides a 24.6% improvement in the total wire capacitance and a 23.6% improvement in the total wire length compared with TG respectively. The running times of both algorithm are shown in the last two columns. As we can see that though our approach is slower, the runtimes are still very practical. For all the test cases, the running times of our approach are within seconds. On average, the major path expansion algorithm, the topology refinement step and the wire replacement step take 74%, 17% and 9% of the total running time respectively. Note that in some cases, the running time of "Ours1" is even larger than that of "Ours" although "Ours1" does not perform the wire replacement step. This is because the inputs to the topology refinement procedure in "Ours" and "Ours1" are different as "Ours1" does not perform wire replacement. There are thus variations in the running times of the topology refinement step.

If we allow both types of wires on each layer, further reduction in wire capacitance can be obtained and the results are shown in column 6, 10 and 14 of Table 1. As we can see from the result, our approach can make good use of the availability of different wire types to further reduce the capacitance. For example, in test2, the wire capacitance can be reduced significantly by 49% (from 9.68pf to 4.92pf) with the wire replacement step.

5.2 Lowest Achievable Delay

Our approach can actually produce solution with better delay than the TG approach. We have run our algorithm on all the test cases to get the smallest achievable delays. The results are shown in Table 3. For almost all the test cases, we can further reduce the delays generated by TG. Take *test3* as an example, we can significantly reduce the delay from 0.80ps to 0.55ps, which shows an advantage of using our method in satisfying stringent user specified delay limits. In practice, designers may not know whether a delay limit is achievable for a circuit. Our approach can help in

Table 2: Non-tree Algorithm

Test Cases	C (pf)	WL (mm)	D (x ps)	T(s)	No. P.P.	D_{min} (y ps)	Imp. ($\frac{y-x}{y}$%)
ntest1	2.7	10.5	0.45	0.3	3	0.68	33.8
ntest2	10.1	33.6	0.45	4.7	3	0.71	36.6
ntest3	5.6	22.3	0.60	8.7	3	0.51	-18.5
ntest4	4.2	10.0	1.00	2.1	3	1.26	20.8
ntest5	1.2	2.9	1.03	0.1	3	1.29	20.3
ntest6	6.3	16.3	0.66	1.4	3	1.25	47.0
ntest7	7.6	18.1	1.35	1.6	3	2.02	33.3
ntest8	4.6	11.2	1.30	0.5	3	1.98	34.4
ntest9	14.5	33.9	2.00	25.0	3	2.42	17.4
ntest10	8.7	20.4	1.80	15.9	3	2.33	22.6
ntest11	5.1	11.1	0.80	6.8	3	1.24	35.4
ntest12	2.0	4.9	1.70	0.2	3	1.65	-3.0
ntest13	7.3	19.0	1.25	1.0	3	1.35	7.2
ntest14	3.4	8.8	0.58	0.8	3	0.98	40.8
Ave.	5.94	15.9		4.92	3		23.4

Table 3: Lowest Achievable Delays

Test Cases	Capacitance (pf)	Wire Length (mm)	Delay (ps)	Runtime (s)
test1	2.27	10.6	0.45	0.20
test2	6.08	34.6	0.47	4.69
test3	5.06	24.6	0.55	4.56
test4	1.75	10.2	1.00	2.12
test5	0.55	3.0	0.86	0.12
test6	2.83	15.5	0.83	1.01
test7	3.02	18.0	1.35	1.60
test8	1.86	11.2	1.32	0.50
test9	5.45	33.7	1.95	28.22
test10	3.28	20.1	1.67	22.27
test11	1.92	10.7	0.89	2.87
test12	0.91	5.1	1.12	0.22
test13	3.43	19.9	0.90	1.47
test14	1.48	8.4	0.67	0.40

determining the lowest achievable delay by embedding the algorithm in a binary search loop. This is possible since our approach will take the delay limit as an input constraint.

5.3 Results of the Non-tree Extension

To validate the effectiveness of our proposed non-tree algorithm, we further generate 14 test cases from the original ones (the new test cases have their names starting with an "n"). These new test cases are generated as follows. We first sort the ports according to their minimum Elmore delays, which is the delay when a port is connected to its nearest source directly. Then we increase the capacitances of the first three ports in the list so that their minimum delays increase by at least 50%. Detailed results on these new test cases are shown in Table 2.

Total capacitance, total wire length, delay limits, running time and number of problematic ports are shown in column 2-5 respectively. The delay limits D is got empirically for all test cases. The second last column D_{min} in Table 2 shows the minimum delay of the problematic ports when they are connected to the nearest source *directly*. Therefore, these are the lower bound delays achievable using a tree structure. We can see from the comparsion in the last column that our non-tree approach can reduce further the delay by 23.4% on average. For *ntest3* and *ntest12*, our non-tree algorithm does not help much and it automatically degenerates into the original path expansion algorithm (the result is thus a set of trees) because of the high density of the ports especially in the surroundings of the problematic ports. For all the other test cases, our proposed non-tree approach can successfully generate a solution in which the maximum port delay is less than the lower bound delay shown in the second last column. This clearly demonstrates the effectiveness of our proposed non-tree algorithm.

5.4 Simulation Results

We further validate our results using hspice simulation. The slew of the input signals are set to be 10ps. Detailed results are shown in Table 4 and Table 5. As we can see from the simulation results, The delay and slew we calculated is very close to the simulation results. For both tables, the correlation coefficient is over 99% between the simulated delay and calculated delay while it is over 94% between the simulated slew and calculated slew. This verifies the effectiveness of our method.

Table 4: Simulation Results for Tree

Test Cases	Calculated Results		Simulation Results	
	Delay (ps)	Slew (ps)	Delay (ps)	Slew (ps)
test1	0.45	10.05	0.45	10.07
test2	1.14	10.32	1.14	10.24
test3	0.80	10.15	0.80	10.15
test4	1.35	10.43	1.34	10.33
test5	1.09	10.29	1.09	10.25
test6	1.25	10.37	1.25	10.32
test7	1.43	10.50	1.43	10.52
test8	1.78	10.76	1.78	10.90
test9	2.75	11.69	2.70	11.43
test10	1.90	10.84	1.90	11.07
test11	1.05	10.26	1.05	10.23
test12	1.28	10.40	1.28	10.31
test13	1.08	10.29	1.08	10.24
test14	0.95	10.22	0.95	10.20

Table 5: Simulation Results for Non-Tree

Test Cases	Calculated Results		Simulation Results	
	Delay (ps)	Slew (ps)	Delay (ps)	Slew (ps)
ntest1	0.45	10.05	0.45	10.07
ntest2	0.45	10.05	0.45	10.07
ntest3	0.60	10.09	0.60	10.11
ntest4	1.00	10.24	1.00	10.21
ntest5	1.03	10.25	1.03	10.22
ntest6	0.66	10.10	0.66	10.13
ntest7	1.35	10.43	1.35	10.39
ntest8	1.29	10.40	1.29	10.33
ntest9	2.00	10.93	2.00	10.70
ntest10	1.80	10.76	1.80	10.97
ntest11	0.80	10.15	0.80	10.16
ntest12	1.70	10.68	1.70	10.81
ntest13	1.25	10.37	1.25	10.30
ntest14	0.58	10.08	0.58	10.10

6. CONCLUSION

In this paper, we present an efficient algorithm using the heap data structure to construct a post-grid clock network on reserved multi-layer metal tracks. We have compared our approach with the state-of-the-art algorithm and show that our algorithm can significantly improve over the previous work with a 24.6% reduction in wire capacitance and 23.6% reduction in wire length on average while maintaining very practical runtimes. We have also extended the algorithm to allow non-tree structures in order to handle the existence of ports with exceptionally large load capacitances and verified our results using hspice simulation. Our algorithm is expected to bring reduced energy consumption, improve grid-to-port delay in real post-grid clock networks.

7. REFERENCES

[1] *ISPD 2010 High Performance Clock Network Synthesis Contest.* http://www.sigda.org/ispd/contests/10/ispd10cns.html.

[2] P. Chan and K. Karplus. Computing signal delay in general rc networks by tree link partitioning. *IEEE Transactions on Computer-Aided Design of Integrated Circuits and Systems*, 9(8):898–902, Aug 1990.

[3] B. Chazelle. Filtering search: A new approach to query-answering. In *24th Annual Symposium on Foundations of Computer Science*, pages 122–132, 1983.

[4] B. Chazelle and H. Edelsbrunner. An optimal algorithm for intersecting line segments in the plane. In *29th Annual Symposium on Foundations of Computer Science*, pages 590–600, 1988.

[5] C. Kashyap, C. Alpert, F. Liu, and A. Devgan. Closed-form expressions for extending step delay and slew metrics to ramp inputs for rc trees. *IEEE Transactions on Computer-Aided Design of Integrated Circuits and Systems*, 23(4):509–516, April 2004.

[6] M. Mori, H. Chen, B. Yao, and C.-K. Cheng. A mulitple level network approach for clock skew minimization with process variations. In *Proceedings of Asia and South Pacific Design Automation Conference*, pages 184–187, 2000.

[7] D. Pham, T. Aipperspach, D. Boerstler, M. Bolliger, R. Chaudhry, D. Cox, P. Harvey, P. Harvey, H. Hofstee, C. Johns, et al. Overview of the architecture, circuit design, and physical implementation of a first-generation cell processor. *IEEE Journal of Solid-State Circuits*, 41(1):179–196, Jan. 2006.

[8] P. Restle, T. McNamara, D. Webber, P. Camporese, K. Eng, K. Jenkins, D. Allen, M. Rohn, M. Quaranta, D. Boerstler, et al. A clock distribution network for microprocessors. In *IEEE Journal of Solid-State Circuits*, pages 184–187, 2000.

[9] R. Shelar. An algorithm for routing with capacitance/distance constraints for clock distribution in microprocessors. In *Proceedings of the 2009 international symposium on Physical design*, pages 141–148, 2009.

[10] R. Shelar. Routing with constraints for post-grid clock distribution in microprocessors. *IEEE Transactions on Computer-Aided Design of Integrated Circuits and Systems*, 29(2):245–249, Feb. 2010.

Cross Link Insertion for Improving Tolerance to Variations in Clock Network Synthesis

Tarun Mittal
ECE Department, Purdue University, West
Lafayette, IN 47906
tmittal@purdue.edu

Cheng-Kok Koh
ECE Department, Purdue University, West
Lafayette, IN 47906
chengkok@purdue.edu

ABSTRACT

Cross links have been used to reduce skew variations in clock trees. In earlier studies cross links are inserted between the sinks of DC-connected trees. In this paper, we propose a link insertion scheme that inserts cross links between internal nodes of a clock tree. In addition to reducing the skew variability, the proposed approach also reduces the total cross link length. Our work also improves the correlation of sink delays for those sinks within a subtree that have similar path lengths to the cross link. Monte-Carlo (MC) simulations on the ISPD-2010 benchmarks showed that our work could handle variations effectively. In addition to meeting all the design constraints, the solutions produced by our approach have on the average 32% lower capacitance than the least capacitance obtained by the top three teams in the ISPD-2010 design contest [1].

Categories and Subject Descriptors

B.7.2 [**Hardware**]: Integrated Circuits-Design Aids

General Terms

Algorithms, Design, Performance.

Keywords

VLSI CAD, Physical Design, Clock Network, Non-tree Clocks, Cross Links

1. INTRODUCTION

The synthesis of a high quality clock network is an important step in the design of a synchronous VLSI system, as all the data transfer is coordinated by the clock signal. Considering the importance of such networks, many studies have been done on the clock network synthesis (CNS) problem [2, 3, 4, 5, 6, 7]. These studies focus on traditional parameters such as clock skew, buffer insertion, and power reduction. However, with the scaling of the VLSI technology, the power supply and wire width variations tend to

Permission to make digital or hard copies of all or part of this work for personal or classroom use is granted without fee provided that copies are not made or distributed for profit or commercial advantage and that copies bear this notice and the full citation on the first page. To copy otherwise, to republish, to post on servers or to redistribute to lists, requires prior specific permission and/or a fee.
ISPD'11, March 27–30, 2011, Santa Barbara, California, USA.
Copyright 2011 ACM 978-1-4503-0550-1/11/03 ...$10.00.

have a significant impact on the performance of the clock distribution networks. It has been observed that a non-tree structure performs better in reducing the skew variations as compared to a tree structure. Although clock meshes [8, 9, 10] may be effective in reducing skew variability, they have larger area and power overheads than tree structures.

Clock trees with cross links inserted have been proposed in [11, 12, 13, 14] as a cost effective alternative for reducing skew variability. [13, 14] address only unbuffered clock tree. The work of [11, 12] attempts to solve the problem by constructing a buffered clock tree with links inserted in it. The work of [11, 12] uses algorithms proposed in the work of [14] for the link insertion. Our studies in Section 2 show that the link insertion strategy in these studies has certain drawbacks. First, as the buffer levels increases in a clock tree, [11] might perform poorly as links at the sinks may not be sufficient to reduce the variations for all the buffer levels. Second, [11, 12] insert links at the sinks or the DC-connected sinks of a clock tree, which result in different path lengths from the sinks to the cross link.

In this work, we propose a different link insertion scheme for synthesizing a variation tolerant clock tree. This work attempts to overcome the above mentioned drawbacks of the link insertion strategy in [11, 12]. The important contributions of this work are:

- Our link insertion scheme inserts cross links between internal nodes in a clock tree. The end points of a link are selected in order to maximize the reduction of the skew variability on the sinks below that link.

- Our work improves the correlation of the sinks delays, as sinks in a subtree have similar path lengths to the cross link.

- Our link insertion process is integrated in the bottom-up phase and thus avoids the reconstruction of tree.

NGSPICE [15] based MC simulations show that our link insertion methodology is able to satisy all the constraints on the benchmarks released by the ISPD-2010 design contest [1]. The solutions produced by our approach have on the average 32% lower capacitance than the least capacitance obtained by the top three teams in the ISPD-2010 design contest [1]. We also implemented the link insertion strategy proposed in the work of [11] in order to compare it with our link insertion methodology. Our link insertion scheme performs better than the link insertion scheme of [11] in terms of reducing the skew variability.

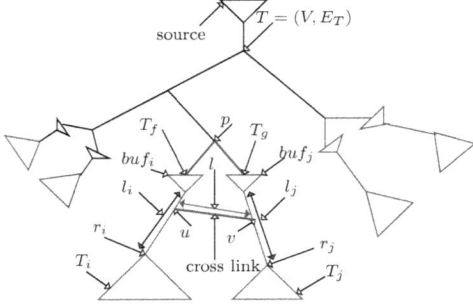

Figure 1: Link Inserted between nodes u and v.

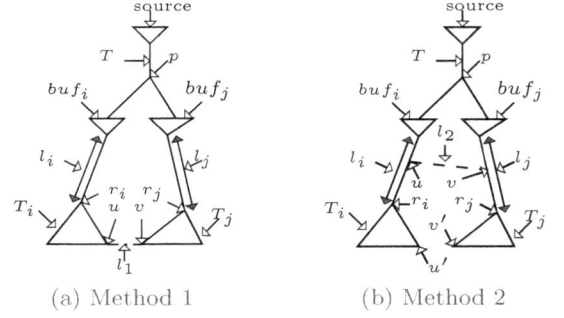

(a) Method 1   (b) Method 2

Figure 2: Clock trees for Method 1 and Method 2.

2. MOTIVATION

In this section we will first perform a qualitative analysis of our link insertion methodology. Next we will conduct experiments to validate our claim. Together, they provide us the motivation for a new link insertion methodology.

2.1 Qualitative Analysis

We first review the link insertion proposed in the work of [13, 14]. Consider a tree $T = (V, E_T)$, where node u is in the subtree $T_f \subset T$, and node v in the subtree $T_g \subset T$ as shown in Figure 1. The root nodes of subtrees T_i and T_j are r_i and r_j, respectively. Node p is called the nearest common ancestor (NCA) of nodes u and v. If a link is inserted between any two arbitrary nodes u and v as shown in Figure 1, the skew between u and v after the link insertion is given by [13]:

$$\hat{q}_{u,v} = \frac{R_l}{R_l + r_u - r_v} \left(q_{u,v} + \frac{C_l}{2}(R_{u,u} - R_{v,v}) \right), \quad (1)$$

where $\hat{q}_{u,v}$ is the final skew after the link insertion between the nodes u and v; $q_{u,v}$ is the original skew before the link insertion; R_l the link resistance; C_l the link capacitance; $R_{u,u}$ and $R_{v,v}$ are the sums of the resistances along the paths from the source to the nodes u and v, respectively; r_u and r_v are equal to Elmore delays at u and v when the node capacitances of u and v are 1 and -1, respectively, and all the other node capacitances are set to zero.

Let $R_{loop} = (R_l + r_u - r_v) = (R_l + r_{p,u} + r_{p,v})$, where $r_{p,u}$ and $r_{p,v}$ denote the sums of the resistances along the paths $p \rightsquigarrow u$ and $p \rightsquigarrow v$, respectively. Hence, R_{loop} is the total resistance along the loop $p \rightsquigarrow u \rightsquigarrow v \rightsquigarrow p$. Therefore, we can rewrite (1) as

$$\hat{q}_{u,v} = (\alpha q_{u,v} + \alpha \beta), \quad (2)$$

where $\alpha = \frac{R_l}{R_{loop}}$, and $\beta = \frac{C_l}{2}(R_{u,u} - R_{v,v})$.

According to (2), a link is more effective in reducing the skew variability when the corresponding α and β are smaller. Assume that we have zero-skew subtrees T_i and T_j with the root nodes r_i and r_j, respectively, as shown in Figure 1. Let us consider two methods of link insertion:

Method 1: Link l_1, with parameters $R_l = R_{l1}$, $\alpha = \alpha_1$, $\beta = \beta_1$, $r_{p,u} = r_{pu_1}$, and $r_{p,v} = r_{pv_1}$, is inserted between the two sink nodes u and v. Therefore, u is assumed to be a sink node in T_i, and v a sink node in T_j, as shown in Figure 2a. This method is used for the link insertion between

the sinks in [13, 14].

Method 2: Link l_2, with parameters $R_l = R_{l2}$, $\alpha = \alpha_2$, $\beta = \beta_2$, $r_{p,u} = r_{pu_2}$, and $r_{p,v} = r_{pv_2}$, is inserted between the two internal nodes u and v, as shown in Figure 2b. The nominal delay from the node u to all the sinks in the subtree T_i is the same as the nominal delay from the node v to all the sinks in the subtree T_j. This method will be used in our proposed link insertion methodology.

We will show that one can choose a link l_2 that performs better than l_1 in reducing the skew $\hat{q}_{u,v}$. Let us define $\gamma_1 = \frac{R_{l1}}{R_{l2}}$, $\gamma_2 = \frac{r_{pu_1}}{r_{pu_2}}$, and $\gamma_3 = \frac{r_{pv_1}}{r_{pv_2}}$. For a link l_2 to be more effective than a link l_1 in reducing the skew variability, we have to satisy $\alpha_2 \leq \alpha_1$, which according to (2), means

$$\frac{R_{l2}}{R_{l2} + r_{pu_2} + r_{pv_2}} \leq \frac{\gamma_1 R_{l2}}{\gamma_1 R_{l2} + \gamma_2 r_{pu_2} + \gamma_3 r_{pv_2}}.$$

We can rewrite the preceding equation as:

$$\frac{R_{l2}}{R_{l2} + r_{pu_2} + r_{pv_2}} \leq \frac{R_{l2}}{R_{l2} + (\frac{\gamma_2}{\gamma_1})r_{pu_2} + (\frac{\gamma_3}{\gamma_1})r_{pv_2}}. \quad (3)$$

Therefore, if we choose $\gamma_1 \geq \gamma_2$ and $\gamma_1 \geq \gamma_3$, we will be able to satisfy (3). In other words, l_2 can be made more effective than l_1 by selecting a much smaller link than the minimum length link l_1 between all sink pairs of subtrees T_i and T_j. Selecting a smaller link l_2 also makes $\beta_2 \leq \beta_1$, as β is proportional to the link length.

Let m and n be the two sink nodes with $q_{m,n}$ as the skew between them. The effect of inserting a link between the nodes u and v on the skew $q_{m,n}$ is given by [13]:

$$\hat{q}_{m,n} = q_{m,n} - \frac{r_m - r_n}{R_{loop}} q_{u,v}.$$

Similar to [13], we present the following scenarios to show the deviation of $\hat{q}_{m,n}$ from $q_{m,n}$ when a link is inserted between the nodes u and v.

Scenario 1: Both m and n are in the same subtree T_i or T_j. Assuming both m and n are in the subtree T_i. Let $PL_{m,u}$ and $PL_{n,u}$ denote the path lengths from the sinks m and n to u, respectively. In Method 2, $PL_{m,u}$ is similar to $PL_{n,u}$, implying that the link will have similar influence on the delays of the sinks m and n. In Method 1, however, depending on the path lengths of sinks m and n to sink u, $PL_{m,u}$ can differ greatly from $PL_{n,u}$ and therefore, the inserted link will have a greater influence on one sink than the other. Thus, link l_2 inserted in Method 2 will reduce the skew variability more as compared to link l_1 inserted in

Table 1: Effect of α on the skew variations.

TC	$l_{1_{min}}$ (nm)	α_1	$skew_1$ (ps)	$l_{2_{min}}$ (nm)	α_2	$skew_2$ (ps)
1	43844	0.06	2.39	11012	0.02	1.32
2	165121	0.22	1.62	73868	0.17	1.35
3	38550	0.12	2.43	8935	0.02	1.50

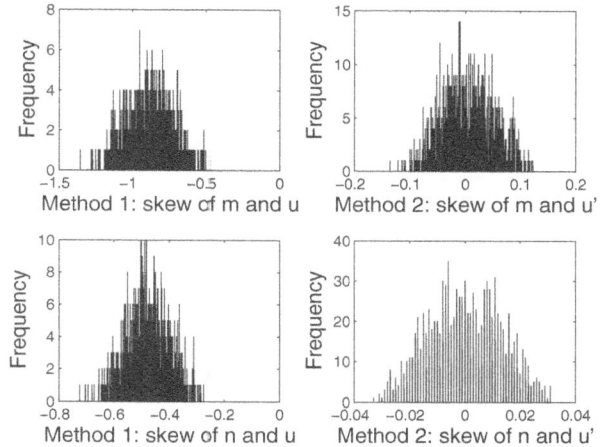

Figure 3: skew variability in sink nodes within a subtree.

Method 1.

Scenario 2: One of the nodes m and n is in subtree T_i and other is in subtree T_j. Assume that $m \in T_i$ and $n \in T_j$. In this scenario, adding a cross link between the subtrees T_i and T_j introduces a correlation between the delays of the sinks m and n, thus reducing the skew variability between them. In Method 2, all the sinks in T_i have similar delays and all the sinks in T_j have similar delays as shown in Scenario 1. Thus, adding the cross link will introduce a uniform correlation between any two sink pairs m and n in subtrees T_i and T_j, respectively. In Method 1, since the delays within the same subtree T_i and T_j are different to start with, the correlation due to the cross link will be different for each sink pair m and n, respectively. Hence, the link inserted in Method 2 will be more effective in reducing the skew variability as compared to the link inserted in Method 1.

Scenario 3: One of the m and n is in subnetwork T_p rooted at NCA p, and other node is not in T_p. For example, m is in T_p, and n is not in T_p. Since there is no overlap between the source-to-n path and T_p, there is no predictable correlation between the delays of nodes m and n. This holds true for both Method 1 and Method 2.

We hereby summarize the important conclusions drawn by us based on the analysis done above. Method 2 allows us to insert small cross links between internal nodes in a clock tree. By selecting a small cross link in Method 2, one can make α_2 smaller than the minimum α_1 of Method 1. This also makes β_2 smaller than the β_1, thereby making link l_2 more effective than link l_1 in reducing the skew variability. Moreover, l_2 has a uniform correlation with all the sink delays below that link as opposed to l_1, where correlation is dependent on the path length of sink node to the cross link l_1. Hence, the link inserted in Method 2 should be better at reducing the skew variability as compared to the link inserted in Method 1.

2.2 Quantitative Validation

Here we will present some experimental results to validate the qualitative analysis presented in Section 2.1.

Experiment 1: This experiment is designed to illustrate the effect of α on the skew variability. Initially two clusters of sink nodes are chosen for two subtrees T_i and T_j, and are zero-skew merged using Elmore delay model to form the root nodes r_i and r_j, respectively, as shown in Figure 2. Clustering of sink nodes into two groups is based on the local skew distance from the ISPD-10 benchmarks [1].

For Method 1, l_{1-min} is selected by choosing the closest sink node pairs between the two subtrees T_i and T_j. Both T_i and T_j are then reconstructed in order to account for the capacitance of link l_{-min} in zero-skew merging [13]. For both Method 1 and Method 2, assume a stem wire w_i of length l_i is placed between the root node r_i and the buffer buf_i as shown in Figure 2. Similarly, assume a stem wire w_j of length l_j is placed between the root node r_j and buffer buf_j.

The basis of the stem wire [16] insertion is explained later in Section 3.2.2, which is on buffer insertion. For link insertion in Method 2, nodes u and v are searched on stem wires w_i and w_j, respectively, such that the nominal delay from the node u to the sinks in subtree T_i is same as the nominal delay from the node v to sinks in subtree T_j. We pick u and v on wires w_i and w_j such that they satisfy (3). Let $l_{2_{min}}$ be the length of the cross link in Method 2. Buffers buf_i and buf_j are zero-skew merged to obtain a clock tree T as shown in Figure 2b. The results of this experiment are shown in Table 1. Variations in the form of wire width and power supply are applied and the results are obtained by running 1000 MC simulations. One can observe from Table 1 that we are able to find link $l_2 \ll l_1$, because of which α_2 is less than α_1 and so are the skew variations.

Experiment 2: The aim of this experiment is to observe the effect of the link on the delays of the sinks within a subtree, for example, sink delays in subtree T_i. In other words, this experiment demonstrates the Scenario 1 in the qualitative analysis. Let m and n be the two sink nodes in the subtree T_i with the largest and the smallest path lengths $PL_{m,u}$ and $PL_{n,u}$, respectively, as defined in Scenario 1 for Method 1. For the Method 2, these path lengths to the link end point u are similar. As node u denotes internal node in Method 2, let us define u' to be the sink node in Method 2 corresponding to sink u in Method 1, as shown in Figure 2b. Using the experimental results from Experiment 1, we plot the skew of m and n w.r.t. sink u in Method 1, and skew of sinks m and n w.r.t. sink u' in Method 2, as shown in Figure 3. Figure 3 shows that for Method 2, the spread of the skews of m and n are nearly the same, suggesting that both m and n have nearly the same skew variability w.r.t. sink u'. However, for Method 1, m has a skew distribution that is more spread out as compared to n, suggesting that m has a higher skew variability than n. Moreover, the skew distributions for Method 2 are narrower as compared to Method 1, meaning that Method 2 has a better skew variability than Method 1. These results are consistent with the qualitative analysis presented in Scenario 1.

Experiment 3: The objective of this experiment is to observe the effect of *change* in α on the skew variability. Start-

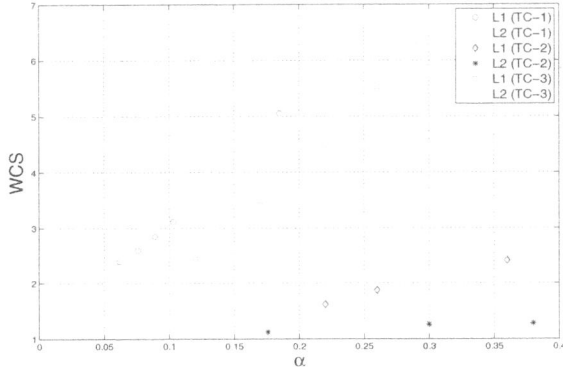

Figure 4: WCS variability with change in α for Method 1 and Method 2.

Table 2: Comparison of link above and below the buffer.

TC	Method	max total current (mA)	max buf_i current (mA)	max buf_j current (mA)	skew (ps)
1	no link	28.50	7.07	7.07	4.281
	link above	28.89	7.37	7.01	4.174
	link below	29.13	7.07	7.45	1.473
2	no link	28.02	7.29	6.86	4.455
	link above	28.68	7.20	7.06	4.090
	link below	28.41	7.18	7.27	1.823
3	no link	28.82	7.22	7.00	4.163
	link above	29.38	7.01	7.11	3.855
	link below	28.76	7.07	7.70	1.382

ing with the setup of Experiment 1, we modified it by adding detour lengths to $l_{1_{min}}$ and $l_{2_{min}}$ in order to vary α in both the cases. The results of this experiment are plotted in Figure 4. Worst case skew (WCS) refers to the maximum skew obtained for clock tree T after running MC simulations. We can observe that the gradient of curve for Method 1 is greater than that for Method 2. This means that even if we have a larger α in Method 2, it may not allow the skew to vary much. However having a high α in Method 1 is not favorable for the skew variability.

We also observed that the skew variability between nodes u and v remains minimal due to the direct effect of cross link in both Method 1 and Method 2. However, the effect of link on the skew variability in Method 1 is dependent on path lengths, as shown in Experiment 2. Therefore, in order to maintain a low WCS, we may have to insert several such links between the sub trees T_i and T_j in Method 1, thereby increasing the cost of link insertion.

Experiment 4 This experiment is designed to compare cross link inserted above the buffers, as in [12], versus cross link inserted below the buffers, as shown in Figure 2. The experimental setup is same as that for Experiment 1. The only change is that instead of inserting the links at the sink, we insert it above the buffers as in [12]. The results of skew variations are shown in Table 2. Since the total number of buffers in the clock trees for the three test cases are same, the total input current is nearly the same. The main issue with our link insertion strategy is the short circuit current through the buffers buf_i and buf_j, which may increase the overall power dissipation and degrade the network. Table 2 lists the maximum value of the currents drawn by the clock tree, buffer buf_i, and buffer buf_j on running trails of MC simulations. Clock tree with no links is considered as the base case for measuring the short circuit current. One can observe that the maximum current drawn by the buffers buf_i and buf_j are nearly the same for links inserted above the buffers and links inserted below the buffers.

3. CLOCK NETWORK SYNTHESIS

First, we will present formulation of the clock synthesis problem. Next, based on the qualitative analysis done in Section 2.1, we propose to construct clock tree with link insertion using Method 2 exclusively to address the problem.

3.1 Problem Formulation

The high performance CNS problem formulated by the ISPD-2010 design contest [1] serves as a good starting point for the problem formulation. Consider a layout of a chip having a simple synchronous circuit using edge-triggered flip flops (FFs), all having the same polarity for latching, as the sequential elements. We use $S = \{s_1, s_2, \ldots, s_n\}$ to denote the set of the clock pins of the flip flops in the circuit with s_i denoting the clock pin of FF_i, and S_o to denote the clock source. A valid clock network T connects S_o to S. In order to maintain the strength of the signal, a maximum slew constraint of M has been imposed on the signal reaching the FF_i. Further all the sinks pairs that have distance between them less than the local distance LD must satisfy the local clock skew constraint LCS.

The layout also has placement blockages denoted by $B = \{b_1, \ldots, b_m\}$, which may be in the form of other macros in the chip. Wire types denoted by $W = \{w_1, \ldots, w_k\}$, and inverters library $I = \{i_1, \ldots, i_l\}$ are the same as those given in [1]. For a solution to be considered valid, clock buffers cannot be placed within blockages, although clock wires can route over the blockages because the clock typically uses a dedicated metal layer. The total capacitance of the clock network is used as a measure for the total power dissipation.

3.2 Our Approach

Our approach comprises of three main steps: Merging, Buffer Insertion, and Link Insertion.

3.2.1 Merging

The general framework of CNS adopted by us is based on the deferred merge embedding approach, i.e. a bottom-up phase for zero-skew merging followed by a top-down phase for embedding [17, 2, 7].

We focus on the bottom-up phase, as the top-down phase is identical to [17]. Nearest Neighbour Graph (NNG) is constructed using root nodes of subtrees as the vertices. Initially, the vertices represent the clock sinks. The weight of an edge between any two vertices is equal to the length of the wire required to zero-skew merge those two nodes using Elmore delay. The clock tree is constructed iteratively. In each iteration, the two nodes connected with the least weight edge are merged together. NGSPICE simulations are used to fine tune the merging process, in order to reduce the nominal skew. After merging, NGSPICE is used to check whether the resulting subtree meets the slew constraints (explained in the next section). NNG is updated to remove the edges incident on those two nodes. If the slew constraints are met, new vertex and new edges corresponding to the parent node

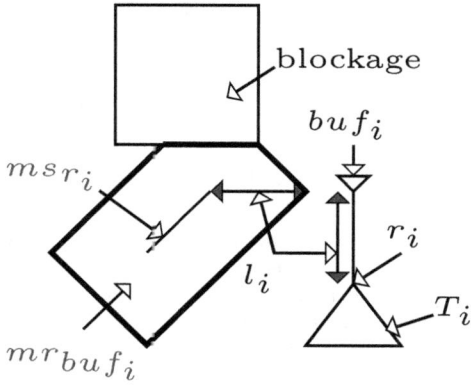

Figure 5: Merging region of buffer buf_i.

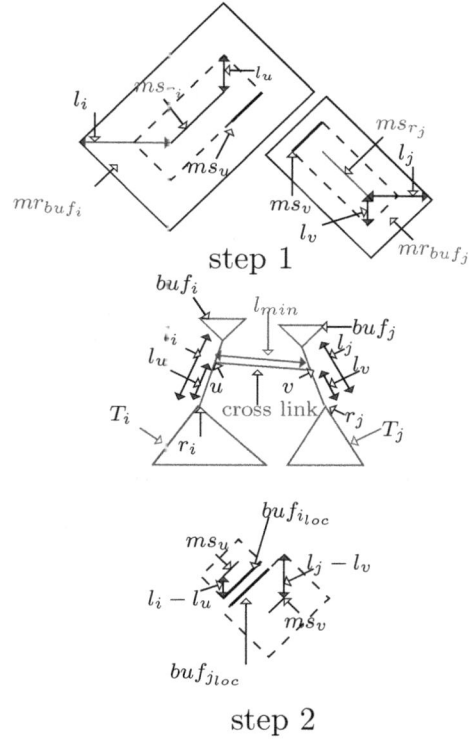

Figure 6: Link insertion strategy between node u and node v.

of the last two merged nodes are added to the NNG. Otherwise, buffer insertion is done, and the node is locked for further merging i.e. buffer inserted node is removed from the NNG. This process is repeated until all the nodes are locked. When that happens, all nodes are unlocked and a new NNG is constructed based by the buffered subtrees, and the iterative process is repeated until a clock tree is constructed.

3.2.2 Buffer Insertion

From now onwards, a subtree between any two consecutive levels of buffers is referred to as a DC-connected subtree. In order to keep the slew at the DC-connected sinks below the slew limit M, we may have to insert buffers, resulting in many levels of buffers in T. We will use buffers and inverters interchangeably unless mentioned explicitly. We use the same buffer size in our implementation of the clock tree.

We determine whether a buffer should be inserted similar to the buffer insertion strategy proposed in [11]. When two nodes a and b are merged to form a parent node q, we perform a NGSPICE simulation as follows: A buffer having an input slew of $slew_i < M$ is placed on top of the node q. If the slew rates at the sinks of the DC-connected subtree rooted at node q is greater than $slew_i$, nodes a and b are unmerged and buffers are inserted on top of nodes a and b using stem wires [16]. The length of stem wire is chosen such that the maximum slew at the sinks of the DC-connected subtrees a and b is $slew_i$.

The input slew $slew_i$ is chosen such that under variations it does not degrade to be more than the slew limit M. One can perform experiments on a DC-connected tree and obtain the value of safe $slew_i$. Once $slew_i$ is chosen, it should be kept constant for the entire CNS. For our CNS, we have chosen $slew_i$ as 70ps.

Now, we will describe the approach used to find the length of the stem wire. Let r_i with the merging segment ms_{r_i} be the root of a DC-connected subtree T_i. In order to satisfy $slew_i < M$ at the DC-connected sinks of T_i, we insert a stem wire w_i of length l_i on top of r_i before inserting a buffer buf_i as shown in Figure 5. It also shows the merging region mr_{buf_i} of buffer buf_i, which is a tilted rectangular region (trr) constructed on ms_{r_i} with merging radius l_i. In order to find the length l_i of stem wire w_i, we start with an initial guess that is an upper bound of the stem wire length.

Our initial guess is the length required by a buffer buf_m to drive another buffer buf_n, such that the input slews at both the buffers buf_m and buf_n are $slew_i$. This length can be found empirically depending upon the buffer size used for the CNS. Then, we perform a binary search for stem wire length, using NGSPICE simulation to guide the search process. Blockages are considered by chopping off the part of merging region mr_{buf_i} that lies within the blockage as shown in Figure 5.

3.2.3 Link Insertion

Our objective is to design a variation tolerant CNS. As discussed in [13], a link is beneficial only when it is inserted between any two zero skew or near zero skew nodes. Based on our analysis in Section 2.1, we propose to insert a link between the two zero skew nodes u and v on the stem wires w_i and w_j, connecting the root nodes r_i to buf_i and r_j to buf_j, respectively, as shown in Figure 1. According to Figure 5, buffers buf_i and buf_j have merging regions mr_{buf_i} and mr_{buf_j} associated with them depending on the inserted stem wire lengths l_i and l_j, respectively. The insertion of link is divided into 2 steps as shown in Figure 6. Let l_u and l_v denote the distances of nodes u and v from the root nodes r_i and r_j, respectively. Initially, we start with $l_u = 0.5l_i$.

Step 1: This step involves finding the merging segments of the nodes u and v, between which, the cross link is inserted. Starting with the initial l_u for node u, we perform a binary search on the length l_j of the stem wire w_j in order to find a node v, such that the delays from the node u to sinks in T_i is same as the delays from the node v to sinks in T_j. This process is guided by NGSPICE simulation. Based on the lengths l_u and l_v, merging segments of nodes u and v are chosen respectively as shown in step 1 of Figure 6, such that the length l_{min} of the cross link is minimum.

If a link is far too long, it has higher α, which is not favorable as discussed in Section 2. Therefore, we try a longer l_u, obtained by performing a binary search between the current l_u as minimum and the l_i as maximum length. Then, we find the new corresponding location for node v (with binary search again). Similarly, if a link is far too short, than based on our experience we have observed that it is difficult to achieve a low nominal skew at NCA node p, shown in Figure 1. Therefore, we try a shorter l_u, obtained by performing a binary search between the 0 as minimum, and the current l_u as maximum length. Then, we find the new corresponding location for node v (with binary search again). Links shorter than 5μm are treated as short links. Similarly, links larger than 900μm are treated as long links in our CNS. Both these values are obtained empirically based on the ISPD'10 wire and inverter parameters.

Step 2: Using the corresponding merging segments ms_u and ms_v for nodes u and v computed in step 1, we find the buffer locations $buf_{i_{loc}}$ and $buf_{j_{loc}}$ for buffers buf_i and buf_j, respectively, in this step.

Based on the merging radius $l_i - l_u$ for node u and $l_j - l_v$ for node v, $trrs$ are constructed on the merging segments ms_u and ms_v, respectively, as shown in step 2 of Figure 6. Solution is made obstacle-aware by chopping off the $trrs$ that lie within blockages, as discussed in the Section 3.2.2. We select the buffer locations $buf_{i_{loc}}$ and $buf_{j_{loc}}$ for buffers buf_i and buf_j, respectively, such that the distance between them is minimum. Other criteria, such as minimizing the detour between buffer nodes i and j on later merging, can also be used for selecting the buffer locations $buf_{i_{loc}}$ and $buf_{j_{loc}}$, respectively.

3.3 Overall Design Flow

This section summarizes the entire design flow discussed earlier.

1. NNG is constructed using roots of subtrees, as discussed in the Section 3.2.1.

2. Closest nodes are chosen to merge with each other and edges incident on those two nodes are deleted from NNG.

3. After merging, a test buffer is placed on the parent node and NGSPICE is used to check whether the slews at the inputs of DC-connected sinks are no greater than $slew_i$.

4. (a) If the slews at the inputs of DC-connected sinks are no greater than $slew_i$:
      i. NNG is updated to include the parent node.
      ii. Return to 2.
   (b) Else:
      i. The two nodes are unmerged and buffers are placed on top of these two nodes using stem wires as explained in Section 3.2.2.
      ii. Both these buffer inserted nodes are locked for any future merging. The nodes are removed from NNG.

5. 2 – 4 are continued till we have all the nodes locked.

6. Cross links are inserted between these locked nodes using the process described in Section 3.2.3.

7. All the nodes are unlocked.

8. 1 – 7 are repeated again till there is a single root node.

3.4 Merits of our Link Insertion Flow

1. One obvious advantage is that similar to [12], our link insertion flow allows us to control the link length, which is not possible in the other link insertion strategies. However in [12], cross links are inserted right at the buffers input, and as shown in Experiment 4 of Section 2.2, inserting a link below the buffer helps more in reducing the variations effects of buffer as compared to inserting above it. Moreover, our scheme works with ordinary buffer cells as opposed to the work of [12], which requires special tunable buffers.

2. A less obvious advantage of the proposed link insertion flow is that it helps in reducing the total capacitance of the clock tree. In our proposed link insertion flow, we choose to place the buffers closer to each other. This helps in reducing the buffer levels in the design. Consider a DC-connected subtree where the sinks are placed far apart to start with, because of the slew constraint M, the height of the root node of that DC-connected subtree will be less as compared to the case where the sinks are placed closer to each other.

3. Our proposed link insertion flow is integrated in the bottom-up phase and does not involves reconstruction of Tree T as is currently the case of [11, 12].

4. EXPERIMENTAL RESULTS

In this section, we compare the skew variability results of the solutions obtained through our proposed link insertion scheme with the top 3 teams of the ISPD-2010 design contest [1]. The proposed CNS flow is implemented as a C program, and the experiments are performed on the Linux machine with $2.2GHz$ Intel microprocessor and $2.9GB$ memory. The benchmark circuits are downloaded from [1]. In our experiments, we followed exactly the same as ISPD-2010 design contest [1]. Power supply has $\pm 7.5\%$, and wire width has $\pm 5\%$ variations. Inverter inv-2 [1], with input capacitance of 4.2fF; output capacitance of 6.1fF; and output resistance of 440Ω is used for the buffer sizing in our CNS. The buffer size used in our experiments consists of 10 inv-2 in the first layer, driving 40 inv-2 in the second layer. Each buffer has an input slew of 70ps at the nominal voltage. Wire type wire-1 [1], with $0.0001(\frac{\Omega}{nm})$ as unit wire resistance, and $0.0002(\frac{fF}{nm})$ as unit wire capacitance, is used for the interconnects in our CNS. For each solution, 500 trails of NGSPICE [15] based MC simulations are performed to obtain the 95% local clock skew. Table 3 compares our solutions with the top 3 teams of [1]. "LCS" refers to the local clock skew constraint set on the benchmark. The entries "med" and "95%" refers to the median and 95th percentile local clock skew, respectively, obtained on performing the MC simulations. The original names of benchmarks has prefix $ispd10cns$ attached to them, for example, the actual name of 01 is $ispd10cns01$. Table 3 shows that our solutions has been able to satisfy the LCS for all the benchmarks. The total

Table 3: Experimental results for the 8 benchmarks of the ISPD-2010 design contest [1]. For Contango and CNSRouter the reported results are from [18] and [19] respectively. Results for *med* were not available for [20].

BM	#sinks	LCS (ps)	Method	*med* (ps)	95% (ps)	Cap (fF)	Cap Ratio	CPU (s)
01	1107	7.5	Contango [1, 18]	5.16	7.01	198337	1.39	12015
			CNSRouter [1, 19]	5.07	7.23	1168104	8.19	675
			NTUclock [1]	6.64	8.66	293887	2.06	15
			work in [20]	-	7.16	445331	3.12	0.40
			our solution	5.06	7.32	142644	1.00	1092
02	2249	7.5	Contango [1, 18]	5.58	7.34	375863	1.41	25006
			CNSRouter [1, 19]	5.43	7.35	2099811	7.91	2140
			NTUclock [1]	8.10	10.73	832483	3.13	176
			work in [20]	-	7.33	933574	3.52	2.42
			our solution	5.81	7.42	265207	1.00	4314
03	1200	4.9	Contango [1, 18]	3.03	4.18	55861	1.52	3840
			CNSRouter [1, 19]	2.89	3.95	93965	2.56	21
			NTUclock [1]	6.76	8.63	167062	4.56	6
			work in [20]	-	4.88	183702	5.01	1.57
			our solution	2.74	4.49	36609	1.00	383
04	1845	7.5	Contango [1, 18]	3.26	4.46	71843	1.40	6075
			CNSRouter [1, 19]	5.27	7.25	125333	2.45	22
			NTUclock [1]	7.36	9.55	325206	6.36	58
			work in [20]	-	4.09	196337	3.84	0.27
			our solution	3.96	6.70	51070	1.00	934
05	1016	7.5	Contango [1, 18]	3.01	4.41	37690	1.49	2406
			CNSRouter [1, 19]	4.34	7.27	74084	2.94	10
			NTUclock [1]	5.17	6.98	130389	5.18	11
			work in [20]	-	3.81	89094	3.54	0.10
			our solution	2.16	4.78	25129	1.00	278
06	981	7.5	Contango [1, 18]	5.03	6.05	47810	1.46	2660
			CNSRouter [1, 19]	5.75	6.79	87390	2.67	46
			NTUclock [1]	408.72	416.62	2E+06	61.19	1
			work in [20]	-	7.49	160447	4.90	0.28
			our solution	4.73	6.41	32680	1.00	285
07	1915	7.5	Contango [1, 18]	3.41	4.58	72664	1.50	2351
			CNSRouter [1, 19]	4.18	5.97	128351	2.65	27
			NTUclock [1]	6.07	8.12	275597	5.70	66
			work in [20]	-	6.24	228243	4.72	0.30
			our solution	4.04	5.86	48316	1.00	818
08	1134	7.5	Contango [1, 18]	4.15	5.15	52490	1.60	1987
			CNSRouter [1, 19]	3.58	5.37	97421	2.97	19
			NTUclock [1]	5.85	7.64	165883	5.07	7
			work in [20]	-	5.47	228243	6.90	0.28
			our solution	3.41	5.07	32699	1.00	327

capacitance of the solutions obtained through our proposed approach lies within $62.5\% - 71.9\%$ of the least capacitance obtained by the top three teams of the ISPD-2010 design contest [1].

There was no easy way for us to compare our proposed approach directly with the link insertion methodology proposed in [11] as we did not have information about the buffer and wire models used in [11]. In order to circumvent this problem, we implemented the link insertion methodology proposed in [11] using the same buffer size and wire model as used in our CNS This enabled us to compare both link insertion schemes, it may not be exactly fair. Table 4 compares the link insertion strategy of our work with our implementation of the link insertion strategy proposed in [11]. One can observe that the link length of our work on an average is 57% less than the link length inserted using the strategy in [11], and still have lower skew variability. Further, one can also observe that the links inserted using the strategy in [11] perform poorly on benchmarks 01 and 02 as compared to other benchmarks. These benchmarks have relatively larger number of buffer levels in them. This suggests that as buffer levels increases, inserting links at the

Table 4: Comparison of our link insertion scheme with our implementation of the link insertion scheme proposed in work of [11].

BM	Method	Total link length (mm)	95% (ps)	Ratio of total link length to total wire length
01	work in [11]	52.34	10.92	0.1948
	our work	16.63	7.32	0.0516
02	work in [11]	67.10	9.81	0.1079
	our work	29.41	7.42	0.0505
03	work in [11]	3.079	5.42	0.0764
	our work	2.14	4.49	0.0545
04	work in [11]	5.25	5.06	0.0614
	our work	5.00	6.70	0.0419
05	work in [11]	2.42	5.25	0.0483
	our work	1.44	4.78	0.0298
06	work in [11]	6.596	6.91	0.1415
	our work	3.02	6.41	0.0710
07	work in [11]	7.934	7.15	0.1020
	our work	3.83	5.86	0.0491
08	work in [11]	3.930	7.24	0.0709
	our work	2.60	5.07	0.0534

sinks may not be sufficient for skew reduction under variations. We speculate that adding the cross links at the sinks of DC-connected subtrees [12] may help in that situation.

5. CONCLUSIONS

A new link insertion methodology, where links are inserted between pairs of internal nodes in a clock tree, is proposed in this work. The proposed methodology improves the correlation of sink delays for sinks that have similar path lengths to an inserted cross link. The effectiveness of the proposed link insertion methodology is validated using NGSPICE based Monte-Carlo simulations. Experimental results showed that our new link insertion strategy can handle the variations effectively.

6. ACKNOWLEDGMENT

The authors would like to thank Dr. Cliff Sze of the IBM Austin Research Laboratory for the evaluation of the results in the review process.

7. REFERENCES

[1] ISPD 2010 High Performance Clock Network Synthesis Contest website[Available Online] http://www.ispd.cc/contests/.

[2] R.-S. Tsay. "Exact zero skew". In *Proceedings of the IEEE/ACM ICCAD, Santa Clara, CA*, pages 336–339, November 1991.

[3] Y. P. Chen and D.F. Wong. "An Algorithm for Zero-Skew Clock Tree Routing with Buffer Insertion". In *Proceedings of the ED and TC, Pairs, France*, pages 230–236, March 1996.

[4] S. Pullela, N. Menezes, and L. T. Pillage. "Low power IC clock tree design". In *Proceedings of the CICC*, pages 263–266, May 1995.

[5] A. Vittal and M. Marek-Sadowska. "Low-power Buffered Clock Tree Design". In *IEEE Transactions on CAD*, volume 16, pages 965–975, September 1997.

[6] K. Wang and M. Marek-Sadowska. "Clock network sizing via sequential linear programming with time-domain analysis". In *Proceedings of the International Symposium of Physical Design, Monterey, CA*, pages 182–189, April 2003.

[7] M. Edahiro. "A clustering-based optimization algorithm in zero-skew routings". In *Proceedings of the ACM/IEEE Design Automation Conference*, pages 612–616, 1993.

[8] P.J. Restle, T.G. McNamara, D.A. Webber, P.J. Camporese, K.F. Eng, K.A. Jenkins, D.H. Allen, M.J. Rohn, M.P. Quaranta, D.W. Boerstler, C.J. Alpert, C.A. Carter, R.N. Bailey, J.G. Petrovick, B.L. Krauter, and B.D. McCredie. "A clock distribution network for microprocessors". In *IEEE Journal of Solid-State Circuits,*, pages 792–799, May 2001.

[9] N. A. Kurd, J. S. Barkatullah, R. O. Dizon, T. D. Fletcher, and P. D. Madland. "A multigigahertz clocking scheme for the pentium 4 microprocessor". In *IEEE Journal of Solid-State Circuits,*, volume 36, pages 1647–1653, November 2001.

[10] N. Bindal, T. Kelly, N. Velastegui, and K. L. Wong. "Scalable sub-10ps skew global clock distribution for a 90nm multi-GHz IA microprocessor". In *Proceedings of*

the IEEE International Solid-State Circuits Conference,*, pages 346–355, 2003.

[11] Anand Rajaram and David Z. Pan. "Variation Tolerant Buffered Clock Network Synthesis with Cross Links". *Proceedings of the ACM International Symposium of Physical Design*, pages 157–164, November 2006.

[12] G. Venkataraman, N. Jayakumar, J. Hu, P. Li, S. Khatri, A. Rajaram, P. McGuinness, and C. Albert. "Practical Techniques for Minimizing Skew and its Variation in Buffered Clock Networks". In *Proc. of the ICCAD, San Jose, CA*, pages 592–596, November 2005.

[13] Anand Rajaram, Jiang Hu, and Rabi Mahapatra. "Reducing Clock Skew Variability via Cross Links". *Proceedings of the ACM/IEEE Design Automation Conference*, pages 18–23, June 2004.

[14] A. Rajaram, D.Z. Pan, and J. Hu. "Improved Algorithms for Link-Based Non-Tree Clock Networks for Skew Variability". *Proceedings of the ACM International Symposium of Physical Design*, pages 55–62, April 2005.

[15] ngspice [Available Online] http://ngspice.sourceforge.net.

[16] Y.P. Chen and D.F. Wong. "An Algorithm for Zero-Skew Clock Tree Routing with Buffer Insertion". In *Proceedings of the 1996 European conference on Design and Test,*, 1996.

[17] T.-H. Chao, Y.-C. Hsu, J.-M. Ho, K. D. Boese, and A. B. Kahng. "Zero skew clock routing with minimum wire-length". In *IEEE Transactions on CS-ADSP*, volume 39, pages 799–814, November 1992.

[18] D. Lee, M.C. Kim, and I.L. Markov. "Low-power Clock Trees for CPUs". In *Proc. of the ICCAD, San Jose, CA*, November 2010.

[19] Linfu Xiao, Zigang Xiao, Zaichen Qian, Yan Jiang, Tao Huang, Haitong Tian, and Evangeline F.Y. Young. "Local Clock Skew Minimization Using Blockage-Aware Mixed Tree-Mesh Clock Network". In *Proc. of the ICCAD, San Jose, CA*, November 2010.

[20] X.W. Shih, H.C. Lee, K.H. Ho, and Y.W. Chang. "High Variation-Tolerant Obstacle-Avoiding Clock Mesh Synthesis with Symmetrical Driving Trees". In *Proc. of the ICCAD, San Jose, CA*, November 2010.

Synthesis of Low Power Clock Trees for Handling Power-supply Variations

Shashank Bujimalla
School of Electrical and Computer Engineering,
Purdue University, W. Lafayette, IN 47907, USA
soujimal@purdue.edu

Cheng-Kok Koh
School of Electrical and Computer Engineering,
Purdue University, W. Lafayette, IN 47907, USA
chengkok@purdue.edu

ABSTRACT

The International Symposium on Physical Design (ISPD) 2010 contest presents the challenge of synthesizing clock distribution networks that are tolerant to severe power-supply and wire-width variations. In particular, a robust clock network should satisfy the local clock skew (LCS) constraint, i.e., the clock skew between any pair of sequential elements that are closer than a user-specified distance is below a user-specified limit, even in the presence of variations. In this paper, we identify a few factors that help in tolerating these variations in clock trees. Our proposed clock tree router uses a two-stage flow to construct low-power clock trees for which the LCS constraints are met. Our clock tree router has been tested on ISPD'10 contest benchmark circuits. Extensive Monte-Carlo simulations showed that low power clock tree solutions could effectively handle variations, even when we imposed more stringent conditions in the experimental setup.

Categories and Subject Descriptors

B.7.2 [**INTEGRATED CIRCUITS**]: Design Aids

General Terms

Algorithms, Performance, Reliability

Keywords

VLSI CAD, Physical design, Clock trees, Variations

1. INTRODUCTION

Because of technology scaling and shrinking cycle time, the synthesis of clock distribution networks (CDNs) has become a challenging and important problem for high performance circuits. The clock skew between a pair of sequential elements is defined as the difference in the arrival times of clock signal at the two elements. The clock skews in a circuit have to satisfy certain constraints for the correct operation of the circuit. It becomes all the more difficult to satisfy the constraints in the presence of process and power-supply variations. The observed clock skew could deviate significantly from the nominal clock skew if these variations are not taken into account in the process of clock network synthesis (CNS). Buffers are typically inserted along the path from the source to a sink in order to improve the slew (generally defined as the time difference between 10 and 90 percent voltage levels) of the clock signal. However, the performance of buffers depends on process and voltage variations, making the CNS problem even more challenging.

CDNs are actively switching circuits and they dissipate a significant amount of power in VLSI design. The capacitance of CDN is generally used as a metric to measure the power it dissipates [1]. Non-tree solutions (e.g., [13], [12]) use redundant paths from clock source to sinks to handle variations. However, they typically dissipate more power than clock tree structures, which are relatively less tolerant to variations.

It is evident that clock tree structures can potentially be more economical than non-tree structures. This paper deals with the problem of clock tree construction, which we refer to as clock tree synthesis (CTS). In particular, we address the following important questions that arise in CTS: (1) What are the factors that can affect the robustness of the clock tree solutions; (2) Can clock tree structures meet the skew requirements in the presence of variations?

The skew requirement that we aim to satisfy is the local clock skew constraint, which is introduced in International Symposium on Physical Design (ISPD) 2010 contest. The clock skew between any pair of sequential elements that are closer than a user-specified distance is defined as the local clock skew (LCS). A valid CDN must have all LCSs below a user-specified limit [15]. This LCS constraint is imposed on a circuit that is subject to severe power-supply and interconnect variations. The interconnect variations are modeled using wire-width variations. In addition to this, blockage and slew constraints are also imposed on the circuit. We identify a few factors that have high impact on clock skew and could help in handling these variations in clock trees. We propose a two-stage blockage-avoiding clock tree router (CTR) to construct clock trees that meet the LCS and slew constraints.

Benchmark circuits from ISPD'10 contest [1], which are based on Intel and IBM microprocessor designs (scaled to 45nm technology), have been used to evaluate our CTR. Monte-Carlo (MC) simulations based on NGSPICE [2] are run to evaluate the performance of our clock trees. Our

Permission to make digital or hard copies of all or part of this work for personal or classroom use is granted without fee provided that copies are not made or distributed for profit or commercial advantage and that copies bear this notice and the full citation on the first page. To copy otherwise, to republish, to post on servers or to redistribute to lists, requires prior specific permission and/or a fee.
ISPD'11, March 27–30, 2011, Santa Barbara, California, USA.
Copyright 2011 ACM 978-1-4503-0550-1/11/03 ...$10.00.

CTR could satisfy the stringent LCS constraints of all the benchmark circuits for the MC simulations used in the contest [1], while having lower capacitance values than [10], [16] and [14], which are published by the top three teams of the contest. Moreover, we introduce more stringent conditions in the MC simulations and analyze the performance of clock trees constructed by our CTR under these conditions.

The remainder of this paper is organized as follows. In Section 2, we give the problem description and definition. In Section 3, we analyze and identify the factors that have high impact on clock skew and LCS. In Section 4, we describe the techniques used in the proposed CTR. We present the experimental set-up and results in Section 5 and draw our conclusions in Section 6.

2. PROBLEM DEFINITION

Consider a synchronous circuit, as in the ISPD'10 contest, using edge-triggered Flip Flops (FFs) as sequential elements. We use $S = \{s_1, s_2, \ldots, s_n\}$ to denote the set of clock pins of FFs in the circuit, with s_i denoting the clock pin of FF_i. A clock tree T connects clock source, denoted as s_0, to all clock pins in S. The clock tree may be synthesized using wires of 2 different widths $W = \{w_1, w_2\}$ and a set of 2 inverters $I = \{i_1, i_2\}$ of different drive-strengths. (A buffer is formed by a chain of two inverters. Unless otherwise specified, inverters and buffers are used interchangeably in the rest of the paper.) The clock signal slew should be below a certain slew limit M. The layout also has m placement blockages denoted by $K = \{b_1, b_2, \ldots, b_m\}$. The inverters used in the CTS cannot be placed in the regions occupied by K although wires can be routed over K. The capacitance of the clock tree is used as the metric to measure the power it dissipates.

The path delay from s_0 to any sink in S is also affected by variations in power-supply (ΔV) around the nominal power supply (V) and variations in wire-width (Δw) around the nominal wire-width (w_i). These variations are assumed to be uniformly distributed. If the Manhattan distance between s_i and s_j in S, denoted by $MD(i,j)$, is less than a user-specified value L, s_i and s_j form a local clock skew (LCS) pair. If the clock signal arrives at LCS pair s_i and s_j at times t_i and t_j respectively, the local clock skew (LCS) between s_i and s_j is defined as $LCS(i,j) = |t_i - t_j|$. The maximum local clock skew (MLCS) of a clock tree T, denoted as $MLCS(T)$, is defined to be the largest LCS in the entire tree:

$$MLCS(T) = \max_{MD(i,j) \leq L} |t_i - t_j|. \quad (1)$$

In particular, the ISPD'10 contest requires that the 95th percentile of $MLCS(T)$, denoted as $95\%MLCS(T)$, be kept below a user-specified LCS constraint C under power-supply variations and wire-width variations. For a more detailed description of the ISPD'10 contest, please refer to [1]. We define the 95th percentile of a random variable X (denoted as $95\%X$) as:

$$P(X < 95\%X) = 0.95. \quad (2)$$

3. ANALYSIS

In this section, we first derive an a model for estimating the $95\%MLCS(T)$. Based on the $95\%MLCS(T)$ model, we identify a few factors that have high impact on the clock skew under power-supply and wire-width variations.

3.1 Upper bound on $95\%MLCS(T)$

In a buffered clock tree, we refer to the interconnect load that a buffer directly drives, as a DC-connected subtree. The buffer and the DC-connected subtree that it drives, is called a buffer stage. Let T_D be a metric-free clock tree [8], i.e., all path delays from source to sinks are identical and all buffer stages have identical and independent delay distributions, with possibly overlapping paths from source to sinks. It has N sinks and all the path delays follow normal distribution with the same mean and variance. Let T_I be another metric-free clock tree that is identical to T_D, but with the assumption that these overlapping paths are considered to be independent. Let R_I be the variable denoting the clock skew in T_I and R_D be the variable denoting the clock skew in T_D. It was shown in [7] that

$$P(R_D < z) \geq P(R_I < z). \quad (3)$$

Consequently, the expected clock skew in T_D, denoted as $E(R_D)$, is no greater than the expected clock skew in T_I, denoted as $E(R_I)$. If z in (3) is taken as $95\%R_I$, the 95th percentile of R_I, we can show that $95\%R_I$ is no less than $95\%R_D$, the 95th percentile of R_D. Therefore,

$$95\%R_I \geq 95\%R_D, \quad (4)$$

which can be also written as:

$$95\%R_D = \alpha \cdot 95\%R_I, \quad (5)$$

where α takes a value between 0 and 1. Consider a clock tree T similar to T_D, except that T is not a perfectly zero skew clock tree like T_D, at nominal parameters. Let the nominal clock skew (defined as the clock skew at the nominal parameters) of T be NCS and the random variable denoting clock skew in T be R. The 95th percentile of R, denoted as $95\%R$, can be approximated as:

$$95\%R \simeq NCS + \alpha \cdot 95\%R_I \quad (6)$$

The asymptotic formulae for $E(R_I)$ and variance of R_I, $Var(R_I)$ are given in [8] when the number of stages in the clock tree is $\log_2 N$ and the variance of delay per buffer stage is σ_0^2. Although, these formulae apply only for metric-free tree, we believe that it is a fair approximation for balanced clock trees where the buffers in various parts of the tree drive similar loads. If the number of buffer levels is assumed to be B, the formulae are:

$$E(R_I) = \sigma_0\sqrt{B}\left[\frac{4\ln N - \ln\ln N - \ln 4\pi + 2C}{(2\ln N)^{1/2}} + O\left(\frac{1}{\log N}\right)\right] \quad (7)$$

$$Var(R_I) = \frac{\sigma_0^2 B}{\ln N}\frac{\pi^2}{6} + O\left(\frac{1}{\log^2 N}\right), \quad (8)$$

where $C (= 0.5772\ldots)$ is Euler's constant. If the sample set is large, we can approximate the distribution of R_I as a normal distribution. Then, $95\%R_I$ is given by:

$$95\%R_I = E(R_I) + 2\sqrt{Var(R_I)}. \quad (9)$$

By substituting (9) into (6), we obtain an estimate of $95\%R$ of clock tree, which would depend on the parameters N, B and σ_0. The estimation of α will be elaborated in Section 4.2. Eqn. (6) expresses the intuition that $95\%MLCS(T)$ would be low if the nominal clock skew NCS, number of buffer levels B, number of sinks N and delay variation σ_0 are all low.

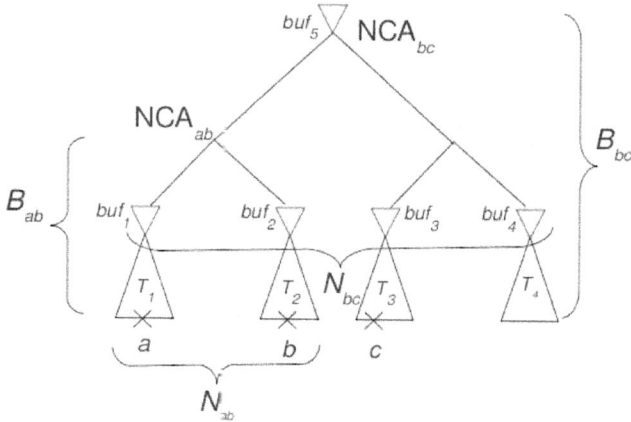

Figure 1: N and B **parameters for LCS.**

If we consider LCS instead of clock skew R, in Eqn. (6), the parameters N, B and σ_0 change as follows. Consider a LCS pair (e.g., sinks a and b in Fig. 1). The LCS between these sinks comes only from the subtree that is under their nearest common ancestor (NCA). The parameters, N and B, would respectively refer to the number of sinks and the number of buffer levels in this subtree. We assume that all the paths connecting the clock source to the sinks have the same number of buffer levels. This is a popular technique that is used for handling variations in buffered clock trees [6], which we use in the proposed CTR.

When a good (i.e., low) signal slew is to be maintained in a buffered clock tree, we would have short DC-connected subtrees. The impact of variations in these interconnects on the clock skew is relatively short compared to the impact of power supply variations. We validate this in Section 5.1. Hence, this paper handles interconnect variations by building in a small margin (e.g., 0.5 ps) when constructing a clock tree, i.e., the clock tree is constructed to satisfy the LCS constraint less the margin. The delay variations in a buffer stage, due to power-supply variations, comes mostly from the delay variations in the buffer. Therefore, we revise the definition of σ_0 as the standard deviation of buffer delay. Hence, the delay variations can be considered to be minimal in the subtrees below the first-level buffers (counting the buffer levels bottom-up, starting from the sinks). These first-level buffer outputs (like N_{bc} in Fig. 1) could be considered as the sinks since subtrees below them do not have any buffers. N is redefined as the number of such first-level buffers. NCS is the nominal LCS between the LCS pair under consideration.

Using the revised definitions of N, B and σ_0, we can estimate the 95th percentile of LCS, denoted as $95\%LCS$, between any LCS pair. We can then estimate the $95\%MLCS(T)$, which is the largest $95\%LCS$ among all the LCS pairs. N and B would then be determined by the LCS pair that has the highest $95\%LCS$.

3.2 Delay variations in buffer

PMOS and NMOS transistor characteristics change becuase of power-supply variations and this affects the CMOS inverter delay. We denote the delay of the inverter for rising and falling input transitions as $t_{pHL}(inv)$ and $t_{pLH}(inv)$

respectively. V_{dd} and V_{ss} are used to denote the voltages of power-supply and ground rail, respectively. If one of V_{ss} or V_{dd} increases, keeping the other constant, $t_{pHL}(inv)$ increases and $t_{pLH}(inv)$ decreases. Therefore, the delay of an inverter varies monotonically with variations in V_{ss} or V_{dd}. Consider a buffer (called buf) which is a chain of two inverters where $inv1$ is the driving inverter and $inv2$ is the second inverter. For a rising input signal, the delay of the buffer is denoted by $t_{pLH}(buf)$. It is given by:

$$t_{pLH}(buf) = t_{pHL}(inv1) + t_{pLH}(inv2). \qquad (10)$$

If one of V_{ss} or V_{dd} increases, keeping the other constant, $t_{pHL}(inv1)$ increases and $t_{pLH}(inv2)$ decreases. $t_{pLH}(buf)$ is determined by the resultant of these two delay changes which are of opposite signs. This way, a buffer delay variation is less than an inverter delay variation. Similar logic can be applied for a falling transition. Our first order analysis here shows that if both the inverters of a buffer get identical power-supply voltages, buffer delay would be more tolerant to power-supply variations than an inverter delay. In [4], a similar observation is made using MOSFET equations and it is empirically validated.

By intuition, if we want the buffer in a clock tree to have minimum σ_0 under power-supply variations, the function to minimize is

$$F = 0.5\Delta t_{pHL} + 0.5\Delta t_{pLH}, \qquad (11)$$

where Δt_{pHL} and Δt_{pLH} represent the maximum delay variations in t_{pHL} and t_{pLH}, respectively, in the presence of power-supply variations, for any given load.

F depends on the input slew and load of a buffer. It also depends on the size of the buffer and the choice of whether an inverter or a buffer (chain of two inverters) is used. We use F_{inv} and F_{buf} to denote F for inverter and buffer respectively. We performed NGSPICE simulations on the inverter library used in the ISPD'10 contest [1]. We make the following key observations, which are used to handle power-supply variations in our clock trees. (a) F_{buf} is generally less than F_{inv} for reasons discussed earlier. (b) F_{buf} and F_{inv} increase when input signal slew increases. The intuitive explanation for this observation is that changes in power supply have smaller effect on delays when signal slew is lower. (c) At low input slews, F_{buf} and F_{inv} do not vary significantly with buffer or inverter size. (d) At low input slews, F_{buf} does not vary significantly with load (typical loads at which buffers are inserted without resulting in slew violation) for typical buffer sizes.

Based on these observations, we cannot get any significant improvement in σ_0 of buffer by varying the buffer size. Moreover, inserting a specific buffer size at a specific load (where its F is minimum), may affect the input slew of subsequent buffer stage. This may degrade the σ_0 of subsequent buffer stage. For simplicity, we therefore try to use a single buffer size in our clock tree (see blockage avoidance in Section 4.2 for exception). However, different buffers have different drive strengths and input gate capacitances. Consequently, the selection of this buffer size has a huge impact on the parameters N and B. We also try to maintain a low signal slew in our clock trees in order to reduce σ_0 of buffers.

4. METHODOLOGY

In this section, we give an overview of the proposed CTR and discuss the various CTS techniques we use.

4.1 Overview of the proposed CTR

We observe that the selection of buffer size is the most important step that determines $95\%MLCS(T)$. Different buffer sizes may result in clock trees with different values of $95\%MLCS(T)$. An appropriate buffer size has to be chosen such that the clock tree meets the LCS constraint. We refer to such a buffer size as the desired buffer size. It is difficult to find the desired buffer size without constructing the clock tree, because the layout would generally have non-uniform sink distribution and blockages, and different clock tree topologies may arise when we use different buffer sizes. Since we want to find the desired buffer size from a buffer library, the CTR should be fast. It should also be accurate enough to achieve a low NCS. Therefore, we propose the following two-stage CTS flow.

We pre-characterize the buffer library to find σ_0 values for different buffer sizes. Given an inverter library, in general, finding the σ_0 values for various buffer sizes is a one-time effort. In the first stage, we use a fast but possibly inaccurate (in terms of NCS) clock tree router, called Fast Clock Tree Router (FCTR), to find the desired buffer size. In FCTR, we use Eqn. (6) to estimate the $95\%MLCS(T)$ of the constructed clock tree, when a particular buffer size is used in our clock tree. We perform linear search with various buffer sizes, starting with a typical buffer size (with reasonable drive strength), until a desired buffer size is found. In the second stage, we use a relatively slow but accurate clock tree router called Timing-Accurate Clock Tree Router (TACTR) and construct a low-NCS clock tree that uses the buffer size determined in the first stage.

4.2 CTS techniques

In this section, we discuss the CTS techniques that are used in both TACTR and FCTR. Other than buffer modeling and fine-tuning clock skew, FCTR and TACTR use the same CTS techniques. Therefore, FCTR provides a good estimate of the N, B and $95\%MLCS(T)$ of a clock tree T constructed using TACTR. We construct a zero skew clock tree at nominal parameters in order to minimize the NCS. We use a DME-based CTS paradigm [3], with certain modifications in the buffer insertion strategy and buffer modeling, in FCTR and TACTR. We highlight only these modifications in the remainder of this section.

Figure 2: Buffer insertion strategy.

We use the following buffer insertion strategy. Individual sub-trees are allowed to merge until slew constraint is violated (i.e., slew is greater than slew limit M) at the sinks of DC-connected sub-trees. When slew constraint is violated, the new sub-tree that is formed by merging is locked and is prevented from participating in further merging. When all the sub-trees in the circuit are locked, the locked nodes are removed. Buffers are then inserted at a certain distance (call stem-length) from the children of these locked nodes

where slew violations are imminent. Consider two nodes u and v with sub-trees T_u and T_v respectively that are merged to form node w with sub-tree T_w (see Fig. 2). When slew violation occurs at node w, T_w is locked. Node w is removed and buffers buf_1 and buf_2 are inserted on new nodes u' and v' above nodes u and v respectively. The locations of nodes u' and v' are chosen such that the distance between merging segments of u' and v' is shorter than that between u and v. Since u and v were merged previously to form w, this strategy increases the chances of merging of u' and v'. This helps in the reduction of wire-length, hence reducing the capacitance of the circuit. (The use of a wire between a buffer and a sub-tree has been used earlier in [5] for balancing the delays of all buffered subtrees). These buffer-inserted nodes are the new roots of the subtrees below them. This creates a new level of buffers and the subtrees are unlocked for further merging.

As discussed in Section 3.2, we try to maintain a low signal slew in our clock tree. However, the number of buffer levels B should not be increased while trying to maintain a low slew. We assume M as the buffer input slew and find a stem-length such that the clock slews at the sinks of DC-connected subtree are similar (e.g., differ by less than 10ps), and the worst case clock slew (among various combinations of power-supply voltages and input transitions) at these sinks is below the slew constraint M. We observe that this strategy helps in maintaining sufficiently low nominal signal slews at buffer inputs in the clock tree and the signal slews under variations are generally close to these slews. This also helps us to reduce B while meeting the slew constraint. The actual input slew of the buffer is not known while the clock tree is being constructed. We assume a particular buffer input slew (e.g., M) to estimate the buffer delay. If the actual buffer input slew is different from this, buffer delays across all buffers at same buffer level would change similarly since their input slews are similar. Therefore, a low nominal clock skew is still maintained.

Placement blockages K are taken into consideration while inserting buffers. If a Tilted Rectangular Region (TRR) [6], where a buffer has to be inserted, lies partially in K, then the portion of TRR that is inside K is chopped off and not considered for merging anymore (like the highlighted part in TRR_1 of Fig. 3). If a TRR lies completely in K, a buffer with higher drive strength (i.e., a bigger buffer) is used so that the TRR radius increases and it does not lie entirely inside K (TRR_2 changes to TRR_2' of Fig. 3). However, we observe that a bigger buffer is rarely used in our CTR. Our CTR can be easily modified to incorporate more rigorous blockage avoidance techniques like the ones described in [9].

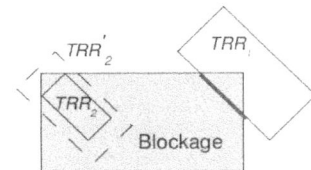

Figure 3: Blockage avoidance.

4.3 TACTR

In TACTR, we use NGSPICE [2] to model a buffer. However, if we simulate the entire subtree that a buffer drives using NGSPICE, the run-time of TACTR would be very high. Instead, we simulate, using NGSPICE, only the DC-connected subtree that the buffer directly drives. The sinks of these DC-connected subtrees could be buffers driving the next stage in the clock tree. These buffers are modeled using their input gate capacitance, so as to decrease the run-time of TACTR.

After merging any two sub-trees, we run NGSPICE to check if the clock skew is less than a specified limit (e.g., 0.20 ps). If the clock skew exceeds the limit, the edge-lengths connecting the two sub-trees to their parent are adjusted to reduce the clock skew. This adjustment is done for a specific number of iterations (e.g., 30) or until the clock skew comes entirely from either of these sub-trees. This is generally necessary when wire snaking has to be done for zero skew merge. The number of such fine-tunes and the number of iterations required to fine-tune NCS have been observed to be small. This is mainly because our buffer insertion strategy typically results in a balanced clock tree.

4.4 FCTR

In FCTR, we use an approach similar to [11], to estimate the delay and output slew of a buffer that drives a DC-connected subtree. [11] describes a fast iterative approach to estimate the delays and slews of various nodes in an interconnect tree that is driven by a buffer. This is done by modeling the interconnect tree that the buffer drives, as an effective load capacitance of the buffer. In order to model a buffer operated at voltage V, we maintain a look-up table (LUT) of buffer delay and output slew for different output load capacitances and input slews of concern. The values in the LUT are obtained by performing NGSPICE simulations. FCTR uses linear interpolation to obtain the buffer delay and output slew for general values of output load capacitance. Since the buffer model is not accurate, the accuracy of clock signal delays is not high when we use FCTR. Moreover, we do not fine tune the clock skew in FCTR. Therefore, NCS of clock trees constructed using FCTR is not low enough to satisfy the LCS constraints.

For a given σ_0, α in Eqn. (6) depends on N and B. We empirically estimated the value of α by performing MC simulations on clock trees having typical values of N and B for benchmark circuits from [1]. In general, an H-tree [6] can be used to build a look-up table of α for general values of N and B.

5. EXPERIMENTAL RESULTS

Table 1: Benchmark circuits of ISPD'10 contest [1].

BM	$|S|$	$|K|$	C (ps)	L (nm)	W (nm)	H (nm)
01	1107	4	7.50	600000	8000000	8000000
02	2249	1	7.50	600000	13000000	7000000
03	1200	2	4.99	370000	3071928	492989
04	1848	2	7.50	600000	2130492	2689554
05	1016	1	7.50	600000	2318787	2545448
06	981	0	7.50	600000	1949600	890880
07	1915	0	7.50	600000	2536640	1447680
08	1134	0	7.50	600000	1837440	1628160

The proposed CTS flow has been implemented as a C program. It has been evaluated with the ISPD'10 benchmark circuits [1] (Table 1), which are based on real Intel and IBM microprocessor designs, scaled to 45nm technology, on a 2GHz Intel CPU Linux workstation with 3GB memory. The inverter library (Table 2) and interconnect library used in CTS are the same as those used in the contest [1]. We use the wire with lower resistance-capacitance product per unit length in our CTR because it would help in decreasing the number of buffer levels in the clock tree. NGSPICE has been used to simulate the constructed clock trees and Monte-Carlo (MC) simulations have been performed to evaluate their performance. Wire-width variations upto $\pm5\%$ and power-supply variations upto $\pm7.5\%$, have been introduced into the above mentioned MC simulations. The magnitudes of these variations are the same as the ones in the contest. Uniform distribution has been used for both of them.

Table 2: Inverter library.

Name	Input cap	Output cap	Output res
inv-1	35 (fF)	80 (fF)	61.2 (Ω)
inv-2	4.2 (fF)	6.1 (fF)	440 (Ω)

Power-supply variations: Similar to the contest [1], placement of inverter is formulated as a "point" and more than one inverter can be placed at a single location. Although this is not physically realizable, the performance of a clock tree, which has such an inverter placement, would not be significantly different from the physical realization, if the variations in these inverters are modeled apppropriately. The power-supply variations of such inverters or buffers, which are placed at a single location, can be modeled in a couple of ways. All such inverters could get same or different power-supply variations. The Monte-carlo (MC) simulations of [1], wherein all the inverters placed at same location could have different power-supply voltages, are referred to as "ISPD MC simulations".

However, all such closely placed inverters would be expected to have similar power-supply voltages in reality. We therefore introduce another set of MC simulations called the "Single location - single voltage (SLSV) MC simulations", wherein all inverters placed at same location have the same power-supply voltage. When power-supply variations in the problem definition of Section 2 are modeled in the above ways, the problems are referred to as "ISPD problem" and "SLSV problem" respectively, from here onwards.

The MC simulations of [1] provide the entire power supply variation to power rail V_{dd}. However, power-supply is generally measured as the V_{dd} referenced relative to ground V_{ss}, which we call as effective V_{dd}. We could expect the power-supply variations to come from either of these rails. We perform various MC simulations in terms of how the power-supply variation is distributed to these rails. Type-1 MC simulations are the same as in [1], where V_{dd} gets all the power supply variation while V_{ss} is kept at a constant nominal value zero. Using the same seed of power-supply variations obtained from Type-1 MC, Type-2 MC simulations divide the variation equally between V_{dd} and V_{ss} whereas Type-3 MC simulations divide the variation randomly between V_{dd} and V_{ss}. Since the same seed is given to Type-1, Type-2 and Type-3 MC simulations, the effective V_{dd} for each inverter or buffer is the same in all these simulations. These 3 types of MC simulations could be performed for each of the earlier mentioned sets of MC simulations, i.e., ISPD MC and SLSV MC. Table 3 tabulates the various types of MC simulations. Note that SLSV MC simulations are still not realistic because inverters that are not placed very far from each other get different voltages.

Table 3: Different types of MC simulations.

Set	Type of MC	Name of MC simulation
ISPD MC	Type-1	Type-1 ISPD MC
	Type-2	Type-2 ISPD MC
	Type-3	Type-3 ISPD MC
SLSV MC	Type-1	Type-1 SLSV MC
	Type-2	Type-2 SLSV MC
	Type-3	Type-3 SLSV MC

Note: In the following tables, we represent the benchmarks (BM) by their number (e.g., ispd10cns01 is denoted as 01). "nom", "mean" and "max" denote the nominal, mean and maximum values respectively. 95th percentile is denoted using "95%". All clock skew values are measured in picoseconds (ps). "Cap" is the capacitance of the CDN, measured in picoFarad. Time represents the run-time of the tool in seconds. (*) is used to denote LCS constraint violation. 100 MC simulations have been run on the clock trees that are constructed using our work to get the values in tables of this section. Around 60 MC simulations are run for Table 8 and BM 02 in Table 6.

5.1 Wire-width variations in CTS

Table 4 shows the MC simulation results of our sample clock tree solution on BM circuit 05. The results of (a) MC with only wire-width variations, (b) Type-1 ISPD MC with only power-supply variations, (c) Type-1 ISPD MC with power-supply and wire-width variations, (d) Type-1 SLSV MC with only power-supply variations and (e) Type-1 SLSV MC with power-supply and wire-width variations, have been presented. We can see that the effect of wire-width variations on $MLCS(T)$ (and clock skew in general) is relatively small when compared to that of power-supply variations, in buffered clock trees.

Table 4: Wire-width vs power-supply variations.

	(a)	(b)	(c)	(d)	(e)
nom MLCS	1.63	1.63	1.63	1.63	1.63
95% MLCS	2.21	4.46	4.90	6.06	6.56

5.2 ISPD problem

The inverters in a buffer (chain of two inverters) do not get the same power supply voltage in the ISPD problem. Therefore, buffers may not have lower σ_0 than inverters. Consequently, inserting a buffer instead of inverter may not be effective in solving the ISPD problem. When many parallel inverters with possibly different power-supply voltages are placed at a single location, the delay distribution of the set of parallel inverters peaks at about the nominal delay and its σ_0 would decrease. For example, the delay distribution of 30 parallel inverters has a peak around 12ps as shown in Figure 4, while that of a single inverter would be uniform (not shown here). This figure was obtained with 1000 Type-1 ISPD MC simulation runs.

We use this observation to solve the ISPD problem. We insert 30 "inv-2" parallel inverters, as one single entity, in our CTS, to solve the ISPD problem. The first stage of our CTS flow estimates that a buffer size of 30 "inv-2" parallel inverters ($\sigma_0 \simeq 0.5$ps) would solve the ISPD problem for all

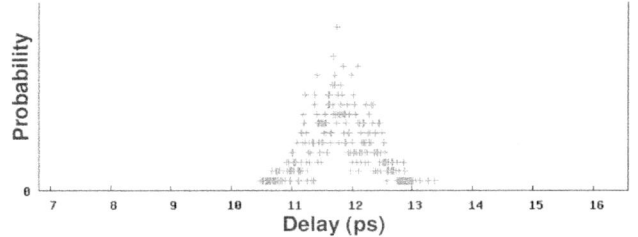

Figure 4: Delay distribution of 30 parallel inverters.

the BM circuits. In Table 5, the results of our CTR on ISPD problem are presented. The clock trees constructed from our work satisfy LCS constraints for all the BM circuits, while the run-time of our tool is less than 12 hours that was set in [1].

The top three teams of the ISPD'10 contest have presented their results in [10], [16] and [14]. Contango 2.0 [10] uses clock tree structrures while [16] and [14] use mesh structures. All three of them satisfy LCS constraints on ISPD'10 BM circuits. On the average, the capacitances of CDNs of [16] and [14] are atleast 2 times as much as the capacitances of clock trees generated by Contango 2.0. Therefore, we make a direct comparison of our clock trees with those generated by Contango 2.0 [10] in Table 5. On the average, the capacitances of clock trees of Contango 2.0 are 1.22 times as much as the capacitances of clock trees constructed using our work. We attribute these smaller capacitance values to the simple and effective techniques we used to construct balanced clock trees. Although it may not be a fair comparison, the runtime of Contango 2.0 is atleast 2 times (on an average) as much as the runtime of our tool.

5.3 SLSV problem

We observe that the SLSV problem could not be solved using inverters. This is because σ_0 of inverters is not low enough to meet the LCS constraint. The first stage of our CTS flow also estimates the same. In Table 5, we present the Type-1 SLSV MC simulation results on our clock trees that satisfied the LCS constraint for ISPD problem. We show here that using parallel inverters may not solve the SLSV problem, although it solves the ISPD problem. The results from [10], [14] on SLSV MC simulations are not available. If they have used parallel inverters to solve the ISPD problem, they may also fail when SLSV MC simulations are used. It is not necessary that such CDNs would satisfy the LCS constraint for SLSV MC simulations, unless their CTS process ensures to do so.

We observe that buffers could help in solving the SLSV problem because σ_0 of buffers is lower than that of inverters, in general. In fact, σ_0 of buffers could be low enough to meet the LCS constraint. We assume a single σ_0 value (found using NGSPICE simulations for a typical buffer input slew) for all buffer sizes and loads since it is observed to be nearly constant for low input slews. We use one set of parallel inverters driving another set of parallel inverters to form a buffer (e.g., 10 inverters driving 40 inverters). Here, parallel inverters are used only to size the buffer, so that N and B values of the clock tree could be reduced. In Table 6, the results of our CTR on SLSV problem are presented. It can be seen that our clock trees satisfy LCS constraints for

Table 5: Type-1 MC simulations on clock trees constructed to solve the ISPD problem using inverters.

BM	Contango 2.0 [10]					Our work								
	MLCS						MLCS							
	Type-1 ISPD MC			Cap	Time	nom	Type-1 ISPD MC			Type-1 SLSV MC			Cap	Time
	mean	max	95%				mean	max	95%	mean	max	95%		
01	2.51	5.16	7.01	198.33	12015	2.13	4.01	7.45	5.79	17.47	31.30	*25.76	177.46	2790
02	2.99	5.58	7.34	375.86	25006	2.67	4.98	7.50	6.69	20.29	29.54	*27.83	329.92	7787
03	1.50	3.03	4.18	55.86	3840	1.41	2.44	4.24	3.46	10.40	16.66	*14.54	50.81	2094
04	2.07	3.26	4.46	71.84	6075	1.54	2.84	4.21	3.79	12.18	23.41	*18.13	57.44	2763
05	1.50	3.01	4.41	37.69	2406	1.99	2.72	4.69	3.68	8.94	16.37	*13.35	28.93	1100
06	4.29	5.03	6.05	47.81	2660	2.32	3.03	4.69	4.01	11.19	19.63	*15.28	36.12	1142
07	2.22	3.41	4.58	72.66	2351	2.83	3.81	5.91	5.65	12.12	18.80	*16.46	57.93	2968
08	3.42	4.15	5.15	52.49	1987	1.73	2.89	5.13	4.24	12.12	19.09	*16.34	40.43	1498
Avg.				1.22×									1.00	

Table 6: Type-1 MC simulations on clock trees constructed to solve the SLSV problem using buffers.

BM	[16]					Our work								
	MLCS						MLCS							
	mean	max	95%	Cap	Time	nom	Type-1 ISPD MC			Type-1 SLSV MC			Cap	Time
							mean	max	95%	mean	max	95%		
01	5.07	9.43	7.23	1168.10	675	1.47	4.21	8.60	5.58	6.96	11.40	*10.29	189.06	2324
02	5.43	8.99	7.35	2099.81	2140	1.42	4.79	7.06	6.77	8.30	14.19	*12.30	341.08	6723
03	2.89	4.23	3.95	93.96	21	0.64	1.96	3.42	2.96	3.47	5.80	4.95	69.15	1269
04	5.27	7.64	7.25	125.33	22	0.81	3.38	7.34	5.69	5.27	8.32	7.17	56.59	2711
05	4.34	9.40	7.27	74.08	10	0.81	2.32	5.27	3.67	3.64	5.64	5.00	26.25	1057
06	5.75	8.04	6.79	87.39	46	0.66	2.80	5.94	4.58	4.25	6.40	5.97	32.57	1027
07	4.18	6.67	5.97	128.35	27	1.09	3.20	6.29	4.91	5.03	8.99	7.07	56.13	2917
08	3.58	6.62	5.37	97.42	18	0.94	3.10	5.29	4.83	4.60	7.39	6.53	37.40	1427
Avg.				2.33×									1.00	

6 out of the 8 BM circuits, on SLSV MC simulations. The first stage of our CTS flow could not find a buffer size that could solve SLSV problem on BM circuits 01 and 02. The clock trees for these 2 BM circuits could not satisfy LCS constraint on SLSV MC simulations. The first stage of our CTS flow predicted this without performing expensive MC simulations. We may have to use more rigorous CTS techniques to solve the SLSV problem on these 2 BM circuits. We compare our SLSV MC simulation results with [16]. [16] add only a single inverter at any specified location, which would mean that their ISPD and SLSV MC simulation results would be the same. The difference here is that [16] add only a single inverter while we add multiple inverters (having same power-supply voltages) at any specified location. On the average, The capacitances of CDNs of [16] are 2.33 times as much as the capacitances of our clock trees, when we consider only those trees that satisfy LCS constraints (i.e., BM 03–08).

We observe that the clock trees required to solve the SLSV problem could have higher or lower capacitances than those required to solve the ISPD problem using inverters in our CTS (i.e., the ones in Table 5). The MLCS statistics of these clock trees on ISPD MC simulations are also presented in Table 6, which show that buffers can also be used to solve the ISPD problem, for all the BM circuits.

5.4 Model validation

In Table 7, we empirically validate the model proposed in Section 3.1. F95% denotes the $95\% MLCS(T)$ obtained using Eqn. (6). 95% denotes the observed $95\% MLCS(T)$ obtained

Table 7: Model validation.

BM	MLCS					
	Type-1 ISPD MC using inverters			Type-1 SLSV MC using buffers		
	nom	F95%	95%	nom	F95%	95%
01	2.13	5.67	5.79	1.47	10.41	10.29
02	2.67	6.84	6.69	1.42	12.17	12.30
03	1.41	3.45	3.46	0.64	4.76	4.95
04	1.54	3.83	3.79	0.81	5.98	7.17
05	1.99	4.11	3.68	0.81	4.87	5.00
06	2.32	4.15	4.01	0.66	4.07	5.97
07	2.83	5.13	5.65	1.09	6.21	7.07
08	1.73	3.77	4.24	0.94	5.17	6.53

by performing MC simulations on clock trees constructed using our work. We can see that Eqn. (6) gives a good estimate of the observed $95\% MLCS(T)$.

5.5 Different types of MC simualations

In Table 8, we show the results of different types of MC simulations on BM circuits 05 and 06 when we use inverters or buffers (chain of two inverters) in our CTS. When ISPD MC or SLSV MC simulations are run on clock trees that have inverters, $95\% MLCS(T)$ for Type-1 MC is higher than that for Type-2 and Type-3 MC. This trend is also observed when ISPD MC simulations are run on clock trees that have buffers. However, when SLSV MC simulations are run on clock trees that have buffers, $95\% MLCS(T)$ for Type-1 MC is similar to that for Type-2 and Type-3 MC. The reason for this is as follows.

Table 8: 95% MLCS for various MC simulations.

BM	Use	Set	Type-1	Type-2	Type-3
05	Inv	ISPD MC	4.86	3.65	3.54
	Inv	SLSV MC	15.06	5.52	9.58
	Buf	ISPD MC	4.36	2.19	3.09
	Buf	SLSV MC	6.64	6.01	6.12
06	Inv	ISPD MC	5.75	4.53	4.84
	Inv	SLSV MC	16.44	6.17	11.43
	Buf	ISPD MC	3.83	1.96	2.52
	Buf	SLSV MC	5.62	5.12	5.22

Consider an inverter, with rising input signal, that is subject to variation in V_{dd}, like in Type-1 MC. Keeping the same effective V_{dd}, distribute the variation to both V_{dd} and V_{ss}. For example, an increase in effective V_{dd} is distributed as a smaller increase in V_{dd} and decrease in V_{ss}. $t_{pHL}(inv)$ increases with increase in one of V_{ss} or V_{dd}. As a result, an increase in V_{dd} and decrease in V_{ss} would result in a smaller change in $t_{pHL}(inv)$ than when V_{dd} gets all the variation. This would lead to a smaller σ_0 and hence a smaller $95\% MLCS(T)$. The inverters in a buffer get different power-supply voltages in ISPD MC simulations. They would behave like separate inverters and hence the same explanation applies to them. Now, consider a buffer instead of an inverter and assume that the same power-supply voltage is given to both its inverters, like in SLSV MC. $t_{pLH}(buf)$ is the sum of $t_{pHL}(inv1)$ and $t_{pLH}(inv2)$ (see Eqn. (10)). Using a similar explanation that was given for an inverter, the variation in $t_{pHL}(inv1)$ or $t_{pLH}(inv2)$ gets smaller when we distribute the variations between V_{dd} and V_{ss}. But these changes have opposite polarities (because the two inverters have different transitions) and they negate each other. As a result, σ_0 of buffer remains nearly the same as that when V_{dd} gets all the variation, and hence a similar $95\% MLCS(T)$ is observed.

In general, we observe that the $95\% MLCS(T)$ of Type-1 MC is the highest among all the types of MC simulations. Our clock trees that satisfy the LCS constraints for Type-1 MC, would therefore satisfy LCS constraints for Type-2 MC and Type-3 MC simulations as well.

6. CONCLUSION

In this paper, we have identified a few factors that help in tolerating power-supply variations in clock trees. We proposed a two-stage flow to construct clock trees that meet the LCS constraints, while handling blockage and slew constraints. We constructed low power clock trees that solved the ISPD problem. We also introduced more stringent conditions into MC simulations and analyzed the performance of our clock trees, thereby showing that low power clock tree solutions could be used to handle variations even when stringent LCS constraints and conditions are imposed.

Although the techniques proposed in this work are based on power-supply and wire-width variations, we could extend them for other types of variations. We could use the technique (used to solve ISPD problem) of inserting parallel inverters to handle variations that are random or uncorrelated in nature. Similar to power-supply variations, we can analyze other sources of variations like transistor length, threshold voltage and their impact on clock skew. We can incorporate the delay variations that are caused by them into the

σ_0 parameter, and can estimate the clock tree performance without performing expensive MC simulations.

7. REFERENCES

[1] ISPD 2010 High Performance Clock Network Synthesis Contest website. [Available Online] http://www.sigda.org/ispd/contests/10/ispd10cns.html.

[2] NGSPICE http://ngspice.sourceforge.net/.

[3] K. Boese and A. B. Kahng. Zero-skew clock routing trees with minimum wirelength. In *ASIC'92*, pages 1–1.

[4] L. H. Chen, M. Marek-sadowska, and F. Brewer. Coping with buffer delay change due to power and ground noise. In *Proc. DAC'02*, pages 860–865.

[5] Y. P. Chen and D. F. Wong. An algorithm for zero-skew clock tree routing with buffer insertion. In *EDTC'96*, page 230.

[6] C.-K. Koh, J. Jain, and S. F. Cauley. Synthesis of clock and power/ground networks. In L.-T. Wang, Y.-W. Chang, and K.-T. Cheng, editors, *Electronic Design Automation: Synthesis, Verification, and Test*, chapter 13. Morgan Kauffman, 2009.

[7] S. Kugelmass and K. Steiglitz. A probabilistic model for clock skew. In *Proc. Int. Conf. Systolic Arrays '88*, pages 545–554.

[8] S. Kugelmass and K. Steiglitz. An upper bound on expected clock skew in synchronous systems. *Computers, IEEE Trans. on '90*, 39(12):1475–1477, Dec.

[9] D. Lee and I. Markov. Contango: Integrated optimization of SoC clock networks. In *DATE'10*, pages 1468–1473.

[10] D.-J. Lee, M.-C. Kim, and I. Markov. Low-power clock trees for CPUs. In *ICCAD'10*, pages 444–451.

[11] R. Puri, D. S. Kung, and A. D. Drumm. Fast and accurate wire delay estimation for physical synthesis of large ASICs. In *GLSVLSI'02*, pages 30–36.

[12] A. Rajaram, J. Hu, and R. Mahapatra. Reducing clock skew variability via cross links. In *Proc. DAC'04*, pages 18–23.

[13] P. J. Restle, T. G. Mcnamara, D. A. Webber, P. J. Camporese, K. F. Eng, K. A. Jenkins, S. Member, D. H. Allen, M. J. Rohn, M. P. Quaranta, D. W. Boerstler, C. J. Alpert, C. A. Carter, R. N. Bailey, J. G. Petrovick, B. L. Krauter, and B. D. Mccredie. A clock distribution network for microprocessors. *IEEE Journal of Solid-State Circuits*, 36:792–799, 2001.

[14] X.-W. Shih, H.-C. Lee, K.-H. Ho, and Y.-W. Chang. High variation-tolerant obstacle-avoiding clock mesh synthesis with symmetrical driving trees. In *ICCAD'10*, pages 452–457.

[15] C. N. Sze. ISPD 2010 high performance clock network synthesis contest: benchmark suite and results. In *ISPD'10.*, pages 143–143.

[16] L. Xiao, Z. Xiao, Z. Qian, Y. Jiang, T. Huang, H. Tian, and E. Young. Local clock skew minimization using blockage-aware mixed tree-mesh clock network. In *ICCAD'10*, pages 458–462.

RegularRoute: An Efficient Detailed Router with Regular Routing Patterns

Yanheng Zhang and Chris Chu
Iowa State University
Dept. of Electrical and Computer Engineering
Ames, Iowa, U.S.A.
zyh@iastate.edu, cnchu@iastate.edu

ABSTRACT

Detailed routing is an important phase of realizing exact routing paths for optimizing various design objectives and satisfying increasingly complicated design rules. In this paper, we propose RegularRoute, a fast detailed router trying to use regular routing patterns in a correct-by-construction strategy for better routability and design rule satisfaction. Given a 2-D global routing solution and the underlying routing tracks, we generate a detailed routing solution in a bottom-up layer-by-layer manner. For each layer, the routing tracks are partitioned into a number of panels. We formulate the problem of assigning global segments into different tracks of each panel as a Maximum Weighted Independent Set (MWIS) problem. We propose a fast and effective heuristic to solve the MWIS problem. Then unassigned segments after MWIS are partially routed by a greedy technique. For the unrouted portion of each segment, its terminals are promoted so that the assignment is deferred to upper layers. At top layers, we apply panel merging and maze routing techniques to achieve better routability. Due to the unavailability of academic detailed routing benchmarks, we proposed two sets of detailed routing testcases derived from ISPD98 [1] and ISPD05 [2] placement benchmark suites respectively. The experimental results demonstrate the effectiveness and efficiency of RegularRoute.

Categories and Subject Descriptors

B.7.2 [**Hardware**]: Integrated Circuits—*Design Aids*

General Terms

Algorithms

Keywords

VLSI CAD, Detailed Routing

1. INTRODUCTION

Because of the problem complexity, VLSI routing is usually divided into global routing and detailed routing. In the global routing stage, rough routing decisions are made based

Permission to make digital or hard copies of all or part of this work for personal or classroom use is granted without fee provided that copies are not made or distributed for profit or commercial advantage and that copies bear this notice and the full citation on the first page. To copy otherwise, to republish, to post on servers or to redistribute to lists, requires prior specific permission and/or a fee.

ISPD'11, March 27–30, 2011, Santa Barbara, California, USA.
Copyright 2011 ACM 978-1-4503-0550-1/11/03. .$10.00.

on G-Cell-to-G-Cell (e.g., global routing cell) connection on a global routing grid graph. Detailed Routing, on the other hand, realizes exact routing paths considering geometrical constraints based on the global routing solution. Detailed routing is an important stage in the sense that it is directly related to the routing completion and design rule satisfaction. It also impacts many design metrics such as timing, manufacturability, etc.

Detailed routing has been extensively studied since 70's (e.g., [3,4]) but the topic is not frequently seen in recent publications. For modern designs in which over-the-cell routing is applied, the most common technique for detailed routing is rip-up and reroute such as the one in Mighty [5]. However, such a sequential net-by-net approach is ineffective in handling congested designs and it usually creates unnecessary detour. DUNE [6] and MR [7] proposed to handle full chip gridless routing by similar multilevel approaches, in which the routing undergoes a coarsening phase and an uncoarsening phase. But these multilevel routers still rely on the sequential rip-up and reroute technique and nets at the upper levels of the hierarchy are routed based on inaccurate information. There are several attempts which consider nets in a more simultaneous manner during detailed routing. Nam et al. [8] proposed a detailed router for FPGA based on Boolean satisfiability. Though this approach achieves good solution quality, the runtime is extremely long. Zhou et al. [9] introduced track assignment as an intermediate step between global and detailed routing. In track assignment, segments extracted from global routing solution are assigned to routing tracks. This problem is NP-complete and is solved by a weighted bipartite matching based heuristic. However, the connections of a segment to pins or segments in other layers are not completed during track assignment. They are postponed to detailed routing, which may fails to connect different parts of a net. Mustafa [10] presented an insightful technique to perform escape routing for dense pin clusters, which is a bottleneck of detailed routing. A multi-commodity flow based optimal solution and a Lagrangian relaxation based heuristic are proposed. Nevertheless, the technique is not proposed to solve whole-chip scale detailed routing.

With diminishing feature size, many complex design rules are imposed to ensure manufacturability. It has been reported that for 32nm process, the number of rules reaches several thousands [11] and the design rule manual has roughly a thousand pages [12]. The dramatic increase in the number and complexity of the design rules makes detailed routing progressively complicated and time-consuming. We notice that many of those complex rules are triggered only by non-trivial routing patterns. Here we define regular patterns as those avoiding jogs and unnecessary detours as much as possible. Figure 1 illustrates two routing solutions for the same problem. The top one is irregular routing with many jogs and detours while the bottom one is regular routing which only uses simple patterns. If only regular patterns are used, it is not even necessary to check many design rules and the routing solution will be correct by construction. On the other hand, if a routing solution is irregular, even though it may not violate any design rule, it is likely to be detrimental for both yield and routability. Moreover, regular routing introduces less vias, jogs and wirelength, and hence is better in terms of timing, signal integrity and power consumption. In order to reduce the implementation complexity and runtime of detailed routers and to improve the electrical properties, yield and routability of circuits, we propose to perform detailed routing based on regular patterns. Note that this approach is along the lines of the restrictive design rule approach that the industry has started applying to the device layers to enhance manufacturability. In this paper, we extend it to the interconnect layers.

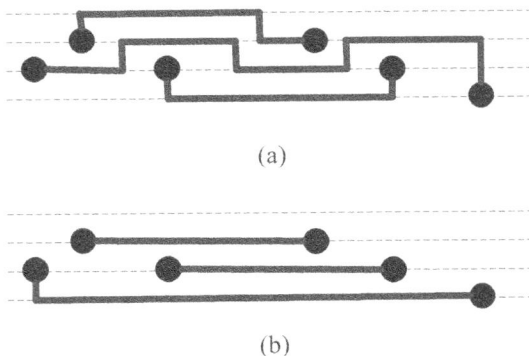

(a)

(b)

Figure 1: (a) Non-trivial routing patterns. (b) Regular routing patterns.

Potentially, regular routing may adversely affect routability because it is more restrictive and may be less effective in resolving congestion. This paper shows that with an appropriate algorithm, regular routing can be effective since the solution space can be explored much more effectively and efficiently. On the contrary, for routing with general patterns, the solution space is much larger. But it is also much harder to be explored. The best known approach is to route the nets one by one using maze routing together with rip-up and reroute. Such an approach is very time-consuming (especially if complicated design rules need to be checked repeatedly throughout the routing process) and is prone to getting stuck in local minima.

In this works, we present a fast and effective algorithm called RegularRoute to perform detailed routing with regular patterns. Novel techniques in RegularRoute are listed below:

- We introduce a new bottom-up layer-by-layer framework for detailed routing.

- We propose a single trunk V-Tree topology for routing nets local to a G-Cell.

- We decompose the routing problem of nets spanning multiple G-cells into assignment of global segments into panels. This approach facilitates parallel processing as assignment for different panels are independent of one another.

- We formulate the global segment assignment problem for each panel as a Maximum Weighted Independent Set (MWIS) problem. This formulation enables all segments to be considered simultaneously.

- We present a fast and effective heuristic to solve MWIS.

- We employ a technique to maximize the usage of a panel by partially assigning some of the remaining segments after MWIS.

- We introduce a terminal promotion technique to connect various segments of each net assigned to different layers.

- We present panel merging and maze routing techniques to handle unassigned segments at the top layers.

We implemented RegularRoute and tested its performance on detailed routing testcases derived from ISPD98 [1] and ISPD05 [2] placement benchmarks respectively. Experiments show that RegularRoute performs well in both quality and runtime.

The rest of paper is organized as follows: Section 2 provides the problem formulation and an overview of RegularRoute. Section 3 discusses the routing for local nets. In Section 4, we introduce techniques for handling global segments assignment. Experimental results are shown in Section 5.

2. PRELIMINARIES

In this section, we will present some terminologies, the problem formulation and the algorithm flow of RegularRoute.

2.1 Terminologies and Problem Formulation

In this paper, as regular routing is considered, we model the routing resource as a 3-D regular grid graph. Each grid edge can accommodate one wire except for edges with blockage, which cannot be used. Each layer of the graph has a *preferred routing direction* and the preferred directions of adjacent layers are perpendicular to each other. We assume the preferred direction of lowest layer (metal1) is horizontal. For each layer, the routing usage that is in the preferred direction is called *preferred usage*. Otherwise, the routing usage that is perpendicular to the preferred direction is called *non-preferred usage*. A sequence of unblocked grid edges along the preferred routing direction of each layer is called a *routing track*.

Assume a placed netlist with exact pin locations and a corresponding 2-D global routing solution are given. In this paper, we assume all pins are on metal1. The detailed routing problem is to route all nets on the grid according to the global routing solution such that routes of different nets do not intersect. The primary objective of detailed routing is to complete as many nets as possible. The secondary objectives include minimizing non-preferred usage, via count and wirelength. In industrial applications, there may be many other design metrics such as timing, crosstalk, yield, etc. These metrics can potentially be incorporated into our framework but they will not be handled directly in this paper.

In our framework, the global routing solution of each net is partitioned into a set of *segments* by breaking it at the turning points. Each segment is a horizontal (or vertical) route which spans multiple G-Cells in a row (or column). Then detailed routing of global nets is formulated as assigning the global segments to the routing tracks. Ideally, each segment should be assigned to one track. In order to make routing less restrictive, assigning a segment to more than one tracks connected by short non-preferred usage or via is allowed but discouraged. We define a *panel* to be the collection of all tracks on one layer within one row (for odd layer) or one column (for even layer) of G-Cells. Figure 2 shows the definitions of track, segment and panel. Note that each segment can only be assigned to tracks on a deck of panels that are on different layers but are associated with the same row/column of G-Cells spanned by the segment. In other words, it is natural to perform the global segment assignment in a panel-by-panel manner.

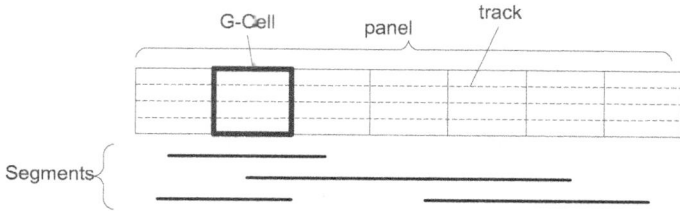

Figure 2: Definitions of track, segment and panel.

2.2 Algorithm Flow

We show the flow of RegularRoute in Figure 3. RegularRoute starts with extracting global segments by breaking the 2-D global routing solution. Then local nets are pre-routed using the single trunk V-Tree topology. In the following global segment assignment, the routed path of local nets are treated as blockages. Next, we perform global segment assignment in a bottom-up layer-by-layer manner. At each layer, the segment assignment of different panels are handled independently. For each panel, we formulate global segment assignment using regular routing patterns as a MWIS problem and solve it by an effective heuristic. After that, we apply a partial assignment technique to increase the utilization of the panel. Then if we have not reached the top horizontal or vertical layers, for the unassigned segments, we promote their terminals and defer their assignment to upper layers. For the unassigned segments at the top layers, we utilize a panel merging technique which provides more flexibility in the assignment by allowing segments to deviate from the global routing solution. Finally, maze routing is applied for the residual unassigned segments.

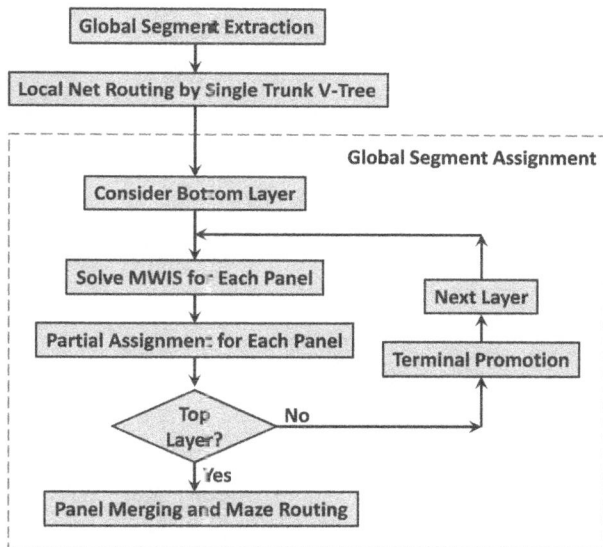

Figure 3: Flow chart for RegularRoute.

3. LOCAL NET ROUTING

In detailed routing, the net or sub-net that resides totally inside one G-Cell is called a local net. In RegularRoute, local nets are routed before assigning global segments. The routing solutions of the local nets are treated as blockages in the following global segment assignment. It is possible to route local nets and global segments simultaneously by integrating local net routing into the MWIS framework in Sec. 4.1. But in order to reduce the size of MWIS problems

and hence the runtime of the whole algorithm, we choose to handle local nets before global segments.

In this section, we first introduce local net routing by single trunk V-Tree topology. We then demonstrate that single trunk V-Tree topology can better preserve routing resources to be used in global segment assignment. It is possible to have conflicts among the single trunk V-Trees of different nets. Hence we also present a branch and trunk shifting technique to resolve the conflicts.

3.1 Single Trunk V-Tree Topology

Single trunk tree has been proposed to predict the routing usage or interconnect properties at early design stages [13]. In here, we use a single trunk tree with the trunk being vertical to route the local nets. We call this topology *single trunk V-tree*. Consider all the pins inside a G-Cell. The x-coordinate of the trunk is set to be the median of the x-coordinates of all pins. The trunk spans from the minimum y-coordinate to the maximum y-coordinate of all pins. The trunk is on metal2. We connect each pin to the trunk with a metal1 (horizontal) connection, which we call a branch, and a via. Figure 4 shows an example of single trunk V-Tree.

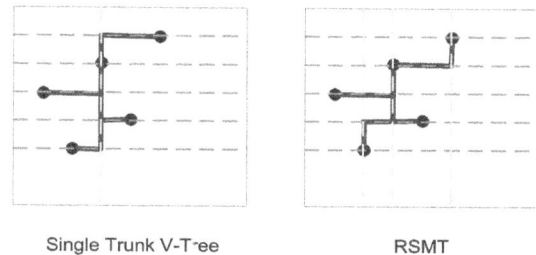

Single Trunk V-Tree   RSMT

Figure 4: Track blockage count for single trunk V-Tree and RSMT.

Single trunk V-Tree can be easily constructed in time linear to the number of pins in a local net. In our testcases, the average pin count is very small (around 3). So the runtime is negligible compared to other steps.

There are many candidate topologies to construct the trees for local nets. For instance, RSMT and RMST are promising candidates. The reason we choose single trunk V-Tree is for the sake of saving routing resources on metal2. In Figure 4, single trunk V-Tree and RSMT are presented for the same 5-pin net. In metal1, five tracks are blocked in both cases. In metal2, only one track is blocked for single trunk V-Tree, but three tracks are blocked for RSMT. As single trunk V-Tree blocks fewer tracks on metal2, global segment assignment will have more tracks to use later on.

3.2 Trunk and Branch Shifting

During the local net routing, we first determine the vertical trunk. If the total pin number is odd, the trunk has only one choice for minimum wirelength. Each branch has only one choice too. If there are multiple local nets inside one G-Cell, there is a risk of conflict among trees of different local nets. To avoid the conflict, we apply trunk and branch shifting by trying neighboring tracks. Any unresolved conflict can be resorted to higher layers (e.g., metal3 and metal4). But in our experiments, all local nets can be routed using only metal1 and metal2. The results will be shown in Section 5.

4. GLOBAL SEGMENT ASSIGNMENT

In this section, we cover the details of global segment assignment. We first present the detailed formulation of assigning global segments to a panel using regular routing patterns. The problem is then converted into a Maximum-Weighted Independent Set (MWIS) problem which is solved by a fast and effective heuristic. We next discuss the technique to perform a partial assignment for increasing the routing resource utilization of current layer. Then we talk about the terminal promotion techniques to defer the unassigned segments to upper layers. For the unassigned segments on the top layers, we develop effective panel merging and maze routing techniques for final routing closure.

4.1 MWIS based Solution for One Panel

In Section 2.1, we have mentioned the global segment assignment problem for each layer is solved by a panel-by-panel strategy. Solving the segment assignment problem of one panel is a fundamental component to the whole algorithm. In this subsection, we will investigate this problem. Without loss of generality, our examples only consider horizontal panels (metal1, metal3, etc.).

A global segment is a horizontal only or vertical only portion of a net extracted from the 2-D global routing solution. The remaining portion of the net in either end of a segment is represented by a *terminal*. When a segment is assigned to a track, each of its ends should be connected to its associated terminal. A terminal can be a pin, a partial assignment of the segment or a neighboring segment. The concept of terminal is illustrated in Figure 5. In this figure, we show two assigned segments. The first segment is incident to a partial segment on the left and a pin on the right. The second segment is incident to a neighboring segment (which has not been assigned and is shown in dotted line) on the left and a pin on the right. The connection between an assigned segment and its terminal is called a terminal connection. Note that this example assumes the pin and partial wire are on the same layer with the segment, but it is not necessary to be true. We will have more discussion on terminals in later subsections.

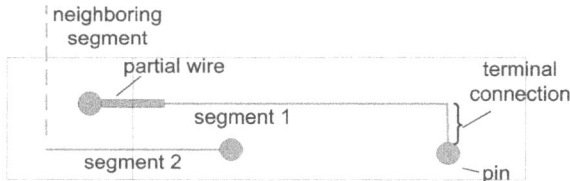

Figure 5: Illustration of terminal.

We introduce the concept of choice for assisting the assignment of a segment. A *choice* is a valid candidate solution to assign a segment using a regular routing pattern when other global segments are ignored. A choice is determined by the track being used and the terminal connections being made. In particular, a choice for a segment can be represented by (t, R). t is a track in the panel that the segment is assigned to, and R is a collection of short wires and/or vias that the assigned segment used to connect to its terminals. A simple example is shown in Figure 6. In this example, segment i has one choice c_i1, and segment j have two choices c_j1 and c_j2. Each choice specifies both the track and the short connections to the terminals. The terminal connections are highlighted as $R1$, $R2$ and $R3$ respectively. When two choices cannot co-exist, we said there is a conflict between the two choices. For instance, c_i1 conflicts with c_j2. Besides, different choices for the same segment mutually conflict with each other. For example, c_j1 conflicts with c_j2.

We formulate the global segment assignment problem for one panel as a Maximum-Weighted Independent Set (MWIS) problem. We can represent the conflicts among the choices by a conflict graph as shown in Figure 6. In the conflict

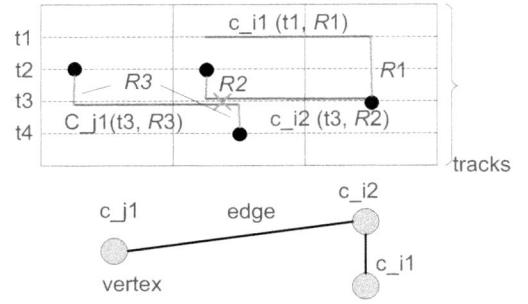

Figure 6: Conflicting choices and conflict graph.

graph, each choice is modeled as a vertex. Each conflict between two choices is modeled as an edge. Each vertex is assigned a weight specifying the benefit of the assignment. Then the problem is to select a set of independent vertices to maximize the total weight.

The weight calculation for each vertex is important in the MWIS problem. It contains several components for differentiating choices and leveraging various objectives. In general, it includes both the *segment differentiation* as well as the *choice differentiation*. The first one differentiates segments and all choices derived from the same segment will share the common weight of this part. The second one differentiates the choices derived from the same segment. We use the following function for weighting a vertex (choice).

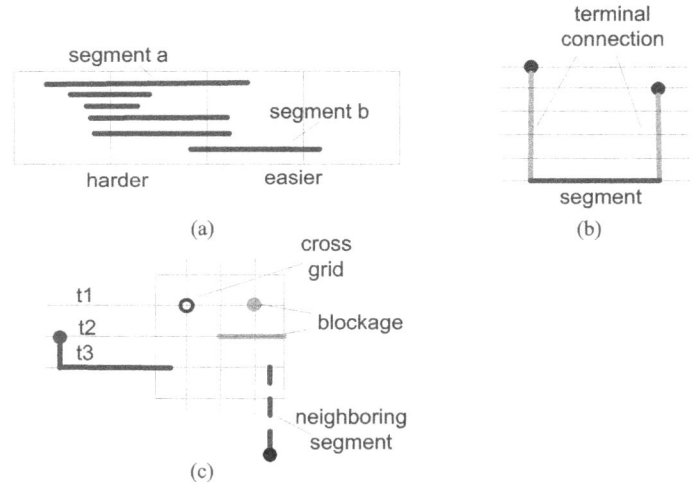

Figure 7: (a) G-Cell boundary density. (b) Terminal connection. (c) Flexibility component

$$W(v) = L + \alpha_1 \times \|R\| + \alpha_2 \times \left(\frac{\sum_{b \in B} (D_b)^2}{\|B\|} \right) + \alpha_3 \times (F1 + F2)$$

(1)

We have four major components for determining the weight of a choice.

1. *Segment Length*
   L is the number of global routing grids that the segment spans. It reflects length of the segment. In the weight calculation, we encourage packing more usage to current layer. This is a component for segment differentiation.

2. *G-Cell boundary density*
   This component is used to increment weight for the

48

segments that cross dense G-Cell boundaries. We use the number of crossing segments to represent the boundary density. Intuitively, the segment passing through denser G-Cell boundaries is harder to assign. They can easily incur conflicts with other segments. We use the average quadratic G-Cell boundary density for this component. In Figure 7(a), segment a is harder to assign than segment b as it passes through denser G-Cell boundaries. In Equation 1, B is the set of G-Cell boundaries the segment passes through. D_b is the density of boundary b. We sum the quadratic value of density and divide it by the number of boundaries. This component is also proposed for differentiating segments.

3. *Terminal Connection*

This component increases the weight for the choices with longer terminal connection route. The longer the terminal connection, the more likely the segment incurs potential conflicts with other segments. We divide the terminal connection usage to be three parts: preferred usage, via count and non-preferred usage. They are adjusted by different coefficients for leveraging their importance. For instance, if via count is critical, then we charge a higher cost for the number of vias in the terminal connection. For the sake of simplifying pool of choices, we generate terminal connection route by maze routing [1]. This component works for differentiating choices for the same segment.

4. *Flexibility Component*

The flexibility component is used to differentiate choices for the segment with one or more ends that are incident to neighboring segment which has not been assigned. The choice of current segment which offers more flexibility for assigning the neighboring segment will be better of routability. Since the direction of the segment and its neighboring segments will be perpendicular, the cross grid is the intersection of tracks between current layer and next layer. we define the flexibility count to be the number of cross grids that have access to upper layer inside the ending G-Cell. It indicates how much freedom of assignment for the neighboring segment. In Equation 1, we use $F1$ and $F2$ to represent the count for both ends. As in Figure 7(c), the flexibility count for track $t1$, $t2$ and $t3$ are 2, 1 and 3 respectively. Thus $t3$ is better in terms of the flexibility for assignment.

In the equation, there are three coefficients (α_1, α_2 and α_3) for tuning the importance of each component. Their exact value are determined by experiment.

The MWIS problem is NP-Complete [14]. Solving it optimally is time-consuming. (Typically there are hundreds of panels for each layer and each panel contains thousands of segments). Instead we develop an efficient and effective heuristic. We first define the cost for each vertex:

$$\text{Cost } C(v) = W(v) - \beta \times SumW_i(v) - \gamma \times SumW_o(v) \quad (2)$$

where W represents the weight of vertex v, $SumW_i$ is the inner sum weight, $SumW_o(v)$ is the outer sum weight, and β and γ are parameters. The inner and outer sum weight are defined with respect to the clique containing all choices for a segment in the conflict graph. The cliques for two segments are illustrated in Figure 8. Vertices A, B, C and D are choices for one segment and they form *clique*1. Similarly, vertices E, F and G are choices for another segment and they

form *clique*2. For each vertex, inner (outer) sum weight is the total weight of its neighboring vertices inside (outside) its corresponding clique. The heuristic to find MWIS selects vertices in the order of decreasing cost. When a vertex is selected, we update the graph by removing all of its neighboring vertices (i.e., vertices conflicting with the selected vertex). The inner/outer sum weight of all vertices incident to the removed vertices are also updated. In order to improve the efficiency of extracting vertex with largest cost and cost update, we utilize a heap to organize the cost for each vertex.

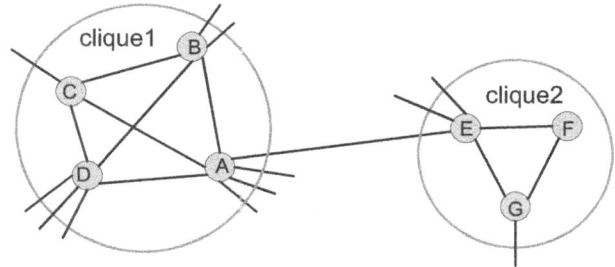

Figure 8: **Cliques for two segments in the conflict graph.**

Basically, vertex with larger weight and smaller inner/outer sum weight is prioritized. First, the vertex with higher weight is preferred since our algorithm tries to optimize the objective defined in Equation (1). Second, larger inner sum weight means more choices for the segment. Its assignment can be deferred so that routing resources are reserved for less flexible segments. Thirdly, the vertex with larger outer sum weight means more conflicts with other segments. Such choice is discouraged. The parameters β and γ are experimentally determined and the same values are used for all testcases.

4.2 Partial Assignment

After solving the MWIS problem by our algorithm, there are potentially large number of remaining unassigned segments in the congested panels. In order to better utilize the routing resources of current layer, we need explore more possibility beyond the choices defined in the MWIS problem. We implement a partial assignment for increasing the utilization of the panel.

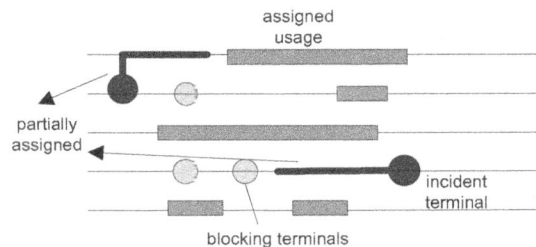

Figure 9: **Partial assignment technique for unassigned segments.**

For each unassigned segment, we will try assigning part of the segment starting from the incident terminals. We rank the unassigned segment based on the grid length (i.e., number of G-Cells passed) and process them in a decreasing order. There are potentially different options to perform partial assignment starting from one terminal by using different tracks. We use the same evaluating function as in Equation 1 to evaluate each option and select the best one. The idea is roughly illustrated in Figure 9. Both terminals of the unassigned segment are on the current layer. They

---

[1] We could save the trouble of maze routing by terminal promotion, which will be discussed later

are highlighted with bigger round shape than other blocking terminals. The assigned usage is treated as blockage as well. The partial assignment of the segment will be the route starting from the terminal to the first blockage met. Please note that in this example we have assumed the terminals are in the current layer. When we discuss the definition of terminal in the beginning of this section, it is not necessary to be true. But it is valid in our algorithm, the reason will be clear after we discuss the next subsection.

We also observe that it is possible to fully assign the segment using two partial assignment originating from each terminal and non-preferred connection to connect the two partial route. Actually the situation is not rare, the routing resources are better utilized and we will have more opportunity to assign more segments. Hence the partial assignment largely improve the efficiency of our algorithm. Besides our technique, an alternative idea is to incorporate the partial assignment into the MWIS problem. In particular, we could enumerate some partial assignment choices associated with each segment. However, it introduces much more vertices and edges and the resulting runtime can be significantly increased. Moreover, it is likely that the routing solution might no be regular routing oriented. So we only apply the partial assignment as postprocessing after solving MWIS problem.

4.3 Terminal Promotion

For the unassigned segments after applying partial assignment, we need defer their assignment to upper layers. However, there will be terminal connection issue when the segment is assigned in upper layer while its terminals are located in lower layers. Let's suppose a horizontal segment is finally assigned to metal5 and one of its terminals is pin which is on metal1. In this case, the terminal connection might not be successful when the routing resource is limited (i.e., nearby routing tracks are taken up), realizing this connection can be headache and may finally results in big effort in rip-up and reroute.

After promoting terminals up, in the assignment in upper layer, we could always treat the terminals as in current layer. The idea could be highly effective for congested panels where the routing resources are limited. Hence the main difference of our algorithm and the track assignment [9] is that ours could guarantee a valid solution after all segments are successfully assigned. Yet track routing may spend a lot of rip-up and reroute effort for correcting failed terminal connection and segments between different layers.

We count the number of tracks that the terminal can access in the upper layer. If there is only one track that the terminal can access. We select the track and promote the terminal accordingly. If the terminal is capable of accessing multiple tracks, we pick the track that is closest to the terminal. If there is no access to upper layer, we rip-up the usage that is small and close to the terminal to guarantee it has at least one access. We use a terminal connection route to promote the terminal. In the upper layer, we create a virtual stripe covering all the tracks in the upper layer that the original terminal can access (virtual means we do not assign actual wire). If the segment is eventually assigned to a track that is not the same as the track we promote the original terminal to, we just rip-up the original terminal connection and correct it.

In Figure 10, we show how we promote the old terminal to upper layer. The horizontal tracks are current layer routing tracks and the dotted tracks are tracks on the upper layer. The vertical short lines in light color are potential via location. The original terminal is located on track $t1$. To promote the old terminal, we extend it with a short wire on

track $t1$ and then add the via. The new terminal is shown on the tracks of upper layer.

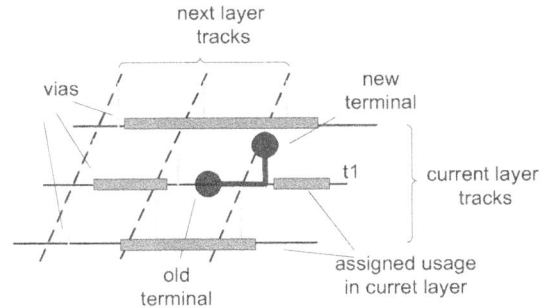

Figure 10: Terminal promotion to avoid terminal connection failure.

Overall, the terminal promotion is a highly effective technique in RegularRoute. A valid detailed routing solution is generated after global segment assignment.

4.4 Unassigned Segments on Top Layers

For unassigned segments on the top two layers (top two layers with horizontal and vertical tracks respectively), there are no upper layers where we can defer the assignment to. For better routability, we apply panel merging and eventually maze routing if the testcase is too hard.

All of our discussion has been based on the assumption that each segment respects the global routing solution. More specifically, the input global routing solution determines which panel each segment must be assigned. This assumption is restrictive in the fact that it forbids the option of trying alternative panel for better routability.

More specifically, for the panel with unassigned segments, we try to merge it with neighboring few panels and redo the assignment. In the case when the neighboring panel has space for holding more segments, it is likely the problem can be resolved. In the experiment, we merge the neighboring one panel (total three panels). This number can be modified depends on the level of hardness and runtime. The panel merging technique is effective for the segments with preferred tracks near the panel boundary. The merging of panels eliminate the boundary and the segment becomes more flexible.

For the very hard testcases, we could apply the panel merging in lower layers instead of waiting till the top layer. Actually, we could run RegularRoute initially for estimation and record the panels that are congested (with unassigned segments). We then run RegularRoute again starting from scratch with the precaution of congested panels. The panels that are predicted congested could be merged with neighboring panels since the lower layers. Again, it is a trade-off between solution quality and runtime and we adopt this idea in the case when testcase is too difficult to handle.

If there are still unassigned segment left, we eventually resort to a line probe maze routing technique or a full 3-D maze routing technique. The maze routing is most flexible technique but detour-prone and time-consuming. We adopt it as the last effort in RegularRoute.

Besides maze routing technique, we could also try academic MWIS solver in order to better solve the problem. However, the near-optimal solver such as [15] will be slow in nature. In order to maintain the fast runtime, we keep the fast heuristic as the main solver.

5. EXPERIMENTAL RESULTS

All our experiments are performed on a machine with 2.67GHz Intel Xeon CPU and 32G memory. We derive two sets of detailed routing testcases from ISPD98 [1] and ISPD05 [2] respectively. In original ISPD98 placement benchmarks, pins are set to be at the center of each standard cell, we develop a program to randomly set the pin coordinate and make sure they satisfy the spacing requirement at the bottom layer. The size of each module in the derived testcases is the same to that of the IBMv2 [16] placement benchmarks. We use Dragon [17] to generate the placed testcases for ISPD98 derived testcases and FastPlace 3.1 [18] for ISPD05 derived testcases. We derive the global routing testcases similar to the format defined by ISPD07/08 global routing benchmarks [19,20]. We then use FastRoute 4.0 [21] to route the global routing testcases and generate the 2-D global routing solution. Both the global routing testcase and the 2-D solution are imported into RegularRoute. Due to the lack of available academic detailed routers, we compared our results with an industrial router - WROUTE. However, WROUTE does not recognize bookshelf placement format or the global routing testcase format we use. We therefore convert the placed testcases by publicly available conversion tool[2] to LEF/DEF format testcases which we can import into WROUTE. Although the testcases are in different formats, we make sure the basic information such as pitch size, module size, routing region, routing layers etc. are identical for both testcases.

5.1 Results of Local Net Routing

We first show RegularRoute's performance on handling local nets based on the single trunk V-Tree topology. We report the final unassigned local nets, total CPU time, final metal2 usage and unassigned global segment count. Here we only use metal1 and metal2. We compare our results with RSMT topology.

In Table 1, the first column lists all experimental testcases. Due to limited space, we only show the results for the ISPD98 derived testcases. The next column shows the total number of local nets for each testcase. The following six columns show results of single trunk V-Tree and RSMT respectively. The RSMT is generated by FLUTE [22] using default settings. First, $\#un.L.$ is the final unassigned local nets. Single trunk V-Tree has no unassigned local nets. But RSMT tree incurs some unassigned nets. Second, CPU is the runtime in seconds. FLUTE runs faster than our algorithm. But the local nets routing runtime is trivial compared with global segment assignment. So the runtime advantage is not important. Third, $metal2\ usage$ is the total usage on metal2 after routing local nets. The single trunk V-Tree introduces 20% to 30% less metal2 usage, which saves more resources on metal2. $\#un.G.$ is the final unassigned global segment if either topology is applied. RSMT may incur some unassigned global segments and it further suggests RSMT is inferior in preserving routing resources.

| | | Single Trunk V-Tree | | | RSMT | | |
Name	#Loc. Nets	#un. L.	CPU (sec.)	metal2 usage	#un. G.	#un. L.	CPU (sec.)	metal2 usage	#un. G.
ibm01	1081	0	0.04	6300	0	0	0.02	9600	0
ibm02	1750	0	0.09	12800	0	0	0.04	15300	0
ibm07	4479	0	0.18	22300	0	7	0.05	32600	5
ibm08	5539	0	0.23	27800	0	0	0.11	39600	0
ibm09	5429	0	0.20	28200	0	9	0.08	37900	0
ibm10	2984	0	0.27	17400	0	0	0.12	29400	1
ibm11	6983	0	0.26	38900	0	4	0.07	50100	7
ibm12	2433	0	0.32	14500	0	0	0.12	26800	0

Table 1: Results for Local Net Routing on ISPD98 Testcases

5.2 Results for ISPD98 Testcases

In Table 2, we show the results for global segment assignment of RegularRoute on the eight ISPD98 testcases. We compare the results with WROUTE (version 3.0.61). The testcase statistics are shown in the first seven columns, for total number of nets ($\#Nets$), G-Cell grids ($Grid$), total number of global segments ($\#Seg.$), total number of local nets ($\#Loc.Nets$), the average net degree for the whole netlist ($Avg.Deg.$), maximum number of segment ($MaxSeg.$) in one panel and maximum number of pin in one panel ($MaxPin$) respectively. These statistics provide an overall idea about the complexity of these testcases. The next column shows the runtime for FastRoute 4.0 [21]. The global routing runtime gives the rough idea of how fast our detailed router compared with the the global router. The following columns show the results of RegularRoute and WROUTE respectively. $\#unassigned$ is the count of segments that cannot be handled by RegularRoute. We develop an internal checker to make sure the assigned segments respect the design rules we have specified (spacing rule). CPU is the runtime in seconds. The WROUTE results are reported with similar metrics except "viol.", the number of design rule violations.

First, RegularRoute is capable of routing through all the eight testcases. WROUTE, nevertheless, can route four testcases without violation. Here the number of violation is the number of spacing rule violation caused by inability to allocate the nets. Second, in terms of runtime, RegularRoute is better compared with WROUTE, which spends a lot of runtime on rip-up and reroute. Please note that it is likely that WROUTE incorporates more design objectives than ours, though we strived to turn off all non-relevant design objectives. However, the point we want to demonstrate is the restrictive regular routing is doable for good routing completion. As we have mentioned in earlier part, we could incorporate more design metrics into our framework. The weight function or cost function for solving MWIS problem can be thus extended to incorporate other design objectives. And we also see better chance for satisfying various design rules, as more and more complicated design rules are triggered by non-trivial routing patterns. The good routing completion rate could also save additional effort during the design rule clean-up stage and thus better manufacturability. Third, we achieve comparable results in terms of wirelength and via count. It is because RegularRoute tries to restrict the usage of detoured routing path. During the routing we also charge additional penalty for the regular routing choices with larger via count.

5.3 Results for ISPD05 Testcases

We show the complete results on six testcases derived from ISPD05 [2] placement benchmarks. They are much bigger in problem size and more challenging in complexity than the ISPD98 derived testcases. We only show the results for six testcases (eight benchmarks in total) because the other two testcases cannot be routed by WROUTE (crashes during execution). As mentioned earlier, these testcases are made following the similar procedure of ISPD98 testcases. We use FastPlace 3.1 [18] to place all the placement benchmarks with default setting. Likewise, the results are also compared with WROUTE. We show the number of unassigned segments($\#unassigned$), the CPU time, the via count and total wirelength for RegularRoute. We additionally show the violation (viol.) count for WROUTE. RegularRoute is capable of routing five testcases without unassigned segments, and WROTUE can route three testcases. WROUTE is likely to incur a number of design violations. Like the experiments for ISPD98 testcases, we tried our best to switch off unrelated design objectives. Based on the results, we show RegularRoute is also doing well for larger designs.

---

[2]We use PlaceUtil executable developed by Umich, http://vlsicad.eecs.umich.edu/BK/PlaceUtils/bin/Sol64

	Testcase Statistics							FR4.0	RegularRoute				WROUTE (Encounter)			
Name	#Nets	Grid	#Seg.	#Loc. Nets	Avg. Deg.	Max Seg.	Max Pin	CPU (sec.)	#unass- igned	CPU (sec.)	via ×10e5	wlen ×10e5	viol.	CPU (sec.)	via ×10e5	wlen ×10e5
ibm01	11507	133×132	42307	1118	3.85	238	463	0.47	0	3.17	0.84	6.9	0	47	0.84	7.1
ibm02	18427	152×151	80891	1616	4.23	366	616	2.71	0	14.4	2.9	15.9	3	155	3.0	16.1
ibm07	44394	229×228	162009	5691	3.7	507	877	8.51	0	34.3	3.8	39.9	12	190	3.8	40.6
ibm08	47944	239×238	198188	6905	4.13	568	1042	10.1	0	54.6	4.4	44.5	0	193	4.4	44.1
ibm09	50393	243×242	179942	6590	3.73	509	1016	6.11	0	43.1	3.9	37.0	0	184	3.9	37.4
ibm10	64227	316×315	282041	4329	4.19	640	1045	8.97	0	66.9	6.0	68.5	0	290	6.2	69.5
ibm11	66994	276×275	230365	8486	3.54	637	1140	15.7	0	68.1	4.8	53.2	23	287	5.1	53.8
ibm12	67739	341×340	336106	3810	4.34	736	1151	25.4	0	112.1	7.0	97.4	9	422	7.2	98.3

Table 2: Results of RegularRoute and WROUTE for ISPD98 Testcases

	Testcase Statistics							FR4.0	RegularRoute				WROUTE (Encounter)			
Name	#Nets	Grid	#Seg.	#Loc. Nets	Avg. Deg.	Max Seg.	Max Pin	CPU (sec.)	#unass- igned	CPU (sec.)	via ×10e6	wlen ×10e7	viol.	CPU (sec.)	via ×10e6	wlen ×10e7
adaptec1	219243	893×892	988418	54374	4.28	1424	2594	141	0	622	1.5	8.4	0	1201	1.5	8.5
adaptec2	257659	1174×1172	1040019	44356	4.09	1533	3065	189	0	558	1.9	10.2	221	1344	2.0	10.4
adaptec3	466293	1935×1946	1887820	44356	4.01	2142	4950	342	0	1176	3.5	21.8	0	3939	3.6	22.1
adaptec4	515300	1933×1945	1812333	85000	3.70	1884	3820	289	4	1330	3.0	19.8	324	4424	3.2	20.4
bigblue1	282399	893×892	1182506	25288	4.02	1410	2534	134	0	911	2.2	9.8	0	1802	2.2	9.7
bigblue2	576618	1560×1568	1826150	92945	3.60	1648	3804	249	0	1177	3.7	21.2	54	2856	3.9	22.0

Table 3: Results of RegularRoute and WROUTE for ISPD05 Testcases

6. ACKNOWLEDGMENT

The authors would like to thank Dr. Hardy Leung for valuable discussions which inspire our work, and the University of Michigan CAD group for the helpful placement utility tools to convert bookshelf placement format to LEF/DEF format.

7. CONCLUSION

In this paper, we propose a detailed router which seeks to route global segments with regular routing patterns. The whole algorithm is based on a bottom-up layer-by-layer processing. The problem for each layer is partitioned into subproblems by panels. Inside each panel, the global segment assignment is formulated as a MWIS problem. An effective heuristic and a few postprocessing techniques are developed. We have shown RegularRoute's performance on detailed routing testcases derived from real circuits. In the future, we would like to further improve RegularRoute's performance and incorporate more design objectives to make our tool more suitable for industrial applications. In addition, we are interested in making a parallel version of our tool for further runtime reduction.

8. REFERENCES

[1] IBM-Place 1.0 benchmark suites. http://er.cs.ucla.edu/benchmarks/ibm-place/.

[2] ISPD05 placement contest benchmarks. http://www.sigda.org/ispd2005/contest.htm.

[3] A. Hashimoto and J. Stevens. Wire routing by optimizing channel assignment within large apertures. In Proc. ACM/IEEE Design Automation Conf., pages 155–169, 1971.

[4] T. Yoshimura and E. Kuh. Efficient algorithms for channel routing. IEEE Trans. on Computer-Aided Design, 1(1):633–647, Jan 1982.

[5] H. Shin and A. Vincentelli. A detailed router based on incremental routing modifications: Mighty. IEEE Trans. on Computer-Aided Design, 6(6):942–955, Nov 1987.

[6] J. Cong, J. Fang, and K. Khoo. DUNE-a multilayer gridless routing system. IEEE Trans. on Computer-Aided Design, 20(5):633–647, May 2001.

[7] Y. Chang and S. Lin. MR: A new framework for multilevel full-chip routing. IEEE Trans. on Computer-Aided Design, 23(5):793–800, May 2004.

[8] G. Nam, K. Sakallah, and R. Rutenbar. A new FPGA detailed routing approach via search-based boolean satisfiability. IEEE Trans. on Computer-Aided Design, 21(6):674–684, Jun 2002.

[9] S. Batterywala, N. Shenoy, W. Nicholls, and H. Zhou. Track assignment: A desirable intermediate step between global routing and detailed routing. In Proc. Intl. Conf. on Computer-Aided Design, pages 59–66, 2002.

[10] M. Ozdal. Detailed-routing algorithms for dense pin clusters in integrated circuits. IEEE Trans. on Computer-Aided Design, 28(3):340–349, March 2009.

[11] Danny Rittman. Nanometer DFM – the tip of the ice. Intelligence: From Science to Industry, Tayden Design newsletter, March 2008.

[12] Luigi Capodieci. Layout printability verification and physical design regularity: Roadmap enablers for the next decade. In edp, 2006.

[13] H. Chen, C. Qiao, F. Zhou, and C. Cheng. Refined single trunk tree: A rectilinear Steiner tree generator for interconnect prediction. In Proc. ACM Intl. Workshop on System Level Interconnect Prediction, pages 85–89, 2002.

[14] M. R. Garey and D. S. Johnson. Computers and Intractability: A Guide to the Theory of NP-Completeness. Freeman, NY, 1979.

[15] M. Halldorsson. Approximations of weighted independent set and hereditary subset problems. Journal of Graph Algorithms and Applications, 1(4):1–16, Apr 2000.

[16] IBM-Place 2.0 benchmark suites. http://er.cs.ucla.edu/benchmarks/ibm-place2/.

[17] X. Yang, B. Choi, and M. Sarrafzadeh. Routability-driven white space allocation for fixed-die standard-cell placement. IEEE Trans. on Computer-Aided Design, 22(4):410–419, April 2003.

[18] N. Visvanathan, M. Pan, and C. Chu. FastPlace 3.0: A fast multilevel quadratic placement algorithm with placement congestion control. In Proc. Asia and South Pacific Design Automation Conf., pages 135–140, 2007.

[19] ISPD07 global routing contest benchmarks. http://www.sigda.org/ispd2007/contest.htm.

[20] ISPD08 global routing contest benchmarks. http://www.sigda.org/ispd2008/contest.htm.

[21] Y. Xu, Y. Zhang, and C. Chu. FastRoute 4.0: Global router with efficient via minimization. In Proc. Asia and South Pacific Design Automation Conf., pages 576–581, 2009.

[22] C. Chu and Y. Wong. FLUTE: Fast lookup table based rectilinear Steiner minimal tree algorithm for VLSI design. IEEE Trans. on Computer-Aided Design, 27(1):70–83, Jan 2008.

An Enhanced Global Router with Consideration of General Layer Directives[*]

Tsung-Hsien Lee[†], Yen-Jung Chang[†], and Ting-Chi Wang[‡]

[†]Department of Electrical and Computer Engineering, University of Texas at Austin, Austin, Texas 78712
[‡]Department of Computer Science, National Tsing Hua University, Hsinchu, Taiwan 30013
tsunghsien.lee@gmail.com, cyenjung@mail.utexas.edu, tcwang@cs.nthu.edu.tw

ABSTRACT

In this paper we study a global routing problem that considers not only overflow and wirelength but also layer directives. A layer directive is often given to a timing-critical net for meeting target performance, and it specifies a range of consecutive layers on which the net should be routed. Unlike a previous work that focuses only on a restricted set of layer ranges, our problem allows arbitrary layer ranges to be specified. We present a global router which enhances an academic router by employing techniques to take general layer directives into account during two-dimensional (2D) routing and layer assignment. The experiment results show that our global router can produce encouraging solutions for all test cases.

Categories and Subject Descriptors

B.7.2 [**Integrated Circuits**]: Design Aids—*Placement and Routing*; J.6 [**Computer-Aided Engineering**]: Computer-Aided Design

General Terms

Algorithms, Design

Keywords

Computer-Aided Design, Global Routing, Layer Directives, VLSI

1. INTRODUCTION

Routing is an important design stage because it determines the physical routes of nets on a chip. To cope with high complexity, routing is typically performed by global routing followed by detailed routing. In 2007 and 2008, International Symposium on Physical Design (ISPD) held global routing contests [1, 2], and many academic global routers, e.g., [5, 6, 7, 9, 11, 15, 17, 21, 22], were developed to produce satisfactory solutions for the benchmarks released in the contests. However, the ISPD contests only focused on minimizing total overflow and wirelength, and could cause these modern academic global routers to ignore other important and practical issues [10, 13, 20].

A new set of global routing benchmarks was released in International Conference on Computer-Aided Design (ICCAD) 2009 [16], and it was produced by making some modifications to the ISPD 2008 benchmarks. The information for considering timing is added, and one of them is to specify layer directives for a subset of nets. A layer directive is often given to a timing-critical net for meeting target performance in a modern physical design flow [3, 14], and it specifies a range of consecutive layers on which the net should be routed. The ICCAD 2009 benchmarks number metal layers as 1, 2, ..., from bottom to top, and each layer range is specified by $[sl : el]$, where sl and el are layer numbers with $sl < el$. Furthermore, sl is an odd number larger than 1, while el is the highest layer and is an even number. Taking a 6-layer design as an example, sl could be 3 or 5, and el is 6; therefore 2 possible layer ranges $[3 : 6]$ and $[5 : 6]$ can be specified in layer directives. It is clear that for any two layer ranges specified in the ICCAD 2009 benchmarks, one is completely contained in the other. For example, the layer range $[5 : 6]$ is completely contained in the layer range $[3 : 6]$ in a 6-layer design.

A global router, called GLADE [4], was recently developed for the ICCAD 2009 benchmarks. Based on extensions of NTHU-Route 2.0 [5]. GLADE not only considers traditional routing objectives such as overflow and wirelength but also focuses on honoring layer directives. The experiments on the ICCAD 2009 benchmarks indicate that GLADE produced encouraging routing solutions that respect all layer directives, and meanwhile maintain similar qualities in overflow and wirelength when compared with NTHU-Route 2.0. Unfortunately, the ICCAD 2009 benchmarks only address a restricted set of layer ranges, and hence GLADE was designed to handle these special layer ranges only and was not able to consider other layer ranges that are also often encountered in real designs. For example, in a 6-layer design, there are totally 20 possible layer ranges (i.e., $[1 : 1], [2 : 2], [3 : 3], [4 : 4], [5 : 5], [6 : 6], [1 : 2], [2 : 3], [3 : 4], [4 : 5], [5 : 6], [1 : 3], [2 : 4], [3 : 5], [4 : 6], [1 : 4], [2 : 5], [3 : 6], [1 : 5], [2 : 6]$) that can be specified in layer directives, but only two of them (i.e., $[3 : 6]$ and $[5 : 6]$) are considered by GLADE. Note that the layer range $[1 : 6]$ is not considered in a layer directive,

[*]This work was partially supported by National Science Council of Taiwan under Grants NSC-99-2220-E-007-008 and NSC-99-2221-E-007-113-MY3.

Permission to make digital or hard copies of all or part of this work for personal or classroom use is granted without fee provided that copies are not made or distributed for profit or commercial advantage and that copies bear this notice and the full citation on the first page. To copy otherwise, to republish, to post on servers or to redistribute to lists, requires prior specific permission and/or a fee
ISPD'11, March 27–30. 2011, Santa Barbara, California, USA.
Copyright 2011 ACM 978-1-4503-0550-1/11/03 ...$10.00.

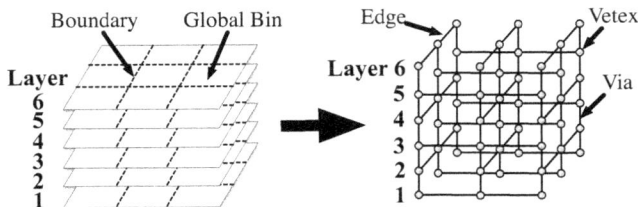

Figure 1: A 6-layer design with preferred routing directions.

because any net with this range can be routed on any layers of the design and hence has no layer restriction.

In this paper, we study a global routing problem that allows arbitrary layer ranges to be specified in layer directives. We present a global router which enhances GLADE by employing techniques to take general layer directives into account during two-dimensional (2D) routing and layer assignment. The experiment results show that our global router can generate encouraging solutions for all test cases.

The rest of the paper is organized as follows. In Section 2, our global routing problem is defined. In Section 3, we briefly review GLADE. We then give an overview and describe the details of our global router in Sections 4, 5 and 6. The experiment results are reported in Section 7 and the conclusion is made in Section 8.

2. PROBLEM FORMULATION

2.1 Preliminaries

A multi-layer global routing problem can be modeled by a three-dimensional (3D) grid graph, where each vertex denotes a global bin and each edge denotes the boundary between two adjacent global bins on the same metal layer. Figure 1 shows an example with preferred routing directions, where horizontal and vertical wires are alternate on different metal layers. Each edge e in the grid graph is given a capacity c_e, denoting the number of available routing tracks that e contains. There is also a set of nets, and each net consists of a set of vertices to be connected. The global routing problem is to find a route for each net such that it connects the corresponding set of vertices of the net using edges and vias. Each edge e in a (partial) global routing solution is associated with a demand d_e that represents the amount of nets passing through e. If the demand d_e exceeds the capacity c_e, edge e has a non-zero overflow which is defined as $(d_e - c_e)$; otherwise, e has no overflow (i.e., overflow is zero). A typical objective for global routing is to minimize both the total overflow of all edges and the total wirelength of all routes.

2.2 General Layer Directives

In this paper, a subset of nets is associated with layer directives in a given global routing instance. Each layer directive specifies a range $[sl : el]$ of consecutive layers on which the associated net should be routed. Here both sl and el can be any arbitrary layer numbers as long as $sl \leq el$. Note that if $TotalLayers$ denotes the total number of layers in the design, then the layer range $[1 : TotalLayers]$ is not considered as a layer directive, because it does not impose any layer restriction.

A net with a layer directive is called a LD net. If a routed LD net passes through an edge on an non-preferred layer (i.e., a layer not in the layer range of the net), then this edge

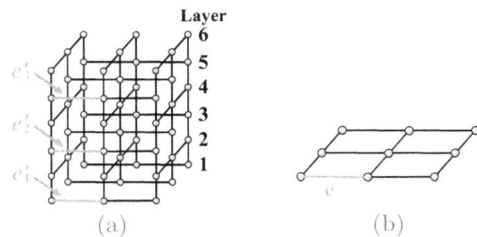

Figure 2: Global routing grid graph model. (a) A 6-layer design with preferred routing directions. (b) The projected 2D routing graph, where e'_1, e'_2 and e'_3 in (a) are projected to e.

induces one unit of LD violation for this net. For example, for a net whose layer range is $[3 : 6]$ in a 6-layer design, if the net passes through an edge on layer 1 (or layer 2), then the net has one unit of LD violation on this edge.

2.3 Objectives

Given a multi-layer design with a subset of nets associated with general layer directives, the primary objective of our global routing problem is to find a routing solution that minimizes the total LD violation as well as the total overflow (TOF) of all edges. The secondary objective is to minimize the total wirelength (TWL) of the routing solution.

3. REVIEW ON GLADE

Since our global router enhances GLADE [4] to take arbitrary layer ranges into account, we review the major ideas of GLADE in this section. GLADE extends NTHU-Route 2.0 [5] to handle a special set of layer ranges specified in the ICCAD 2009 benchmarks, and consists of four stages: initial stage, main stage, refinement stage, and layer assignment. The first three stages perform 2D routing, while the last stage converts a 2D solution into a 3D one.

The initial stage of GLADE is the same as the initial stage of NTHU-Route 2.0. It first projects a multi-layer design onto a 2D plane. Figure 2(a) shows a 3D routing graph for a 6-layer design with preferred routing directions and Figure 2(b) gives the projected 2D plane. To distinguish edges in the two routing graphs, the edges in Figure 2(a) are called 3D edges and those in Figure 2(b) are called 2D edges. Then GLADE generates a wirelength-driven 2D routing solution without considering any layer directives by using FLUTE [8], a probabilistic routing method, an edge shifting technique [18], and a L-shaped pattern routing method.

Since a 2D plane does not have any layer information, the original router (i.e., NTHU-Route 2.0) is unable to consider layer directives when performing 2D global routing. To cope with this shortcoming, GLADE adopts the concepts of *virtual capacity* and *virtual demand*. Suppose there are n different layer ranges given in the design, and hence there are n different types of LD nets. For each 2D edge e, GLADE calculates a virtual capacity $vc_e(t)$ for each LD type t, which is the sum of the capacities of the corresponding 3D edges located in the associated layer range. For example, assume that the capacities of the 3D edges e'_1, e'_2, and e'_3 in Figure 2 are denoted by $c_{e'_1}$, $c_{e'_2}$, and $c_{e'_3}$; for LD types $t_1 = [5 : 6]$ and $t_2 = [3 : 6]$, we have $vc_e(t_1) = c_{e'_3}$ and $vc_e(t_2) = c_{e'_3} + c_{e'_2}$. The $vc_e(t)$ specifies the maximum capacity that e can provide to LD nets of type t without causing any overflow.

For each 2D edge e, GLADE also calculates a virtual de-

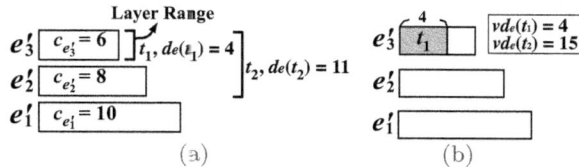

Figure 3: Illustration of how GLADE calculates virtual demand. (a) A 2D routing result on an edge e. (b) Calculating virtual demand by pseudo layer assignment.

mand $vd_e(t)$ for each LD type t. GLADE presumes that LD nets whose layer ranges are completely contained in the layer range of t have higher priorities to be processed during layer assignment. Therefore some certain amount of the virtual capacity $vc_e(t)$ may be consumed by those nets and need to be included as a part of $vd_e(t)$. To calculate this amount, GLADE performs *pseudo layer assignment*. Pseudo layer assignment does not actually assign each individual net to a 3D edge. Instead, for each type of LD nets that are routed through a 2D edge, it calculates how many of them can be assigned to the corresponding 3D edges located in their layer range and do not cause overflow on any of those 3D edges.

Let $d_e(t)$ denote the amount of LD nets of type t that have been routed through e. We use Figure 3 to explain how GLADE calculates $vd_e(t)$. In Figure 3(a), the 3D edges, e'_1, e'_2, and e'_3 are projected to a 2D edge e, and have capacities $c_{e'_1}=10$, $c_{e'_2}=8$, and $c_{e'_3}=6$, respectively; there are 4 LD nets of type t_1 (i.e., $d_e(t_1) = 4$) and 11 LD nets of type t_2 (i.e., $d_e(t_2) = 11$) that have been routed through e; e'_3 is on the preferred layer for LD nets of type t_1, while e'_2 and e'_3 are on the preferred layers for LD nets of type t_2. Since the layer range of t_1 is completely contained in that of t_2, $vd_e(t_1)$ will be calculated first by GLADE. There is no other LD type whose layer range is completely contained in that of t_1, and hence none of the virtual capacity $vc_e(t_1)$ is consumed by any other type of LD nets. Therefore, GLADE sets the virtual demand $vd_e(t_1)$ equal to the actual demand $d_e(t_1)$, i.e., $vd_e(t_1) = d_e(t_1) = 4$. To calculate the virtual demand $vd_e(t_2)$ for type t_2, GLADE performs pseudo layer assignment for LD nets whose layer ranges are completely contained in that of t_2; for our example, those nets are all of type t_1. Therefore, GLADE assigns as many nets of type t_1 to the 3D edges on their preferred layers as possible, subject to the capacity constraint imposed on each 3D edge. Hence 4 LD nets of type t_1 are assigned to e'_3 through pseudo layer assignment (see Figure 3(b)). As a result, the amount of virtual capacity of type t_2 that is consumed by LD nets of type t_1 is 4 due to the pseudo layer assignment result. GLADE then includes this amount into the virtual demand of type t_2, and sets $vd_e(t_2)$ equal to the sum of this amount and the actual demand of type t_2; that is, $vd_e(t_2) = 4 + d_e(t_2) = 4 + 11 = 15$.

During 2D global routing, in order to estimate the total LD violation and additional TOF that a 3D solution might have after actual layer assignment, for each 2D edge e, GLADE calculates a LD overflow, denoted by $LDOF_e(t)$, for each LD type t, which is defined as $max(vd_e(t) - vc_e(t), 0)$. This LDOF (standing for LD overflow) value tells that after pseudo layer assignment, how many LD nets of type t that have been routed through e cannot be assigned to their preferred layers without increasing additional overflow. For the example in Figure 3, we have $LDOF_e(t_1) = max(4-6, 0) =$

0 and $LDOF_e(t_2) = max(15-14, 0) = 1$. The LDOF of each 2D edge e, denoted by $LDOF_e$, is then defined and calculated as the sum of $LDOF_e(t)$'s among all LD types, and the total LDOF of a 2D global routing solution is the sum of $LDOF_e$'s among all edges. Each unit of LDOF implies one unit of LD violation (if overflow must be minimized) or one unit of additional overflow (if layer directive must be honored) after actual layer assignment is done.

We now return to the main stage of GLADE. GLADE improves the solution generated in the initial stage by iteratively ripping-up and rerouting overflowed nets to reduce total overflow as well as the total LDOF. Every ripped-up two-pin net is rerouted by monotonic routing [19] or multi-source multi-sink maze routing. When routing a LD net of type t through a 2D edge e, GLADE calculates the cost of edge e as follows:

$$cost_e = B_e \times GC + H_e \times P_e + V_e \times GC \qquad (1)$$

Here B_e is the wirelength of edge e, GC decreases its value moderately from 1 to 0 as the iteration count increases, H_e is the cost for historical overflow of edge e, and P_e is the congestion cost of e and is calculated below:

$$P_e = \left(\frac{vd_e(t) + 1}{vc_e(t)} \times f \right)^{k_1} \qquad (2)$$

where $vd_e(t)$ and $vc_e(t)$ are the current virtual demand and the virtual capacity of edge e for LD type t, k_1 is a user defined parameter, and f is a penalty amplifier which dramatically raises the value of P_e when e is nearly overflowed; V_e is the expected via cost when using edge e causes a bend. Note that when routing a non-LD net through edge e, the virtual demand $vd_e(t)$ and the virtual capacity $vc_e(t)$ in Eq. (2) are replaced by the actual demand d_e and the actual capacity c_e. In addition, GLADE calculates the total LDOF and the total overflow at the end of each ripup-and-reroute iteration and check if the solution gets converged.

The refinement stage focuses on finding overflow-free and LDOF-free paths for all nets. It simply rips up and reroutes nets if they pass through any overflowed edge. In this stage, each edge cost is determined as follows:

$$cost'_e = \begin{cases} 0 & \text{if } e \text{ has been passed through by the same net} \\ 0 & \text{if } d_e \leq c_e \text{ (for a non-LD net)} \\ 0 & \text{if } vd_e(t) \leq vc_e(t) \text{ (for a LD net of type } t) \\ 1 & \text{otherwise} \end{cases}$$

In the layer assignment stage, GLADE extends the layer assignment algorithm, COLA [12], to map the 2D routing solution from the projected plane to a 3D routing solution on the original multiple layers. To do it, GLADE first sorts nets by the sizes of their target layer ranges: nets with narrow ranges have high priorities. For nets with the same priority, GLADE calculates a score for each of them, which is a tie breaker and based on the pin count and wirelength of a net. Then GLADE follows the net order and applies a single-net layer assignment method, which is based on a dynamic programming technique, to the nets one after the other for finding a result with minimal via count.

4. OVERVIEW OF OUR GLOBAL ROUTER

Our global router enhances both 2D global routing and layer assignment of GLADE to handle general directives.

During 2D global routing, we modify the pseudo layer assignment method for calculating virtual demand and LDOF. During layer assignment, our router first applies COLA to find an initial 3D solution, and then uses a min-cost max-flow based method to iteratively refine the 3D solution for further minimizing the total LD violation and via count. These enhancements are respectively detailed in the next two sections.

5. PSEUDO LAYER ASSIGNMENT FOR GENERAL LAYER DIRECTIVES

For each 2D edge e, the virtual capacity $vc_e(t)$ for each LD type t is calculated in the same way as GLADE. To calculate the virtual demand $vd_e(t)$, we need to modify the pseudo layer assignment method of GLADE. Intuitively, nets with narrower layer ranges have less layers to choose for layer assignment, and therefore our pseudo layer assignment method will first sort the given LD types in a non-decreasing order of their range sizes. When two LD types have the same range size, the one with a smaller start layer (i.e., the one with a smaller sl) appears before the other in the sorted order. Now let us assume the sorted order of the LD types is denoted by $t_1, t_2, ..., t_n$.

To calculate the virtual demand $vd_e(t_i)$ for LD type t_i, our pseudo layer assignment method follows the sorted order to decide how many nets which have been routed through e and have LD types appearing before t_i can be assigned to their preferred layers. It then calculates how much virtual capacity of type t_i, say x, that is consumed by those nets after pseudo layer assignment. Finally the virtual demand $vd_e(t_i)$ is calculated as the sum of x and $d_e(t_i)$, where $d_e(t_i)$ is the amount of LD nets of type t_i that have been routed through e. Clearly our pseudo layer assignment method assumes that nets whose LD types appear before t_i in the sorted order have a higher priority to be processed, because they have smaller or the same range sizes than LD nets of type t_i. Therefore some certain amount of the virtual capacity $vc_e(t_i)$ may be consumed by those nets. Therefore the virtual demand $vd_e(t_i)$ will take into account the demands caused by LD nets whose types appear before t_i in the sorted order.

We use Figure 4 to further explain how to calculate virtual demand. In Figure 4(a), there are 5 LD types $t_1, t_2, ..., t_5$, and for these LD types the amounts of nets that have been routed through the 2D edge e are respectively 2, 12, 12, 3, and 1. In addition, the 2D edge e is the one projected from the 3D edges e'_1, e'_2, e'_3, and e'_4 with respective capacities being 10, 10, 5, and 5. Suppose the virtual demand $vd_e(t_5)$ is the target to calculate. Our pseudo layer assignment method iteratively selects a 3D edge from the top and a 3D edge from the bottom and decides how many nets of each LD type other than t_5 can be assigned to the two 3D edges without exceeding their capacities.[1] The method terminates when no more 3D edge can be selected. At the first iteration, our method chooses e'_4 and e'_1 and follow the order of t_1, t_2, t_3, t_4 to assign nets of different LD types to the two 3D edges. It

---

[1] By assuming each layer range has an equal probability to be specified in a layer directive, layers closer to the middle have higher chances of being within layer ranges of LD nets than layers closer to the top or bottom. Therefore our pseudo layer assignment method considers layers from the top and bottom towards the middle.

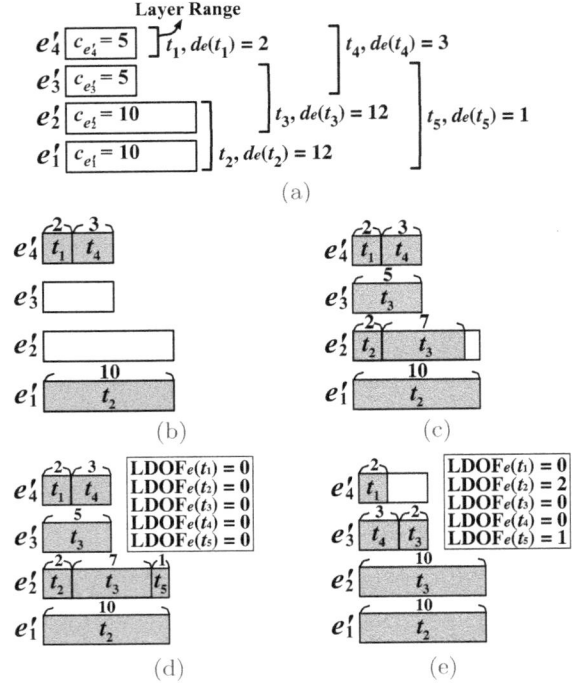

Figure 4: (a) A 2D routing result on an edge e. (b) and (c) Detailed steps of calculating the virtual demand $vd_e(t_5)$. (d) A good pseudo layer assignment result that is generated by our method and the associated LDOF values. (e) A bad pseudo layer assignment result and the associated LDOF values.

first assigns 2 nets of type t_1 and 3 nets of type t_4 to e'_4 (because only these two types of nets have a preferred layer on which e'_4 is located), and then assigns 10 nets of type t_2 to e'_1, under the capacity constraints imposed by e'_4 and e'_1 (see Figure 4(b)). At the second iteration, e'_3 and e'_2 are selected, and this time 5 nets of type t_3 are first assigned to e'_3, and then 2 remaining nets of type t_2 and 7 remaining nets of type t_3 get assigned to e'_2, under the capacity constraints imposed by e'_3 and e'_2 (see Figure 4(c)). No more iteration is needed because all 3D edges have been processed. As a result, for LD type t_5, its virtual capacity is consumed by 24 nets whose LD types appear before t_5 in the sorted order, and hence its virtual demand is $24 + d_e(t_5) = 24 + 1 = 25$.

For each 2D edge e, we also calculate a LDOF for each LD type t_i, which is denoted by $LDOF_e(t_i)$. To calculate $LDOF_e(t_i)$, our pseudo layer assignment method follows the sorted order of LD types and finds a pseudo layer assignment result that tells how many LD nets that have been routed through e can be assigned to their preferred layers without causing any overflow. From the pseudo layer assignment result, we find out the amount of LD nets of type t_i that did not get assigned to their preferred layers, and this amount is defined as $LDOF_e(t_i)$. The LDOF of each 2D edge e, denoted by $LDOF_e$, is calculated as the sum of all $LDOF_e(t_i)$'s, and the total LDOF of a 2D global routing solution is the sum of all $LDOF_e$'s.

Figure 4(d) shows the overall pseudo layer assignment result produced by our method for the example given in Figure 4(a), after assigning the net of type t_5 to e'_2. For this example, it is not hard to see that each type of LDOF is 0. If a different pseudo layer assignment method is used to generate another assignment result as shown in Figure 4(e), then $LDOF_e(t_2)$ becomes 2 and $LDOF_e(t_5)$ becomes 1; as a result, $LDOF_e$ becomes 3. This example shows that different pseudo layer assignment results may produce different LDOF values. Hence it is important for a 2D router to have a pseudo layer assignment method that can reliably find a assignment result with as low LDOF as possible, so that the 2D router can effectively adjust the amount of efforts in fine-tuning a 2D routing solution. Our method actually serves the purpose because it tries to find a assignment result that minimizes the LDOF for each 2D edge.

It is worthy noting that, our router is also able to generate the same 2D solution quality as GLADE on the ICCAD 2009 benchmarks. This fact will also be confirmed by our experimental results in Section 7.1. Nevertheless, our layer assignment algorithm, to be described in the next section, is different from that adopted by GLADE, and hence our final 3D solution may have a different wirelength.

6. LAYER ASSIGNMENT

The main objective of layer assignment in our router is to build a 3D routing result whose total overflow (TOF) is identical to that of a given 2D result. In our problem, layer assignment also needs to minimize LD violation and via count. In order to satisfy those objectives, we develop a two-stage layer assignment algorithm in our router. The first stage adopts COLA [12] to build an initial 3D with via count as small as possible, and the second stage applies a min-cost max-flow based refinement procedure to iteratively refine the 3D result for simultaneous minimization of LD violation and via count. In the following subsections, we describe our two-stage layer assignment algorithm in detail.

6.1 Initial Layer Assignment

At the first stage, we use COLA [12] to generate an initial layer assignment result whose TOF is identical to that of a 2D result. COLA assigns one net at a time for via count minimization but without considering layer directives, and therefore its result may have LD violation even though the LDOF of a given 2D result is 0. However, since we rely on a refinement method which will minimize the LD violation, we do not need to consider LD violation minimization at this stage. Besides, there are several advantages for our router to use COLA. First, COLA is fast such that we can get an initial 3D result in a short time. Second, COLA has notable effects on via count minimization. Third, COLA guarantees that the 3D layer assignment result will have TOF identical to that of a 2D result. These advantages let the following refinement method be able to focus more on LD violation minimization than via count minimization

6.2 Refinement

To fix the LD violation and further reduce the via count from the result generated by COLA in the first stage, we adopt a min-cost max-flow based refinement procedure in the second stage of our layer assignment algorithm. This procedure refines one 2D edge at a time. We discuss this procedure according to the two possible overflow conditions of a chosen 2D edge: overflow-free and overflowed.

6.2.1 Refinement on an Overflow-Free Edge

We use Figure 5 to explain how to use a min-cost max-flow technique to refine the result on an overflow-free 2D

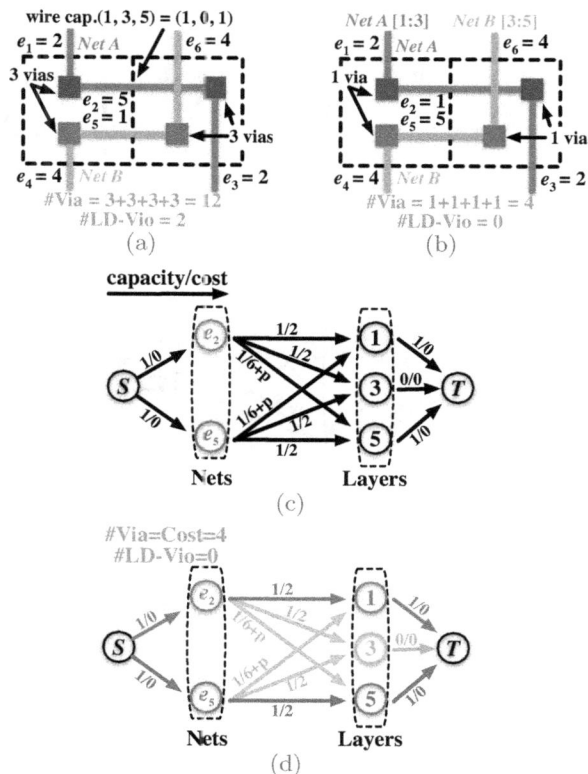

Figure 5: A refinement example on an overflow-free 2D edge by min-cost max-flow. (a) A bad layer assignment result with via count equal to 12 and 2 units of LD violation. (b) A good layer assignment result with via count equal to 4 and 0 unit of LD violation. (c) The flow network for re-assigning edges e_2 and e_5. (d) A min-cost max-flow result of (c) with via count equal to 4 and 0 unit of LD violation.

edge. Figure 5(a) shows the layer assignment results for two LD nets, Net A and Net B. The layer ranges of Net A and Net B are respectively [1 : 3] and [3 : 5]. Net A has three edges e_1, e_2, and e_3 which are assigned to layers 2, 5, and 2, respectively, and Net B has three edges e_4, e_5, and e_6 which are assigned to layers 4, 1, and 4, respectively. The capacities of the edge which e_2 and e_5 pass through are 1, 0, and 1 on layers 1, 3, and 5, respectively; note that this edge is overflow-free; therefore, the layer re-assignment result on this edge shall not cause any overflow. The scenario shown in Figure 5(a) has a via count equal to 12 and 2 units of LD violation due to the inappropriate layer assignments of e_2 and e_5. Figure 5(b) shows another scenario where the via count is reduced to 4, and there is no LD violation if we could re-assign e_2 to layer 1 and e_5 to layer 5. In order to generate this result, we take nets which pass through the same 2D edge and model this layer re-assignment problem of their involved edges as a min-cost max-flow problem. The flow network is shown in Figure 5(c). In the 4-level flow network, from left to right, the first level is source (labeled as S), the second level is composed of the nets that need to be re-assigned, the third level is composed of the assigning layers, and the fourth level is terminal (labeled as T). With this model, we could use capacity to model constraints like the capacity of a layer, net passage, and the layer range of a LD net, and use cost to model via count. For each arc from

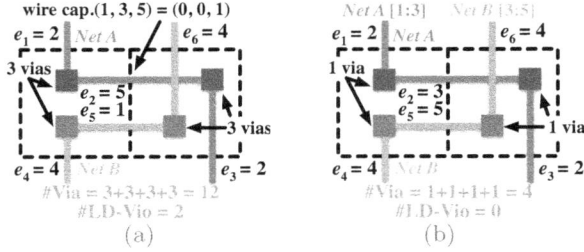

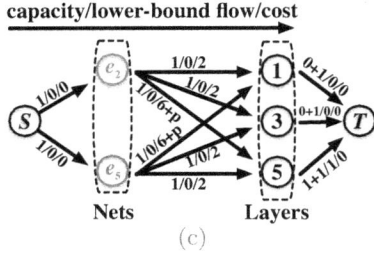

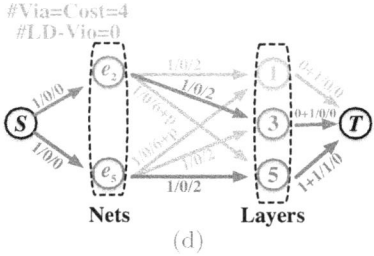

Figure 6: A refinement example on an overflowed 2D edge by min-cost max-flow. (a) A bad layer assignment result with via count equal to 12 and 2 units of LD violation. (b) A good layer assignment result with via count equal to 4 and 0 unit of LD violation. (c) The flow network for re-assigning edges e_2 and e_5. (d) A min-cost max-flow result of (c) with via count equal to 4 and 0 unit of LD violation.

the source to an involved net, its capacity is 1 representing the net passage, and cost is 0. For each arc from an involved net to a layer, if this net is a LD net, its capacity is 1 if the assigning layer is within its layer range, and its cost is the number of vias for connectivity; otherwise, its capacity is 1 and cost is equal to the sum of a *penalty cost* (denoted by p in Figure 5(c) and Figure 5(d)) and the number of vias for connectivity; note that this penalty cost is much larger than the sum of all possible vias that shall be used for re-assignment. On the other hand, if this net is not a LD net, its capacity is 1 representing this net and cost is the required number of vias for connectivity. For each arc from a layer to the terminal, its capacity is equal to the capacity of a layer and cost is 0. Figure5(d) shows a min-cost max-flow result of the flow network in Figure 5(c), which can be used to derive the layer assignment result shown in Figure 5(b).

6.2.2 Refinement on an Overflowed Edge

We explain our refinement method on an overflowed 2D edge by the example shown in Figure 6. Figure 6(a) is similar to Figure 5(a) except the capacity of the edge which e_2 and e_5 pass through is 0 on layer 1. This means this edge has 1 unit of overflow. Figure 6(b) shows the result we anticipate after refinement. Since we want to keep TOF the same, we modify the flow network and show it in Figure 6(c). The modification is that there are an upper and a lower bounds

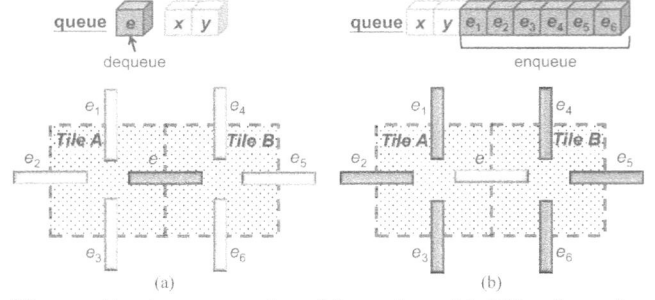

Figure 7: An example of how to add 2D edges for refinement. (a) e is the edge we are refining now, and $e_1 \sim e_6$ are the edges who connect the same tiles with e. (b) If there is any improvement on via count and LD violation from the refinement of e, $e_1 \sim e_6$ are enqueued.

of the flow on each arc now. For each arc except the outgoing arcs from a layer in this model, its upper-bound flow will be identical to the capacity of the corresponding arc in Figure 5(c), and its lower-bound flow is 0. For the outgoing arc of a layer, its upper-bound flow is the sum of its capacity and the overflow value of this chosen edge, and its lower-bound flow is its capacity. The reason we set the flow with upper and lower bounds is to keep TOF the same after refinement. Figure 6(d) shows a min-cost max-flow result of the flow network in Figure 6(c), which can be used to derive the layer assignment result shown in Figure 6(b).

6.2.3 Overall Refinement Procedure

The aforementioned methods illustrate how we refine a given 2D edge. We now describe the overall refinement procedure. We firstly put all 2D edges into a queue, and then perform the min-cost max-flow refinement based for each popped 2D edge due to dequeue. The reason why we firstly enqueue all 2D edges to refine is because that the initial 3D layer assignment result derived from COLA may have LD violation; in order to minimize LD violation which may happen on each 2D edge, it is necessary to refine each edge once. During this process, if the refinement result of a 2D edge has improvement on via count or LD violation, we enqueue the edges which connect the same tiles with this refined 2D edge. We use Figure 7 to illustrate this. In Figure 7(a), there are two adjacent tiles, $TileA$ and $TileB$, which are connected by edge e. $TileA$ has four edges, e_1, e_2, e_3, and e. $TileB$ has four edges e_4, e_5, e_6, and e. Assume that e is the 2D edge popped from the queue, and its refinement result will re-assign those 3D edges on it. If there is any improvement on via count or LD violation, e_1, e_2, and e_3 on $TileA$ and e_4, e_5, and e_6 on $TileB$ may have chances to improve their own layer assignment results. Therefore, in Figure 7(b), we enqueue those edges for further possible improvements. This process will be performed until there is no edge in the queue. Note that there is a hash table to store those 2D edges in the queue to circumvent the duplications of edges for saving run time.

The refinement procedure is applied to minimize LD violation and via count. However, after fixing the LD violation derived from the result of COLA, the via count may increase after refinement. As shall be seen in Section 7.2, it is possible that total wirelength increases due to via count increase after refinement, because LD violation decreases.

6.3 Remarks: Pseudo Layer Assignment

In Section 5, we introduce the pseudo layer assignment method to help us calculate the LDOF of a given 2D edge. Since this method calculates how many wires shall be assigned into each layer, it also can be easily modified and then applied to layer assignment as long as vias are assigned to connect wires. Although we can generate a 3D result with the consideration of minimizing LD violation by applying this method, as shall be seen in Section 7.2, this method suffers from the lack of consideration on via count minimization, and therefore, its performance on total wirelength is not satisfactory.

7. EXPERIMENT RESULTS

In this section, we first compare our router with GLADE [4] on the ICCAD 2009 benchmarks, and then introduce the experiments on the test cases with general layer directives where we compare our router with NTHU-Route 2.0. Our router is implemented in C++ and is run on a Linux machine with an Intel Core Duo 2.2Ghz CPU and 8GB memory.

7.1 Experiments on the ICCAD 2009 Benchmarks

First, we compare our router with GLADE on the ICCAD 2009 benchmarks. The experiment results of GLADE and our router (denoted by "Ours") are detailed in Table 1; for each router, we report the following results: total overflow (denoted by "TOF"), total LDOF of a 2D routing solution (denoted by "LDOF"), total LD violation (denoted by "LD Vio"), total wirelength (denoted by "TWL"), and run time (denoted by "CPU(mins.)"). Note that under the columns of "LD Vio" and "TWL" within major column "Ours," the numbers before the slashes refer to the results before the refinement procedure, and the numbers after the slashes refer to the the results after the refinement procedure.

According to Table 1, GLADE and our router both generate identical TOF, 0 unit of LDOF, and 0 unit of LD violation for each benchmark. Before refinement, since COLA does not consider layer directives, the results have LD violation on each benchmark. After refinement, our router reduces LD violation to 0 for each benchmark and improves TWL by 0.4% on average. This fact clearly shows that our router is capable of generating LD violation free results with high-quality TOF and TWL results for the ICCAD 2009 benchmarks. Note that the increase of run time is mainly because our router spent time running the min-cost max-flow based refinement procedure which did not exist in GLADE.

7.2 Experiments on Test Cases with General Layer Directives

We derived the test cases with general layer directives from the ICCAD 2009 benchmarks. In the ICCAD 2009 benchmarks, there are two types of LD nets (i.e., layer ranges [3 : 6] and [5 : 6]) in 6-layer designs, and three types of LD nets (i.e., layer ranges [3 : 8], [5 : 8], and [7 : 8]) in 8-layer designs. By honoring those timing-critical nets, we generated the new test cases by randomly changing the layer range of each LD net to an arbitrary one, while still keeping each non-LD net intact.

We use these test cases to compare NTHU-Route 2.0 with our router. Besides, to compare with our method, we implement a layer assignment algorithm based on the pseudo layer

assignment method mentioned in Section 5 and also use the min-cost max-flow procedure to refine the result. The experiment results are shown in Table 2 where the major column "NTHU-Route 2.0" lists the results generated by NTHU-Route 2.0, the major column "Ours" lists the results generated by our router, the major column "Ours (pseudo LA)" lists the results generated by our router but with COLA being replaced by pseudo layer assignment. Note that under the columns of "LD Vio" and "TWL" within major columns "Ours" and "Ours (pseudo LA)," the numbers before the slashes refer to the results before the refinement procedure, and the later numbers refer to the the results after the refinement procedure.

From the comparison between the results of "NTHU-Route 2.0" and "Ours," we discuss the results from two aspects: before and after refinement. First, before refinement, our router improves LD violation by 49.8% with only 0.1% TWL increase on average, but still suffers from LD violation even though LDOF is equal to 0 for each test case. After refinement, our router successfully reduces LD violation to 0 for each test case with 0.2% TWL decrease on average. This clearly states the effeteness of the min-cost max-flow based refinement procedure on the minimizations of LD violation and TWL. Note that our router increases TOF for four test cases (i.e., bigblue4, newblue1, newblue4, and newblue7), but we think that they are due to general layer directives that constrain the solution space and hence affects the overflow qualities of our routing results. For the run time increase on each test case, it is because that our router needs time to handle general layer directives in 2D routing, and to run the min-cost max-flow based refinement method which does not exist in NTHU-Route 2.0.

Next, we compare the results of "Ours" and "Ours (pseudo LA)." Both approaches generate results with identical TOF and 0 unit of LD violation for each test case. However, pseudo layer assignment does not consider via minimization. Even with the help of the min-cost max-flow based refinement procedure, its results on TWL still cannot rival with those of our router. Furthermore, "Ours (pseudo LA)" consumes 48.1% more average run time than our router. This is because the layer assignment result generated by pseudo layer assignment leaves too much room for the refinement procedure to improve. As a result, the refinement method costs more time to reach a converged solution.

8. CONCLUSION

In this paper, we have presented a global router for considering general layer directives. The experiment results show that our router successfully honors all layer directives for each test case. Meanwhile, our router can also rival with GLADE, the only router in the literature that was designed specifically for handling a restricted set of layer ranges.

A possible future work is to improve our router such that it can further reduce the overflow values for the test cases that are difficult to route.

9. REFERENCES

[1] ISPD 2007 Global Routing Contest.
[2] ISPD 2008 Global Routing Contest.
[3] C. J. Alpert, C. Chu, and P. G. Villarrubia. The coming of age of physical synthesis. In *Proceedings of International Conference on Computer-Aided Design*, pages 246–249, San Jose, CA, 2007.

Table 1: Comparison between GLADE [4] and our router on the ICCAD 2009 benchmarks [16].

Benchmark	GLADE [4]					Ours				
	TOF	LDOF	LD Vio	TWL	CPU(mins.)	TOF	LDOF	LD Vio	TWL	CPU(mins.)
adaptec1	0	0	0	45.4	7.0	0	0	59849/0	45.2/45.3	10.3
adaptec2	0	0	0	43.9	1.4	0	0	183623/0	43.2/43.8	4.2
adaptec3	0	0	0	115.2	7.2	0	0	210387/0	115.0/114.9	11.3
adaptec4	0	0	0	106.5	1.8	0	0	283214/0	105.9/106.5	3.9
adaptec5	0	0	0	130.1	15.2	0	0	66706/0	129.9/129.6	26.0
bigblue1	0	0	0	48.3	8.7	0	0	53858/0	48.5/48.5	17.1
bigblue2	0	0	0	69.6	7.8	0	0	7248/0	69.6/69.1	10.4
bigblue3	0	0	0	105.9	3.8	0	0	45669/0	105.7/105.5	10.4
bigblue4	188	0	0	178.9	121.0	188	0	71248/0	178.7/177.6	324.8
newblue1	2	0	0	35.6	4.8	2	0	6314/0	35.6/35.5	8.7
newblue2	0	0	0	59.7	0.8	0	0	49218/0	59.5/59.6	2.4
newblue4	140	0	0	108.1	40.1	140	0	45643/0	107.9/107.7	48.6
newblue5	0	0	0	190.7	12.6	0	0	9031/0	190.7/190.3	20.8
newblue6	0	0	0	139.8	11.5	0	0	26887/0	139.8/139.0	23.7
newblue7	78	0	0	281.7	119.9	78	0	113369/0	281.2/279.3	169.9
Comp.	-	-	1.000	1.000	1.000	-	-	-/1.000	0.998/0.996	1.904

Table 2: Comparison between NTHU-Route 2.0 [5] and our router on the test cases with general layer directives. Note that the "LDOF" column is omitted in the major column "Ours (pseudo LA)" since it is identical to the one in the major column "Ours."

Case	NTHU-Route 2.0 [5]				Ours					Ours (pseudo LA)			
	TOF	LD Vio	TWL	CPU(mins.)	TOF	LDOF	LD Vio	TWL	CPU(mins.)	TOF	LD Vio	TWL	CPU(mins.)
adaptec1	0	95066	45.1	5.3	0	0	46284/0	45.1/45.2	11.5	0	0/0	60.5/49.1	11.9
adaptec2	0	289132	43.1	1.3	0	0	145810/0	43.2/43.7	3.3	0	0/0	58.2/46.9	4.3
adaptec3	0	394924	114.9	5.6	0	0	205585/0	115.0/114.9	12.1	0	0/0	142.9/122.9	13.3
adaptec4	0	440412	105.9	1.7	0	0	222726/0	105.9/106.4	4.2	0	0/0	137.2/113.9	5.2
adaptec5	0	120402	129.8	15.7	0	0	59885/0	129.9/129.9	27.3	0	0/0	166.4/139.3	32.1
bigblue1	0	139562	47.8	7.0	0	0	58014/0	48.4/48.5	17.9	0	0/0	62.3/52.7	27.9
bigblue2	0	23070	69.3	6.0	0	0	11067/0	70.3/69.8	17.6	0	0/0	91.5/75.6	25.2
bigblue3	0	101772	105.7	3.7	0	0	54809/0	105.7/105.5	10.6	0	0/0	178.4/117.0	20.7
bigblue4	162	130542	178.7	75.8	236	0	66851/0	178.2/177.1	135.9	236	0/0	297.0/197.4	146.3
newblue1	0	13224	35.6	3.9	6	0	7513/0	35.6/35.5	8.1	6	0/0	50.0/38.7	9.1
newblue2	0	90746	59.4	0.9	0	0	44752/0	59.5/59.5	2.6	0	0/0	84.8/63.8	4.4
newblue4	138	73450	108.3	65.4	156	0	37856/0	107.7/107.5	51.0	156	0/0	145.5/116.9	53.4
newblue5	0	36910	190.7	12.8	0	0	19891/0	190.6/190.2	21.2	0	0/0	258.3/206.3	34.3
newblue6	0	36276	139.8	10.4	0	0	17986/0	139.8/139.0	25.4	0	0/0	193.1/150.7	47.1
newblue7	62	174794	279.8	57.8	82	0	84953/0	280.5/278.4	148.3	82	0/0	453.8/308.4	193.2
Comp.	-	1.000	1.000	1.000	-	-	0.502/0.000	1.001/0.998	1.818	-	0.000/0.000	1.439/1.088	2.299

[4] Y.-J. Chang, T.-H. Lee, and T.-C. Wang. GLADE: A modern global router considering layer directives. In *Proceedings of International Conference on Computer-Aided Design*, pages 319–323, San Jose, CA, 2010.

[5] Y.-J. Chang, Y.-T. Lee, J.-R. Gao, P.-C. Wu, and T.-C. Wang. NTHU-Route 2.0: A robust global router for modern designs. *IEEE Transactions on Computer-Aided Design of Integrated Circuits and Systems*, 29(12):1931–1944, 2010.

[6] H.-Y. Chen, C.-H. Hsu, and Y.-W. Chang. High-performance global routing with fast overflow reduction. In *Proceedings of Asia and South Pacific Design Automation Conference*, pages 582–587, Yokohama, Japan, 2009.

[7] M. Cho, K. Lu, K. Yuan, and D. Z. Pan. Boxrouter 2.0: A hybrid and robust global router with layer assignment for routability. *ACM Transactions on Design Automation of Electronic Systems*, 14(2):1–21, 2009.

[8] C. Chu and Y.-C. Wong. FLUTE: Fast lookup table based rectilinear steiner minimal tree algorithm for VLSI design. *IEEE Transactions on Computer-Aided Design of Integrated Circuits and Systems*, 27(1):70–83, 2008.

[9] K.-R. Dai, W.-H. Liu, and Y.-L. Li. Efficient simulated evolution based rerouting and congestion-relaxed layer assignment on 3-d global routing. In *Proceedings of Asia and South Pacific Design Automation Conference*, pages 570–575, Yokohama, Japan, 2009.

[10] C.-H. Hsu, H.-Y. Chen, and Y.-W. Chang. Multi-layer global routing considering via and wire capacities. In *Proceedings of International Conference on Computer-Aided Design*, pages 350–355, San Jose, CA, 2008.

[11] J. Hu, J. A. Roy, and I. L. Markov. Completing high-quality global routes. In *Proceedings of International Symposium on Physical Design*, pages 35–41, San Francisco, CA, 2010.

[12] T.-H. Lee and T.-C. Wang. Congestion-constrained layer assignment for via minimization in global routing. *IEEE Transactions on Computer-Aided Design of Integrated Circuits and Systems*, 27(9):1643–1656, 2008.

[13] T.-H. Lee and T.-C. Wang. Robust layer assignment for via optimization in multi-layer global routing. In *Proceedings of International Symposium on Physical Design*, pages 159–166, San Diego, CA, 2009.

[14] Z. Li, C. J. Alpert, S. Hu, T. Muhmud, S. T. Quay, and P. G. Villarrubia. Fast interconnect synthesis with layer assignment. In *Proceedings of International Symposium on Physical Design*, pages 71–77, Portland, OR, 2008.

[15] M. D. Moffitt. Maizerouter: Engineering an effective global router. *IEEE Transactions on Computer-Aided Design of Integrated Circuits and Systems*, 27(11):2017–2026, 2008.

[16] M. D. Moffitt. Global routing revisited. In *Proceedings of International Conference on Computer-Aided Design*, pages 805–808, San Jose, CA, 2009.

[17] M. M. Ozdal and M. D. F. Wong. Archer: A history-based global routing algorithm. *IEEE Transactions on Computer-Aided Design of Integrated Circuits and Systems*, 28(4):528–540, 2009.

[18] M. Pan and C. Chu. Fastroute: a step to integrate global routing into placement. In *Proceedings of International Conference on Computer-Aided Design*, pages 464–471, San Jose, CA, 2006.

[19] M. Pan and C. Chu. Fastroute 2.0: A high-quality and efficient global router. In *Proceedings of Asia and South Pacific Design Automation Conference*, pages 250–255, Yokohama, Japan, 2007.

[20] W. Swartz. Issues in global routing. In *Proceedings of International Symposium on Physical Design*, pages 142–147, Portland, USA, 2008.

[21] T.-H. Wu, A. Davoodi, and J. T. Linderoth. A parallel integer programming approach to global routing. In *Proceedings of Design Automation Conference*, pages 194–199, Anaheim, CA, 2010.

[22] Y. Xu and C. Chu. An auction based pre-processing technique to determine detour in global routing. In *Proceedings of International Conference on Computer-Aided Design*, pages 305–311, San Jose, CA, 2010.

Obstacle-Aware Length-Matching Bus Routing

Jin-Tai Yan

Department of Computer Science and Information
Engineering, Chung-Hua University,
Hsinchu, Taiwan, R. O. C.

Zhi-Wei Chen

College of Engineering,
Chung-Hua University,
Hsinchu, Taiwan, R. O. C.

ABSTRACT

As clock frequency increases, signal propagation delay on PCBs is requested to meet the timing specification with very high accuracy. Generally speaking, the net length in a single layer can estimate the routing delay in a single-layer net. In this paper, given a set of r single-layer nets in a bus with their length constraints inside $m \times n$ routing grids with s obstacle grids, based on obstacle-aware region partition inside routing grids, obstacle-aware shortest path generation and two detouring operations, R-flip and C-flip, an efficient $O(mn+s^3)$ algorithm is proposed to generate the length-matching paths for obstacle-aware bus routing. Compared with the published CAFE router, our proposed routing algorithm can save 80.5% of CPU time to complete obstacle-aware length-matching bus routing with no length error for tested examples on the average.

Categories and Subject Descriptors

B.7.2 [**Integrated Circuits**]: Design Aids – *Placement and routing*

General Terms: Algorithms, Design

Keywords:

Length-matching constraint, Single-layer routing, Bus routing

1. INTRODUCTION

As the scale of modern electronic systems grows, the size of advanced printed circuit boards(PCBs) also increases to host thousands of signal nets. As clock frequency increases, signal propagation delay on PCBs is requested to meet the timing specification with very high accuracy. Generally speaking, the net length in a single layer can estimate the routing delay in a single-layer net. To meet the severe timing specification, the length-matching constraint (i.e., all the wires in a bus are expected to be routed with specified length bounds) in bus routing must be satisfied in high-performance PCBs. Clearly, the length-matching constraints make traditional routers not applicable in modern PCBs. Therefore, it is necessary for bus routing to develop a better constrained-driven approach to satisfy the length-matching constraint in high-performance PCBs.

For bus routing with length-matching constraints, some research results[1-3] have been proposed. In [1], based on the direction of the signal nets, the routing area can be partitioned and assigned to each

Permission to make digital or hard copies of all or part of this work for personal or classroom use is granted without fee provided that copies are not made or distributed for profit or commercial advantage and that copies bear this notice and the full citation on the first page. To copy otherwise, or republish, to post on servers or to redistribute to lists, requires prior specific permission and/or a fee.
ISPD'11, March 27–30, 201 , Santa Barbara, California, USA.
Copyright 2011 ACM 978-1-4503-0550-1/11/03...$10.00.

net. Furthermore, the length-matching constraints can be satisfied by using the detouring technique. However, the proposed approach does not consider the obstacles in modern PCB designs. In [2], a Lagrangian relaxation technique is introduced to improve the detouring result as the routing area does not include any obstacle. However, the proposed approach is not guaranteed to obtain a feasible solution for modern PCB designs with obstacles. In [3], based on the structure of bounded-sliceline grid (BSG), the length-matching routing is regarded as an area assignment problem in a gridless routing plane to handle any given topology. However, the proposed approach does not also consider the obstacles in modern PCB designs.

In bus routing for modern PCBs with obstacles, the length controllability of the signal net will be easy if the larger routing area is reserved for routing a signal net. However, the routing area is usually limited and shared by multiple nets. Therefore, the routing area with obstacles must be utilized as much as possible to satisfy the given length-matching constraints of multiple nets simultaneously. For any routing area with obstacles, the longest path of a single net can be obtained within the area. However, it is an NP-hard problem since its decision version is NP-complete[4]. Therefore, two heuristic algorithms[5-6] have been proposed to find the longest path of a single net in a routing area with obstacles. In [5], an upper bound for the wire length of a single net inside routing grids is firstly proposed. The proposed upper bound uses the concept of the biconnected component to exclude useless routing areas efficiently. Furthermore, the proposed algorithm generates the longer path so that the wire length is increased as much as possible. However, the proposed approach is not guaranteed to obtain the longest path for any signal net. In [6], based on the rectangular partition inside routing grids and the analysis of unreachable grids in rectangular pattern detouring, an efficient algorithm is proposed to generate the longest path inside routing grids. Recently, a connectivity-aware algorithm, CAFE[7], for the truck routing problem has been proposed to generate length-matching paths of multiple nets in a routing area with obstacles. However, the proposed approach is not guaranteed to obtain a length-matching solution with no length error.

In this paper, given a set of r single-layer nets in a bus with their length constraints inside $m \times n$ routing grids with s obstacle grids, based on obstacle-aware region partition inside routing grids, obstacle-aware shortest path generation and two detouring operations, R-flip and C-flip, an efficient $O(mn+s^3)$ algorithm is proposed to generate the length-matching paths for obstacle-aware bus routing. Compared with the published CAFE router[7], our proposed routing algorithm can save 80.5% of CPU time to complete obstacle-aware length-matching bus routing with no length error for tested examples on the average.

2. PROBLEM FORMULATION

In general, the routing area in a single layer can be represented by a set of routing grids and the route of a signal net can be represented by a thick path consisting of some connecting horizontal and vertical segments. One grid in the routing area that corresponds to an obstacle can be drawn by black and an arbitrary-shaped routing area can be represented by setting obstacles on the boundary in a rectangular routing area. For any net inside routing grids, two grids are specified as the locations of the start and target terminals in the net. According to the locations of the two terminals in a net, the connecting path of the net connects its terminals by horizontal or vertical segments and passes any available grid at most once. For any connecting path in a net, the *path length* is defined as the number of passed routing grids in the path. According to the locations of the start and target terminals in a net, the path length in the net will be guaranteed to be odd or even.

In [8], the flip operation has been proposed as a routing modification method according to a face on the plain graph. To increase the path length in a net, any path may be modified by using two operations[5][7], *R-flip* and *C-flip*, in detouring path insertion or replacement. For *R*-flip operation as illustrated in Figure 1(a), a partial path of length two can be detoured by inserting two adjacent routing grids and an inserted detouring path, *P'*, can be obtained by recursively using the *R*-flip operation on the path from S to T. Hence, the length of the inserted detouring path, *P'*, must be guaranteed to be even. The *C*-flip operation is a generalization of the *R*-flip operation. For *C*-flip operation as illustrated in Figure 1(b), a partial path, *P*, can be replaced by using another partial path, *P'*, with the same terminals. Hence, the length difference between the replaced path, *P'*, and the original path, *P*, must be guaranteed to be even.

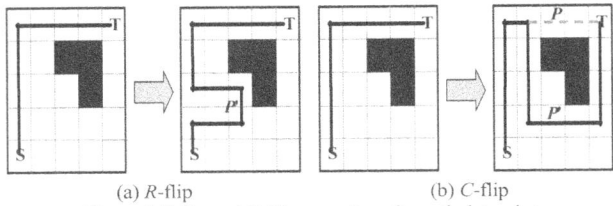

(a) *R*-flip   (b) *C*-flip

Figure 1 *R*-flip and *C*-flip operations for path detouring

Given two sequential sets of start and target terminals for r nets in a bus inside $m \times n$ routing grids with s obstacle grids, it is known that the lower bound of the i-th net for obstacle-aware length-matching bus routing can be obtained as d_i, $1 \leq i \leq r$, where d_i is the connecting length of the shortest path of the i-th net according to the locations of the start and target terminals for the given r nets and the locations of the given obstacles. For obstacle-aware length-matching bus routing inside routing grids, the length constraints, l_1, l_2,...,and l_r, of the r nets are given to satisfy the timing requirement in the bus. In general, the length difference among the given nets in a bus is small. If $l_i < d_i$, $1 \leq i \leq r$, or $\underset{i=1}{\overset{r}{Max}}(l_i) > mn - s$, it is known that the routing problem has no solution. In contrast, if $l_i \geq d_i$, $1 \leq i \leq r$, and $\underset{i=1}{\overset{r}{Max}}(l_i) \leq mn - s$, based on the given obstacles, the routing problem for obstacle-aware length-matching bus routing is to generate the length-matching paths of the r nets inside routing grids without overlapping any obstacle grid. As illustrated in Figure 2(a), two sequential sets of start and target terminals for 6 given nets in a bus are considered to be routed inside 20x18 routing grids with 82 obstacle grids. It is clear that the lower bounds of the 6 nets for obstacle-aware length-matching bus routing are 28, 28, 28, 28, 28 and 34, respectively, and the number of available routing grids is 278 for obstacle-aware length-matching bus routing. If the length constraints of all the 6 nets are set as 46, 6 length-matching paths can be generated for obstacle-aware length-matching bus routing and illustrated in Figure 2(b).

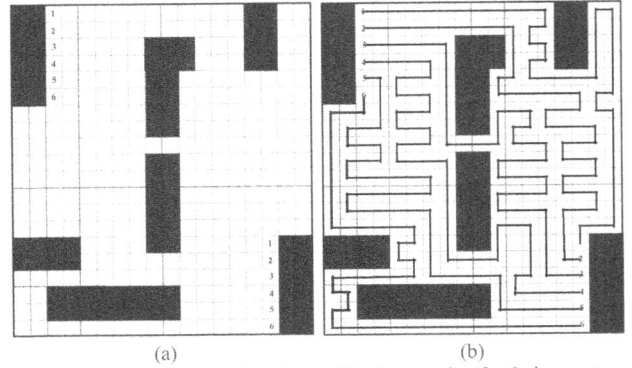

(a)   (b)

Figure 2 Obstacle-aware length-matching bus routing for 6 given nets

3. OBSTACLE-AWARE LENGTH-MATCHING BUS ROUTING

For obstacle-aware length-matching bus routing, an efficient algorithm is proposed to generate the length-matching paths of all the nets in a bus and the design flow is shown in Figure 3. Basically, the routing process can be divided into three sequential phases: *Obstacle-aware region partition inside routing grids*, *Obstacle-aware shortest-path generation* and *Iterative length-matching path generation*.

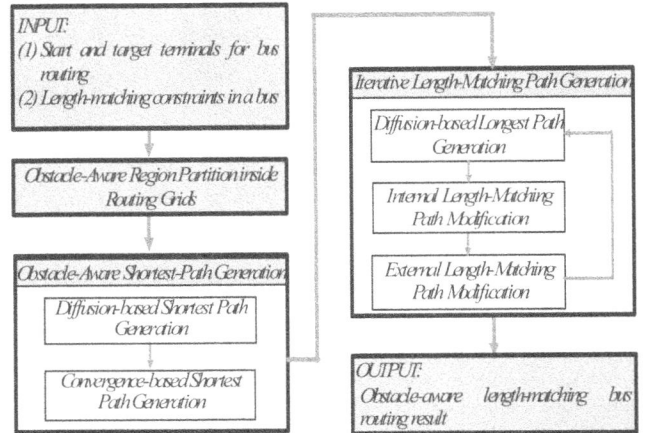

Figure 3 Design flow for obstacle-aware length-matching bus routing

3.1 Obstacle-Aware Region Partition inside Routing Grids

Initially, the minimum covering rectangle of all the routing grids including obstacles can be defined as the *outer boundary* including all the routing grids. Based on the locations of the horizontal and vertical boundaries on all the isolated obstacles and the outer boundary, the extended lines on the horizontal and vertical boundaries can partition all the routing grids into a set of rectangular routing regions. If two horizontal or vertical boundaries on a rectangular routing region are fully located onto the boundaries of two adjacent obstacles or the boundary of one obstacle and the outer boundary, the rectangular routing region will be defined as an *independent region*. In an iterative obstacle-aware region partition, firstly, the original independent regions inside routing grids can be found from the partitioned rectangular routing regions. After assigning the independent regions, the assigned independent regions can be treated as obstacles and a new set of rectangular routing regions can be obtained by extending the horizontal and vertical boundaries to partition the remaining routing grids. Until the remaining routing grid is empty, the iterative assignment process will stop. Hence, all the routing grids including obstacles can be partitioned into a set of assigned independent regions. If an assigned independent region is able to be horizontally and vertically merged with its adjacent independent region, the independent region can be defined as a *corner region*. To reduce the number of the assigned independent regions, all the corner regions in the set of assigned independent regions must be further merged with their adjacent independent regions. Finally, if the start(target) terminals of all the nets are distributed into different regions, the regions including all the start terminals will be merged into a *start(target) region*.

Based on the partitioned regions of all the routing grids and the adjacent relation between two partitioned regions, its corresponding adjacent graph, G(V, E), can be constructed as follows: the adjacent graph, G(V, E), is an undirected edge-weighted graph, where any vertex, V_i, in V represents a partitioned region, R_i, any edge, $E_{i,j}$, in E represents the adjacent relation between two partitioned regions, R_i and R_j, and the edge weight, $w_{i,j}$, represents the allowable net capacity between R_i and R_j. Refer to all the routing grids in Figure 2(a), the routing grids can be firstly partitioned into 53 routing regions as illustrated in Figure 4(a) according to the locations of the obstacle boundaries and the outer boundary. By using an iterative assignment process, 7 independent regions, IR_1, IR_2, IR_3, IR_4, IR_5, IR_6 and IR_7, can be assigned in the first iteration as illustrated Figure 4(b), 4 independent regions, IR_8, IR_9, IR_{10} and IR_{11}, can be assigned in the second iteration as illustrated Figure 4(c), 2 independent regions, IR_{12} and IR_{13}, can be assigned in the third iteration as illustrated Figure 4(d) and one independent region, IR_{14}, can be assigned in the fourth iteration as illustrated Figure 4(e), respectively. Furthermore, the corner region, IR_9, can be merged with its adjacent independent region, IR_2, to reduce the number of the partitioned regions. By renaming the regions, R_S, R_T, R_1, R_2, R_3, R_4, R_5, R_6, R_7, R_8, R_9, R_{10}, R_{11}, the obstacle-aware region partition result can be obtained and illustrated in Figure 4(f). Finally, the corresponding adjacent graph, G(V, E), can be constructed and illustrated in Figure 4(g), where V={ V_S, V_T, V_1, V_2, V_3, V_4, V_5, V_6, V_7, V_8, V_9, V_{10}, V_{11}} and E={ $E_{S,1}(w_{S,1}=2)$, $E_{S,2}(w_{S,2}=6)$, $E_{1,3}(w_{1,3}=2)$, $E_{2,4}(w_{2,4}=1)$, $E_{2,5}(w_{2,5}=5)$, $E_{3,8}(w_{3,8}=3)$, $E_{4,8}(w_{4,8}=1)$, $E_{5,6}(w_{5,6}=1)$, $E_{5,9}(w_{5,9}=2)$, $E_{6,7}(w_{6,7}=1)$, $E_{7,10}(w_{7,10}=1)$, $E_{8,11}(w_{8,11}=2)$, $E_{8,T}(w_{8,T}=6)$, $E_{9,T}(w_{9,T}=2)$, $E_{10,T}(w_{10,T}=1)\}$.

3.2 Obstacle-Aware Shortest-Path Generation

Based on all the adjacent relations and the allowable capacities of all the edges in an adjacent graph, it is known that the number of the maximum routed paths from R_S to R_T can be obtained by running a maximum-flow algorithm[9] from V_S to V_T. If any vertex has only one in-flow edge and one out-flow edge in the adjacent graph and the in-flow of the vertex in the maximum-flow result is equal to the allowable capacity into the corresponding region, the region representing the vertex will be defined as a *critical region*. If the maximum flow in the adjacent graph is smaller than the number of nets in a bus, obstacle-aware bus routing will not be completed. In contrast, if the maximum flow in the adjacent graph is larger than or equal to the number of nets in the bus, all the nets in a bus from R_S to R_T must be assigned onto their region paths in a balanced distribution for obstacle-aware bus routing. It is assumed that M is the maximum flow in the adjacent graph, M flows on the routed paths are sequential numbered from 1 to M and r nets are sequentially numbered from 1 to r. To efficiently use the routing grids to generate length-matching paths for obstacle-aware bus routing, the i-th net in the bus will be assigned onto the f_i-th flow on the routed paths as

$$f_i = 1 + \left\lfloor \frac{(i-1)(M-1)}{r-1} \right\rfloor, \quad if \quad 1 \le i \le r.$$

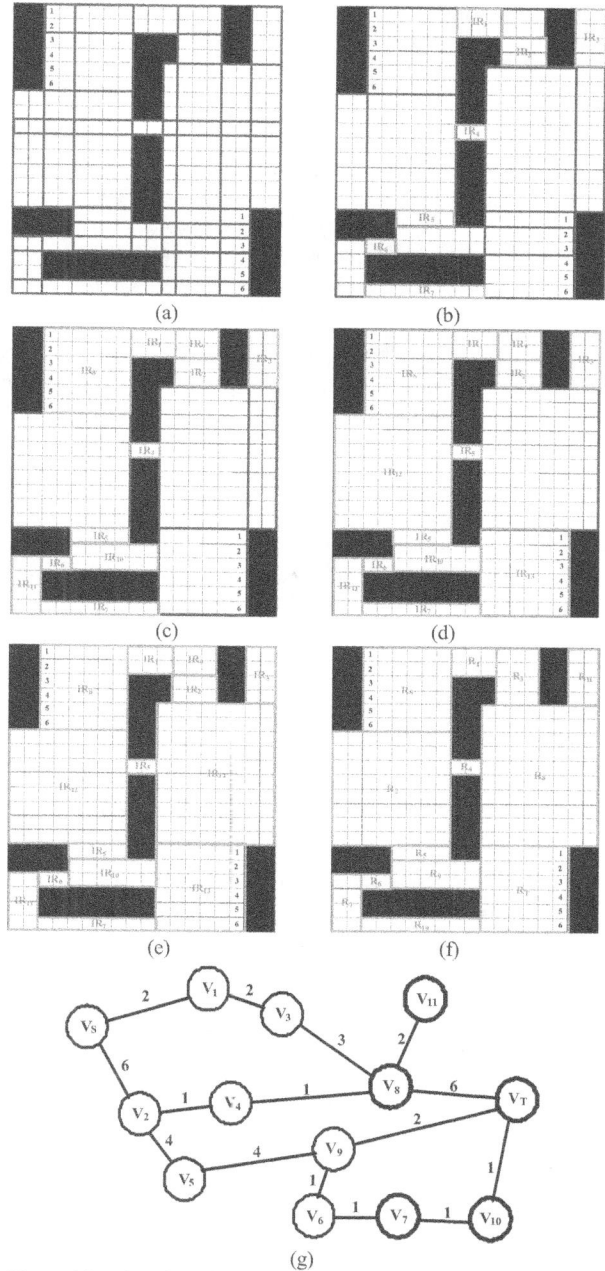

(a)   (b)

(c)   (d)

(e)   (f)

(g)

Figure 4 Iterative obstacle-aware region partition and its corresponding adjacent graph

After all the nets are assigned onto the regions on the region paths, it is known that any region may be only used by some assigned nets. If any region is only used by an assigned net, the region will be defined as a *private region* of the net, that is, all the routing grids inside the region will be only used by the route of the net. Besides that, if any region is not used by any assigned net, the region will be defined as a *share region*, that is, all the routing grids inside the region will be used by the routes of all the possible nets. In contrast, if any region is only used by some assigned nets, the region will be defined as a *public region* of these assigned nets, that is, all the routing grids inside the region will be only used by the routes of these assigned nets. Refer to the adjacent graph in Figure 4(b), the maximum flow from V_S to V_T can be obtained as 6 by running a maximum-flow algorithm as illustrated in Figure 5(a). Based on the pair of the used flow and the allowable capacity on any edge, it is known that the regions, R_1, R_3, R_4, R_6, R_7 and R_{10}, are *critical regions* and the 6 nets in the bus are assigned onto 4 region paths from R_S to R_T. According to the maximum-flow result in Figure 5(a) and the flow assignment of all the nets, nets 1 and 2 will be assigned onto the region path, V_S->V_1->V_3->V_8->V_T, net 3 will be assigned onto the region path, V_S->V_2->V_4->V_8->V_T, net 4 and 5 will be assigned onto the region path, V_S->V_2->V_5->V_9->V_T, and net 6 will be assigned onto the region path, V_S->V_2->V_5->V_9->V_6->V_7->V_{10}->V_T. Clearly, the region, R_4, is a *private region* of net 3, and the regions, R_6, R_7 and R_{10}, are private regions of net 6. The region, R_{11}, is a *share region* of net 1. Besides that, the regions, R_1 and R_3, are *public regions* of net 1 and 2, the region, R_2, is a *public region* of net 3, 4, 5 and 6, the region, R_5, is a public region of net 4, 5 and 6, the region, R_8, is a public region of net 1, 2 and 3, the region, R_9, is a *public region* of net 4, 5 and 6, and R_S and R_T, are *public regions* of net 1, 2, 3, 4, 5 and 6.

In general, the minimum rectangular area including two endpoints can be defined as the routing area of the two-endpoint segment. Based on the minimum-distance maintenance inside its routing area for any two-endpoint segment and the region paths for all the nets in the bus, the shortest paths of all the nets can be assigned onto the routing grids of the regions in the region paths to generate their initial obstacle-aware paths. To guarantee the success of routing the shortest paths of all the nets, the shortest paths of all the nets must be sequentially generated in a diffusion-based style. In diffusion-based shortest path generation, the first and last nets are firstly considered to be routed and their shortest paths pass as the outer routing grids on their region paths as possible. After generating the shortest paths of the first and last nets, the routing grids of the two shortest paths will be treated as obstacles and the shortest paths of the next two unrouted nets will be generated in the diffusion-based process. Until all the nets are routed, the diffusion-based process will stop and the diffusion-based shortest paths of all the nets in the bus will be generated. Furthermore, to guarantee to satisfy the length-matching constraints of all the nets in iterative length-matching path generation, the available routing grids must be integrated together and distributed in the outer routing area. Based on the minimum-distance maintenance inside its routing area for any two-endpoint segment and the routing order of the diffusion-based shortest paths, the convergence-based shortest paths of all the nets in the bus can be reassigned onto the routing grids of the regions in the region paths by using a reverse routing order. Refer to the region paths of the 6 nets in Figure 5(a), the diffusion-based shortest paths of the 6 nets can be assigned onto the routing grids of the regions in their region paths as illustrated in Figure 5(b). It is clear that the lengths of the shortest paths of nets 1, 2, 3, 4 and 5 can be obtained as 28 and the length of the shortest path of net 6 can be obtained as 34. Based on the routing result of the diffusion-based shortest paths for the 6 nets, the convergence-based shortest paths of the 6 nets can be further assigned onto the routing grids of the regions in their region paths as illustrated in Figure 5(c).

3.3 Iterative Length-Matching Path Generation

Based on the convergence-based shortest paths of all the nets in a bus, the length-matching paths of all the nets in the bus can be iteratively generated and the iterative length-matching path generation can be further divided into three steps: *Diffusion-based*

longest path generation, Internal length-matching path modification and *External length-matching path modification.*

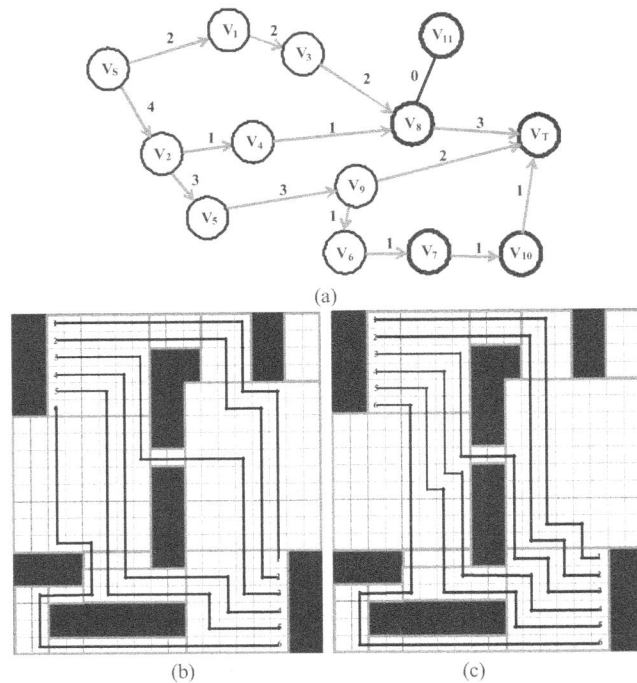

(a)

(b)   (c)

Figure 5 Diffusion-based and convergence-based shortest paths for 6 given nets

In diffusion-based longest path generation, the first and last nets are firstly considered to generate their diffusion-based longest paths. Based on the outer boundary of the routing grids and the locations of the obstacles, the diffusion-based longest path of the first or last net is the path of the routing grids along the outer boundary and the adjacent obstacles. Generally speaking, the diffusion-based longest path of the first or last net goes through the regions in its region path and the available share regions. After the diffusion-based longest paths of the first and last nets are generated, the lengths of two diffusion-based longest paths can be computed and the available routing grids which adjacent to the first and last nets can be obtained for length-matching path generation. Refer to the convergence-based shortest paths of the 6 nets in Figure 5(c), the first and sixth nets are considered in the first iteration for length-matching path generation. As the diffusion-based longest paths of the first and sixth nets are generated as illustrated in Figure 6, it is clear that the diffusion-based longest path of the first net goes through the regions, R_S, R_1, R_3, R_8 and R_T, in its region path and the share region, R_{11}, and the diffusion-based longest path of the sixth net goes through the regions, R_S, R_2, R_5, R_9, R_6, R_7, R_{10} and R_T, in its region path. As a result, the lengths of the longest paths of the first and sixth nets are both computed as 40, 46 available routing grids between the first and second nets can be drawn by yellow and obtained for the length-matching path generation of the first net and 36 available routing grids between the fifth and sixth nets can be drawn by yellow and obtained for the length-matching path generation of the sixth net.

If the length of the diffusion-based longest path of the first or last net is larger than its given constrained length, the diffusion-based longest path of the first or last net must be shortened to satisfy the length-matching constraint. In internal length-matching path modification, the partial or full paths inside the share regions must be shortened to satisfy the length-matching constraint. Generally speaking, if the given length-matching constraint is larger than or equal to its lower length-matching bound, the length-matching path will be obtained by shortening the diffusion-based longest path. As a result, the routing grids in the shortened partial or full paths inside share regions will be defined as *dead grids* and the dead grids will not be used for the routes of the other nets.

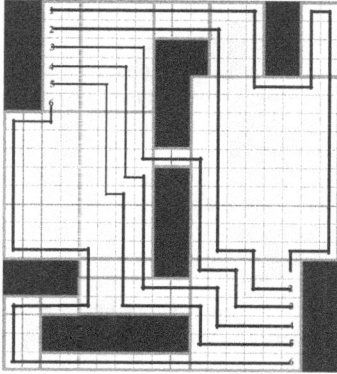

Figure 6 Diffusion-based longest path generation for the first and sixth nets

On the other hand, if the length of the diffusion-based longest path of the first or last net is shorter than its given constrained length, the diffusion-based longest path of the first or last net must be detoured to satisfy the length-matching constraint. In internal length-matching path modification, the length difference between the length of the diffusion-based longest path and its given constrained length must be firstly computed. Furthermore, the routing grids inside its available private and share regions must be considered to meet the length difference. If the number of routing grids inside its available private regions is larger than or equal to the length difference, some adjacent routing grids inside the private regions will be detoured and merged into the diffusion-based longest path to satisfy the length-matching constraint by using some R-flip or C-flip operations. If the number of routing grids inside its available private regions is smaller than the length difference, all the routing grids inside the private regions and some adjacent routing grids inside the share regions will be detoured and merged into the diffusion-based longest path to satisfy the length-matching constraint by using some R-flip or C-flip operations. In contrast, if the number of the total routing grids inside the available private and share regions is smaller than the length difference, all the routing grids inside the private and share regions and some adjacent routing grids inside its public regions will be detoured and merged into the diffusion-based longest path to satisfy the length-matching constraint by using some R-flip or C-flip operations.

Refer to the diffusion-based longest paths of the first and sixth nets in Figure 6, if the constrained lengths of all the nets in the bus are 36, the diffusion-based longest paths of the first and sixth nets must be shortened to satisfy the length-matching constraints and the final shortened paths of the first and sixth nets can be illustrated in Figure 7(a). As a result, one dead region including 4 routing grids is drawn by purple and generated for the first net and two dead regions including 10 routing grids are drawn by purple and generated for the sixth nets in internal length-matching path modification. On the other hand, if the constrained lengths of all the nets in the bus are 46, as illustrated in Figure 7(b), 6 adjacent routing grids inside its public regions drawn by blue must be merged into the diffusion-based longest path of the first net and 6 adjacent routing grids inside its private and public regions drawn by blue must be merged into the diffusion-based longest path of the sixth net to satisfy the length-matching constraints by using 3 R-flip operations. As a result, the

final detoured paths of the first and sixth nets for length-matching path generation can be illustrated in Figure 7(c).

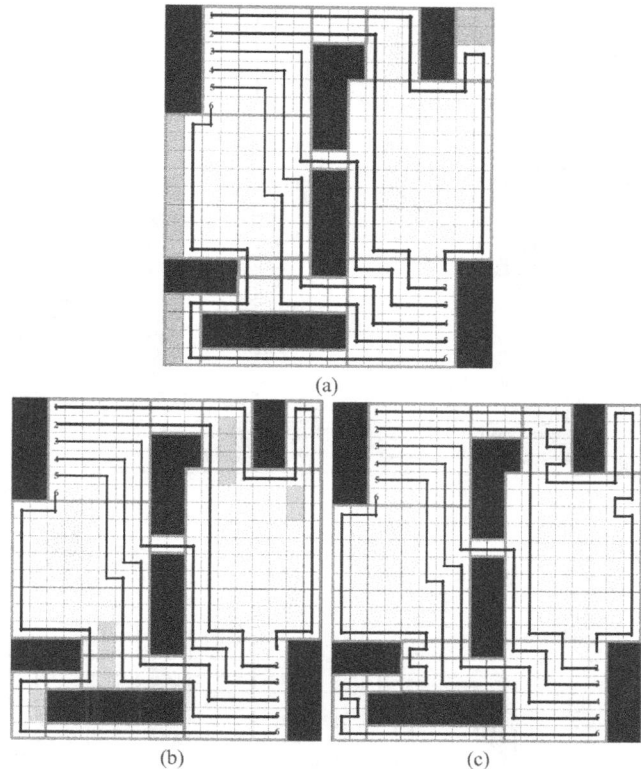

Figure 7 Internal length-matching path modification in the first iteration

After assigning the length-matching paths of the first and sixth nets, the routing grids in the two assigned length-matching paths will be treated as obstacles. Similarly, the second and fifth nets are considered in the second iteration for length-matching path generation. As the constrained lengths of all the nets in the bus are 36 or 46, the length-matching paths of the second and fifth nets will be generated by using some R-flip or C-flip operations and illustrated in Figure 8(a) and 8(b), respectively. After assigning the length-matching paths of the second and fifth nets, the routing grids in the two assigned length-matching paths will be treated as obstacles. Finally, the third and fourth nets are considered in the third iteration for length-matching path generation. As the constrained lengths of all the nets in the bus are 36 or 46, the length-matching paths of the third and fourth nets will be generated by using some R-flip or C-flip operations and illustrated in Figure 8(c) and 8(d), respectively. However, the length of the detoured path of the fourth net in Figure 8(d) does not satisfy its length-matching constraint in internal length-matching path modification.

In internal length-matching path modification, the length-matching paths are limited to only go through the regions in their region paths and the share regions. After running the internal length-matching path modification, the detoured path of any net may not satisfy the given length-matching constraint because the available routing grids inside the regions in their region paths and the share regions are exhaustively used. Hence, some adjacent routing grids in external adjacent regions must be detoured by using some R-flip or C-flip operations to satisfy the given length-matching constraint in external length-matching path modification. Refer to the length-matching paths of the 6 nets in Figure 8(d), it is clear that the detoured path of the fourth net does not satisfy the given length-matching constraint in internal length-matching path modification. Because the length difference is 2, 2 additional routing grids in an external region must be merged into the original detoured path of the fourth net. Hence, a partial path with 1 routing grid in R_T must be replaced with 3 routing grids in R_8 to satisfy the given length-matching constraint by using one C-flip operation in external length-matching detouring as illustrated in Figure 9.

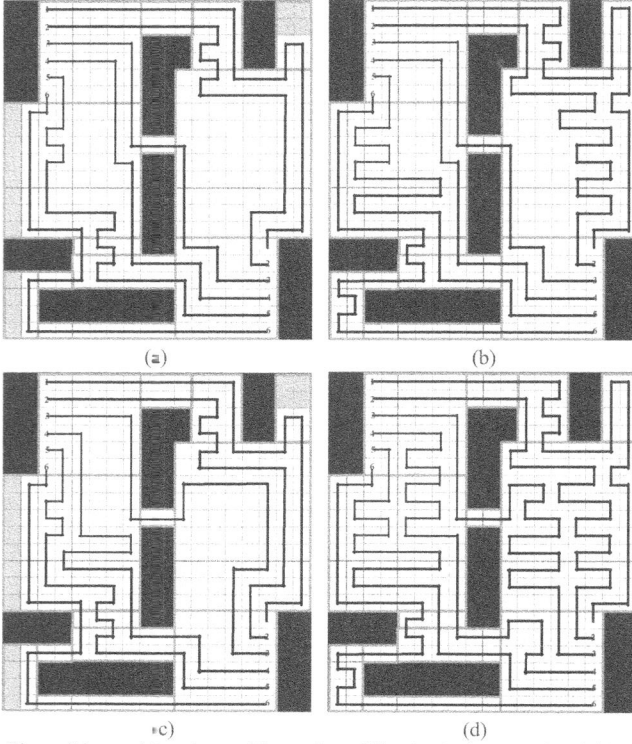

(a) (b)

(c) (d)

Figure 8 Internal length-matching path modification in the second and third iterations

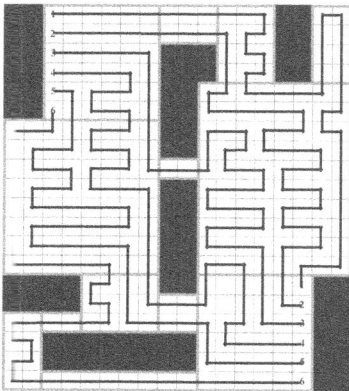

Figure 9 External length-matching path modification for the fourth net

3.4 Analysis of Time Complexity

As mentioned above, the routing process for obstacle-aware length-matching bus routing is divided into three phases: *Obstacle-aware region partition inside routing grids, Obstacle-aware shortest-path generation* and *Iterative length-matching path generation*. For the time analysis in obstacle-aware region partition, based on the iterative assignment process for the independent regions, it is clear that the time complexity of partitioning obstacle-aware

regions is $O(s^2)$. Since an adjacent graph is a planar graph and the number of partitioned regions is in $O(s)$, the time complexity of constructing an adjacent graph is $O(s)$. Furthermore, for the time analysis in obstacle-aware shortest path generation, the time complexity of finding the maximum flow in an adjacent graph is $O(s^3)$ and the time complexity of assigning all the nets onto feasible region paths is $O(rs)$. Based on the assigned region paths of all the nets, the time complexity of generating diffusion-based and convergence-based shortest paths of all the nets is $O(mn)$.

For the time analysis in iterative obstacle-aware length-matching path generation, the time complexity of generating two diffusion-based longest path in each iteration is $O(m+n)$ and the time complexity of performing internal and external length-matching path modification of two diffusion-based longest paths in each iteration is $O(l)$, where l is the given maximum length constraint. Hence, the time complexity of performing iterative obstacle-aware length-matching path generation is $O(r(m+n+l))$. As a result, the time complexity of performing obstacle-aware length-matching bus routing is $O(r(m+n+l)+s^3)$. Since $O(r(m+n+l))$ is equal to $O(mn)$, the time complexity of performing obstacle-aware length-matching bus routing is $O(mn+s^3)$.

4. EXPERIMENTAL RESULTS

For obstacle-aware length-matching bus routing, our proposed algorithm and the CAFE router[7] have been implemented by using standard C++ language and run on a Pentium QuadCore CPU 2.66 GHz machine with 2GB memory. Two tested examples, *data01* and *data02*, are from the previous published paper[7] and listed in Table I. In Table I, "*#Nets*" denotes the number of the given nets in a bus, "*Area*" denotes the dimension of the routing grids for tested examples, "*#Obstale Grids*" denotes the number of obstacle grids for tested examples, "*Length Constraint*" is the required length constraints of all the nets in a bus, and "*Length Error*" is the total length differences of all the nets in a bus based on the given length constraints. In this experiment, the length constraints of all the nets in a bus are set as an length-equivalent constant. Based on obstacle-aware region partition inside routing grids, obstacle-aware shortest path generation and two detouring operations, *R-flip* and *C-flip*, the length-matching routing results of the two tested examples, *data01* and *data02*, have no length error in our proposed routing algorithm. Compared with the CAFE router[7], the experimental results in Table I show that our proposed algorithm generates exact length paths with no length error for obstacle-aware bus routing. Clearly, the experimental results show that our proposed algorithm saves 80.5% of CPU time to generates more accurate length-matching paths for obstacle-aware bus routing than the CAFE router[7] for the tested two examples on the average.

For the tested example, *data01*, all the nets in the bus do not cross each other. In Figure 10, the obstacle-aware bus routing results of the tested example, *data01*, can be illustrated for two length constraints, 100 and 150, respectively. For the tested example, *data02*, some nets in the bus cross each other. In this experiment, the crossing phenomenon can be eliminated by using two-sided untangling[10] and the untangled paths can be routed behind the start and target terminals. After routing the untangled paths, the length constraints of the untangled nets must be modified to satisfy the original length requirements. In Figure 11, the obstacle-aware bus routing results of the tested example, *data02*, can be illustrated for two length-matching constraints, 251 and 301, respectively.

5. CONCLUSIONS

Based on obstacle-aware region partition inside routing grids, obstacle-aware shortest path generation and two detouring operations,

R-flip and *C-flip*, an efficient O($mn+s^3$) algorithm is proposed to generate the length-matching paths with no length error for obstacle-aware bus routing.

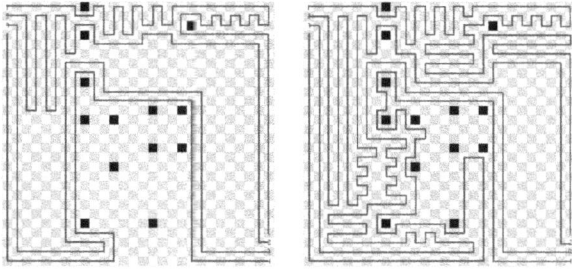

(a) Length constraint=100   (b) Length constraint=150

Figure 10 Experimental result for *data01*

6. REFERENCES

[1] M. M. Ozdal and M. D. F. Wong, "Algorithmic study of single-layer bus routing for high-speed boards," *IEEE Trans. on Computer-Aided Design of Integrated Circuits and Systems*, Vol. 25, no. 3, pp.490–503, 2006.

[2] M. M. Ozdal and M. D. F. Wong, "A length-matching routing algorithm for high-performance printed circuit boards," *IEEE Trans. on Computer-Aided Design of Integrated Circuits and Systems*, Vol. 25, no. 12, pp.2784–2794, 2006.

[3] T. Yan and M. D. F. Wong, "BSG-Route: A length-matching router for general topology," *International Conference on Computer-Aided Design*, pp.499-505, 2008.

[4] A. Itai, C. H. Papadimitriou, and J. L. Szwarcfiter, "Hamiltonian paths in grid graphs," *SIAM, Comp.*, Vol. 11, no. 4, pp.676–686, 1982.

[5] Y. Kohira, S. Suchiro and A. Takahashi, "A fast longer path algorithm for routing grid with obstacles using biconnectivity based length upper bound," *Asia and South Pacific Design Automation Conference*, pp.600–605, 2009.

[6] J. T. Yan, M. C. Jhong and Z. W. Chen, "Obstacle-aware longest path using rectangular pattern detouring in routing grids," *Asia and South Pacific Design Automation Conference*, pp.287-292, 2010.

[7] Y. Kohira and A. Takahashi, "CAFÉ router: A fast connectivity aware multiple nets routing algorithm for routing grid with obstacles," *Asia and South Pacific Design Automation Conference*, pp.281–286, 2010.

[8] Y. Kubo, Y. Takashima, S. Nakatake, and Y. Kajitani, "Self-reforming routing for stochastic search in VLSI interconnection layout," *Asia and South Pacific Design Automation Conference*, pp.87–92, 2000.

[9] T. H. Cormen, C. E. Leiserson and R. L. Rivest, *Introduction to Algorithms*, MIT Press, 2000.

[10] J. T. Yan and Z. W. Chen "Two-sided single-detour untangling for bus routing," *Design Automation Conference*, pp.206-211, 2010.

TABLE I EXPERIMENTAL RESULTS FOR OBSTACLE-AWARE LENGTH-MATCHING BUS ROUTING

Example	#Nets	Area	#Obstacle Grids	Length Constraint	CAFE[7]		Our Bus Routing	
					Length Error	CPU Time(s)	Length Error	CPU Time(s)
data01	4	28x28	13	100	0	0.35(100%)	0	0.08(22.9%)
				150	10	0.42(100%)	0	0.09(21.4%)
data02	13	61x130	3441	251	10	25.63(100%)	0	4.96(19.4%)
				301	6	26.85(100%)	0	5.28(19.7%)
Total					26	53.25(100%)	0	10.41(19.5%)

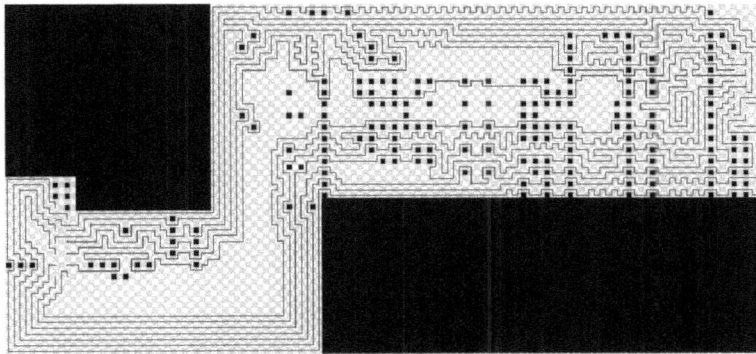

(a)   Length constraint=251

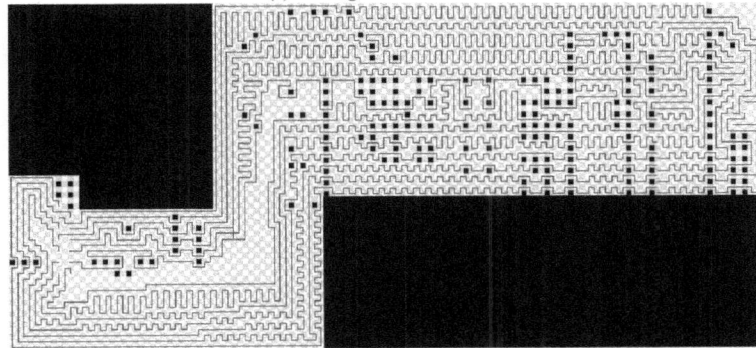

(b) Length constraint=301

Figure 11 Experimental result for *data02*

Co-Optimization of Droplet Routing and Pin Assignment in Disposable Digital Microfluidic Biochips*

Yang Zhao and Krishnendu Chakrabarty

Department of Electrical and Computer Engineering
Duke University, Durham, NC 27708, USA

ABSTRACT

The number of independent input pins used to control the electrodes in digital microfluidic "biochips" is an important cost-driver in the emerging market place, especially for disposable PCB devices that are being developed for clinical and point-of-care diagnostics. However, most prior work on pin-constrained biochip design considers droplet routing and the assignment of pins to electrodes as independent problems. We propose an integer linear programming (ILP)-based optimization method to solve the droplet-routing and the pin-mapping design problems concurrently. The proposed co-optimization method optimizes routing pathways, generates a single pin-assignment, and attempts to minimize the number of control pins. It also overcomes a major limitation of recent work on this problem—the method described in prior work can lead to infeasible designs due to the need to change the mapping of pins to electrodes dynamically during bioassay execution. The effectiveness of the proposed co-optimization method is demonstrated for a commercial biochip that is used to perform n-plex immunoassays, as well as an experimental chip for multiplexed *in-vitro* diagnostics.

Categories and Subject Descriptors

B.7.2 [**Integrated Circuits**]: Design Aids; J.3 [**Computer Applications**]: Life and Medical Sciences—*Biology and genetics, health*

General Terms

Algorithms, Design, Performance

Keywords

Biochip design, digital microfluidics, droplets.

*This research was supported in part by the National Science Foundation under grant CCF-0914895.

Permission to make digital or hard copies of all or part of this work for personal or classroom use is granted without fee provided that copies are not made or distributed for profit or commercial advantage and that copies bear this notice and the full citation on the first page. To copy otherwise, to republish, to post on servers or to redistribute to lists, requires prior specific permission and/or a fee.
ISPD'11, March 27–30, 2011, Santa Barbara, California, USA.
Copyright 2011 ACM 978-1-4503-0550-1/11/03 ...$10.00.

1. INTRODUCTION

Digital microfluidic biochips have now emerged as a promising solution for integrating fluid-handling on a chip [1, 5, 6, 3, 4]. Discrete droplets of nanoliter volumes can be manipulated in a "digital" manner under clock control on a two-dimensional array of electrodes. Compared to traditional bench-top procedures, a microfluidic biochip offers the advantages of low sample and reagent consumption, less likelihood of human error due to automation, high throughput, and high sensitivity.

Applications such as enzymatic assays [14], DNA sequencing [1], and immunoassays [15], protein crystallization [21], and cell manipulation [6, 13] have all been successfully demonstrated on a digital microfluidic chip. These advances in technology and applications serve as a powerful driver for research on tools for the automated design of such chips. Architectural synthesis, physical design, and droplet-routing methods have been developed for the design of chips that can execute laboratory protocols [8, 9, 11, 12, 16, 17, 22, 23, 24].

A number of techniques have been proposed to solve the droplet-routing problem for digital microfluidics [8, 12, 17, 18, 23]. For a specific bioassay, droplet routing is decomposed into a series of sub-problems at different time spans [17]. In each sub-problem, the nets to be routed between the sources and the sinks are first determined. Next the routing method is used to generate paths to efficiently transport all the droplets. A complete droplet-routing solution is obtained for the specific bioassay by solving these sub-problems sequentially.

However, these methods assume a direct-addressing scheme for the control of electrodes, whereby each electrode is connected to a dedicated control pin; it can therefore be activated independently. As more bioassays are concurrently executed on digital microfluidic platforms, system complexity and the number of electrodes is expected to increase steadily. A droplet-based biochip that embeds more than 300,000 20 μm by 20 μm electrodes, and uses dielectrophoresis for droplet manipulation and control, has been demonstrated [13]. A prototype and a soon-to-be announced product for clinical diagnostics include more than 5,000 electrodes [7, 3, 4]. The large number of control pins and the associated interconnect-routing problem significantly add to product cost for disposable PCB-based biochips. These chips need to be disposable because they are targeted for diagnostic tests on human patients, and cleaning and reuse are not always feasible.

Furthermore, a digital microfluidic biochip is controlled by an external micro-controller, which stores the bioassay schedule, resource binding and module placement results, and the resulting pin-actuation sequences in memory [6, 3, 4]. During bioassay execution, the micro-controller reads pin-actuation sequences from memory, translates them into voltage control signals, and activates or deactivates the corresponding control pins of the biochip. However, due to the limited number of output ports of the micro-controller, it is infeasible for a micro-controller to activate a large number of control pins. For example, the micro-controller for a recently developed biochip for n-plex bioassay can only activate 64 control pins [7]. Therefore, it cannot directly control thousands of electrodes.

Pin-constrained design methods for digital microfluidics can utilize a small set of control pins to activate a large number of electrodes [9, 14, 19, 20, 24]. Cross-referencing methods allow control of an $N \times M$ grid array with only $N + M$ control pins [9, 24]. However, the cross-referencing design requires a special electrode structure (i.e., both top and bottom plates contain electrode rows), which results in increased manufacturing cost. The pin-constrained design method in [19] uses array partitioning and careful pin-assignment to reduce the number of control pins. In [20], broadcast addressing is used to design pin-constrained and multi-functional biochips. However, the above solutions require droplet routes to be determined prior to pin assignment, which is clearly sub-optimal since droplet routing and pin assignment are interdependent.

In [11], a two-stage integer linear programming (ILP)-based droplet routing algorithm was presented to solve the droplet-routing and pin-assignment problems concurrently. The proposed method simultaneously minimizes the number of control pins, the number of used cells, and the droplet-transportation time. It generates the mapping of a minimum pins to electrodes in each subproblem such that minimum-length droplet pathways are also obtained.

However, a major problem with [11] is that it can lead to conflicting pin assignments for the different routing subproblems. An electrode in the digital microfluidic array might need to be connected to different input pins for different subproblems. In order to fabricate such a biochip and use it for practical bioassays, we have to integrate multiplexers that can dynamically change the connections between the input pins and the electrodes. Such a solution is prohibitively expensive for low-cost, disposable biochips. Therefore, a fallacy with ILP in [11] is that it overlooks practical constraints and it cannot always be used to generate feasible solutions.

In this paper, we propose a more comprehensive ILP-based optimization method to solve the droplet routing and pin-constrained design problems concurrently. Given the nets to be routed in each subproblem, the proposed method simultaneously co-optimizes the droplet pathways and pin-assignment, in order to minimize the number of control pins. The proposed method generates droplet-routing paths and a single pin-assignment plan that allows droplet-routing in all the subproblems without any conflict in the mapping of pins to electrodes. The effectiveness of the proposed co-optimization method is demonstrated for a commercial biochip that is used to perform n-plex immunoassays, as well as an experimental chip for multiplexed *in-vitro* diagnostics. The first chip and the related design details are not available in the public domain. Therefore, we also report results for the second chip design that can be easily made available to other researchers.

The rest of the paper is organized as follows. In Section 2, we formulate the problem of droplet routing for a pin-constrained design. In Section 3, we describe the proposed ILP-based co-optimization method. Section 4 demonstrates the effectiveness of the proposed co-optimization method for the design of two biochips. Finally, conclusions are drawn in Section 5.

2. PROBLEM DEFINITION

In this paper, we integrate the broadcast-addressing-based design technique of [20] into the proposed ILP-based optimization method. The droplet-routing information is stored in the form of electrode-actuation sequences, where each bit represents the status of the electrode at a specific time-step. The status can be either "1" (activate), "0" (deactivate) or "X" (don't-care). For two electrode-actuation sequences, at every time step, if either the values of two bits are the same, or the value of one bit is "X", we refer to the two sequences as being *compatible*. A group of electrodes whose electrode-actuation sequences are mutually compatible can be controlled using the same pin.

The co-optimization problem for droplet routing and pin assignment can be formulated as follows:

Input: A series of subproblems for the droplet routing in a bioassay. Let D_k be the set of droplets in the k-th subproblem. For the k-th subproblem, we specify the droplets to be routed in D_k, the source electrode (s_x^i, s_y^i) and the sink electrode (t_x^i, t_y^i) for each droplet d_i in D_k, and the maximum completion time $T_{max,k}$ for droplet routing in the k-th subproblem.

Output: (1) Routing paths using which droplets in all the subproblems are transported on the array; (2) A pin-assignment plan that assigns control pins to routing paths, in order to allow droplet-routing in all the subproblems without any pin-actuation conflict.

Objective and constraints: The objective is to minimize the number of pins used for controlling droplet movement, with no increase in droplet-transportation time. Several constraints must be satisfied; these include constraints on how droplets move during routing, static and dynamic fluidic constraints, constraints on electrodes that must be activated or deactivated to control the movement of droplets, as well as the constraints to ensure the compatibility of electrode-actuation sequences. The constraints are explained in detail in Section 3.2.

3. ILP FORMULATION FOR CO-OPTIMIZATION

In this section, we introduce the objective function and constraints in the ILP model. The notation used in the ILP formulation is shown in Table 1. The variables and constraints are not discussed separately in the text due to lack of space. The problems with the ILP model of [11] are highlighted in appropriate places and we also discuss how these problems can be overcome. Each variable in Table 1 has been reformulated using an additional dimension corresponding to the subproblem—a necessary consideration that was overlooked in [11].

3.1 Objective Function

Our objective is to minimize the number of control pins, when the maximum completion time for droplet routing is

Table 1: Notation used in the ILP formulation.

D_k	set of droplets in k-th subproblem
C_k	set of available electrodes in k-th subproblem
B	set of electrodes in microfluidic array
$T_{max,k}$	maximum completion time for droplet routing in k-th subproblem
P_{max}	maximum available control pins
$E_a(x,y)$	set of electrode (x,y) and its non-diagonal adjacent electrodes
$E_b(x,y)$	set of both diagonal and non-diagonal adjacent electrodes of electrode (x,y)
$E_c(x,y)$	set of electrode (x,y) and its both diagonal and non-diagonal adjacent electrodes
(s_x^i, s_y^i)	source electrode of droplet d_i
(t_x^i, t_y^i)	sink electrode of droplet d_i
$c(i,x,y,t,k)$	a 0-1 variable represents that droplet d_i locates at electrode (x,y) at time t in k-th subproblem
$st(i,t,k)$	a 0-1 variable represents that droplet d_i stalls from time $t-1$ to t in k-th subproblem
$uc(x,y)$	a 0-1 variable represents whether electrode (x,y) is used
$a_0(i,x,y,t,k)$	a 0-1 variable represents whether electrode (x,y) must be deactivated to control droplet d_i's movement at time t in k-th subproblem
$a_1(i,x,y,t,k)$	a 0-1 variable represents whether electrode (x,y) must be activated to control droplet d_i's movement at time t in k-th subproblem
$a_X(i,x,y,t,k)$	a 0-1 variable represents whether electrode (x,y) is don't care to control droplet d_i's movement at time t in k-th subproblem
$A_0(x,y,t,k)$	a 0-1 variable represents whether electrode (x,y) must be deactivated to control all droplets' movement at time t in k-th subproblem
$A_1(x,y,t,k)$	a 0-1 variable represents whether electrode (x,y) must be activated to control all droplets' movement at time t in k-th subproblem
$A_X(x,y,t,k)$	a 0-1 variable represents whether electrode (x,y) is don't care to control all droplets' movement at time t in k-th subproblem
$as(x,y,t,k)$	electrode-actuation value of (x,y) at time t in k-th subproblem
$cmp(x_1,y_1,$ $x_2,y_2)$	a 0-1 variable represents the electrode -actuation sequences of (x_1,y_1) and (x_2,y_2) are compatible
$cp(x,y,p)$	a 0-1 variable represents that electrode (x,y) is controlled using control pin p
$up(p)$	a 0-1 variable represents that control pin p is used

given a *priori*. Therefore, the objective function is defined as follows:

$$\text{Minimize}: \sum_{p=1}^{P_{max}} up(p).$$

3.2 Constraints

Assume the droplet routing of the bioassay can be decomposed into N subproblems. Note that t is used to indicate the specific time within each subproblem, therefore we have $0 \leq t \leq T_{max,k}$ for k-th subproblem. The following are the constraints in the ILP formulation. Compared to [11], all the constraints are extended from the case of one subproblem to the complete schedule involving all the subproblems. Moreover, the electrode constraints and broadcast constraints have been completely rewritten to ensure that the resulting chip design is feasible.

1. *Source requirement*: For each subproblem, initially all the droplets within this subproblem must stay at the cor-

responding source electrode at $t = 0$. Therefore, this constraint can be written as: for k-th subproblem ($1 \leq k \leq N$), $\forall\, d_i \in D_k$,

$$c(i, s_x^i, s_y^i, 0, k) = 1 \qquad (1)$$

2. *Sink requirement*: In a specific subproblem (e.g., k-th subproblem), all the droplets will arrive at the corresponding sink nodes within the maximum completion time of droplet routing (e.g., $T_{max,k}$). Once a droplet arrives at the sink node, it remains there until the routing of other droplets is finished. This constraint can be written as: for k-th subproblem ($1 \leq k \leq N$),

$$\sum_{t=0}^{T_{max,k}} c(i, t_x^i, t_y^i, t, k) \geq 1, \forall\, d_i \in D_k \qquad (2)$$

$$c(i, t_x^i, t_y^i, t, k) - c(i, t_x^i, t_y^i, t+1, k) \leq 0, \forall\, d_i \in D_k, \\ 0 \leq t < T_{max,k} \qquad (3)$$

3. *Exclusivity constraint*: In a specific subproblem (e.g., k-th subproblem), each droplet can stay at only one electrode at a specific time. This constraint can be written as: for k-th subproblem ($1 \leq k \leq N$),

$$\sum_{(x,y)\in C_k} c(i,x,y,t,k) = 1, \forall\, d_i \in D_k, 0 \leq t \leq T_{max,k} \qquad (4)$$

4. *Total number of used electrodes*: An electrode (x,y) is used if a droplet is located on it at a specific time within a specific subproblem. Otherwise, if no droplet is located at (x,y) during droplet routing for all the subproblems, (x,y) is not used. We can calculate the total number of used electrodes for droplet routing for all the subproblems using the following constraints: for k-th subproblem ($1 \leq k \leq N$),

$$uc(x,y) \geq c(i,x,y,t,k), \forall\, d_i \in D_k, \\ (x,y) \in B, 0 \leq t \leq T_{max,k} \qquad (5)$$

$$uc(x,y) \leq \sum_{k=1}^{N} \{ \sum_{d_i \in D_k} \{ \sum_{t=0}^{T_{max,k}} c(i,x,y,t,k) \} \}, \\ \forall\, (x,y) \in B \qquad (6)$$

5. *Droplet movement constraint*: As shown in Fig. 1(a), for an electrode (x,y), its adjacent electrodes can be classified into two categories: (1) non-diagonal adjacent electrodes $(x-1,y)$, $(x+1,y)$, $(x,y+1)$ and $(x,y-1)$; (2) diagonal adjacent electrodes $(x-1,y+1)$, $(x+1,y+1)$, $(x-1,y-1)$ and $(x+1,y-1)$.

In a specific subproblem (e.g., k-th subproblem), from current clock cycle t to next clock cycle $t+1$, each droplet can either remain at the same location, or move to its non-diagonal adjacent electrodes. This constraint can be written as: for k-th subproblem ($1 \leq k \leq N$),

$$c(i,x,y,t,k) \leq \sum_{(x',y')\in E_a(x,y)} c(i,x',y',t+1,k), \\ \forall\, d_i \in D_k, (x,y) \in C_k, 0 \leq t < T_{max,k} \qquad (7)$$

Typically, $E_a(x,y)$ includes five electrodes (x,y), $(x+1,y)$, $(x-1,y)$, $(x,y+1)$, $(x,y-1)$. Note that due to the boundary restriction of the microfluidic array, a droplet may not always move to its four non-diagonal adjacent electrodes. For example, the bottom left electrode denoted as (1,1) has only

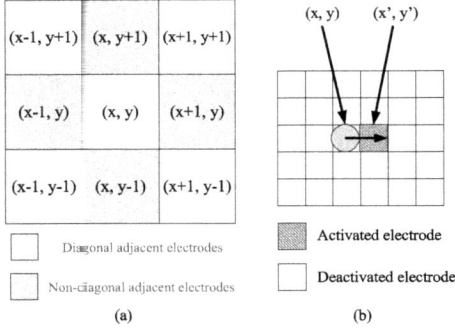

Figure 1: Illustration of constraints in ILP model: (a) Electrode (x, y) and its adjacent electrodes; (b) Electrodes are deactivated at time t when moving d_i from (x, y) to (x', y') from time $t-1$ to t.

two non-diagonal adjacent electrodes (1,2) and (2,1). Therefore, for the electrode (1,1), the constraint can be written as: for k-th subproblem ($1 \leq k \leq N$),

$$c(i, 1, 1, t, k) \leq c(i, 1, 1, t+1, k) + c(i, 1, 2, t+1, k)$$
$$+ c(i, 2, 1, t+1, k), \forall \, d_i \in D_k, 0 \leq t < T_{max,k}$$

6. *Fluidic constraints*: Within a subproblem, both static and dynamic fluidic constraints described in [17] must be met during droplet routing. The static constraint indicates that no droplet (e.g., d_j) can stay at the set of both diagonal and non-diagonal adjacent electrodes of the electrode where a droplet (e.g., d_i) has already existed. The static fluidic constraint is written as: for k-th subproblem ($1 \leq k \leq N$),

$$c(i, x, y, t, k) + \sum_{(x', y') \in E_c(x,y)} c(j, x', y', t, k) \leq 1,$$
$$\forall \, d_i, d_j \in D_k, d_i \neq d_j, (x, y) \in C_k, 0 \leq t \leq T_{max,k} \quad (8)$$

The dynamic fluidic constraint indicates that if a droplet (e.g., d_i) will arrive at electrode (x, y) at clock cycle $t+1$, no droplet (e.g., d_j) can exist in the set of electrode (x, y) and its both diagonal and non-diagonal adjacent electrodes at clock cycle t. The dynamic fluidic constraint can be written as: for k-th subproblem ($1 \leq k \leq N$),

$$c(i, x, y, t+1, k) + \sum_{(x', y') \in E_c(x,y)} c(j, x', y', t, k) \leq 1,$$
$$\forall \, d_i, d_j \in D_k, d_i \neq d_j, (x, y) \in C_k, 0 \leq t < T_{max,k} \quad (9)$$

7. *Electrode constraints*: First, we discuss the constraints for electrodes that must be activated to control the movement of a single droplet (e.g., d_i) during droplet routing within one subproblem. If d_i is at the electrode (x, y) at clock cycle t, exactly one electrode (x, y) must be activated currently. This constraint can be written as: for k-th subproblem ($1 \leq k \leq N$),

$$a_1(i, x, y, t, k) = c(i, x, y, t, k), \forall \, d_i \in D_k, (x, y) \in C_k,$$
$$0 \leq t \leq T_{max,k} \quad (10)$$
$$\sum_{(x,y) \in C_k} a_1(i, x, y, t, k) = 1, \forall \, d_i \in D_k, 0 \leq t \leq T_{max,k} \quad (11)$$

Next we discuss the constraints for electrodes that must be deactivated to control the movement of a single droplet (e.g.,

d_i) during droplet routing within one subproblem. If d_i is at the electrode (x, y) at clock cycle t, the electrode (x, y) must be activated, while its diagonal and non-diagonal adjacent electrodes must be deactivated. We use parameter $K_{(x,y)}$ to denote the number of diagonal and non-diagonal adjacent electrodes of (x, y). This constraint can be written as: for k-th subproblem ($1 \leq k \leq N$),

$$\sum_{(x', y') \in E_b(x,y)} a_0(i, x', y', t, k) \geq K_{(x,y)} \cdot c(i, x, y, t, k)$$
$$\forall \, d_i \in D_k, (x, y) \in C_k, 0 \leq t \leq T_{max,k} \quad (12)$$

Now we consider the electrodes that must be deactivated according to the movement of droplet d_i from time step $t-1$ to t. Assume that d_i stays at the same electrode (x, y) from $t-1$ to t, therefore $K_{(x,y)}$ neighboring electrodes of (x, y) must be deactivated at t. Assume that d_i moves to one of non-diagonal adjacent electrodes of (x, y) from $t-1$ to t, e.g., d_i moves from (x, y) to (x', y'). First, at current time t, both diagonal and non-diagonal adjacent electrodes of (x, y) must be deactivated except for (x', y'), i.e., $K_{(x,y)} - 1$ neighboring electrodes of (x, y) must be deactivated at t. Therefore, if d_i is at (x, y) at $t-1$, no matter it will continue staying at (x, y) or move to (x', y') at t, at least $K_{(x,y)} - 1$ neighboring electrodes of (x, y) must be deactivated at t. This constraint can be written as: for k-th subproblem ($1 \leq k \leq N$),

$$\sum_{(x', y') \in E_b(x,y)} a_0(i, x', y', t, k) \geq$$
$$(K_{(x,y)} - 1) \cdot c(i, x, y, t-1, k),$$
$$\forall \, d_i \in D_k, (x, y) \in C_k, 0 < t \leq T_{max,k} \quad (13)$$

At current time t, if d_i is moved to (x', y'), both diagonal and non-diagonal adjacent electrodes of (x', y') must be deactivated, i.e., $K_{(x',y')}$ neighboring electrodes of (x', y') must be deactivated at t. At current time t, we merge the set for $K_{(x,y)} - 1$ neighboring electrodes of (x, y) and the set for $K_{(x',y')}$ neighboring electrodes of (x', y') into a single set, which denotes electrodes in the entire array that must be deactivated at t, in order to move d_i at (x, y) to (x', y') from $t-1$ to t. We use parameter $K'_{(x,y) \to (x',y')}$ to represent the number of electrodes in the merged set. For example, as shown in Fig. 1(b), when we move d_i from (x, y) to (x', y') from time $t-1$ to t, 7 neighboring electrodes of (x, y) must be deactivated at t, i.e., $K_{(x,y)} - 1$ is equal to 7; 8 neighboring electrodes of (x', y') must be deactivated at t, i.e., $K_{(x',y')}$ is equal to 8; the number of electrodes in the merged set is 11, i.e., $K'_{(x,y) \to (x',y')}$ is equal to 11.

We introduce $st(i, t, k)$ to represent the fact that d_i stalls from $t-1$ to t in k-th subproblem. These constraints can be written as: for k-th subproblem ($1 \leq k \leq N$),

$$st(i, t, k) \geq c(i, x, y, t, k) + c(i, x, y, t-1, k) - 1,$$
$$\forall \, d_i \in D_k, (x, y) \in C_k, 0 < t \leq T_{max,k} \quad (14)$$
$$st(i, 0, k) = 1, \forall \, d_i \in D_k \quad (15)$$
$$\sum_{(x,y) \in C_k} a_0(i, x, y, t, k) = K'_{(x,y) \to (x',y')} \cdot (1 - st(i, t, k))$$
$$+ K_{(x,y)} \cdot st(i, t, k), \forall \, d_i \in D_k, 0 \leq t \leq T_{max,k} \quad (16)$$

8. *Actuation sequence constraints*: Here we consider the constraints that an electrode should be activated, deactivated, or don't care at a specific time t, in order to control the movement of all droplets in D_k of k-th subproblem.

First we consider the constraints for $A_1(x,y,t,k)$. If the electrode (x,y) must be activated at t to control any single droplet (e.g., d_i) in D_k, i.e., $a_1(i,x,y,t,k)$ must be set to 1, then $A_1(x,y,t,k)$ must be set to 1. The constraint can be written as: for k-th subproblem ($1 \leq k \leq N$),

$$A_1(x,y,t,k) \geq a_1(i,x,y,t,k), \forall\, d_i \in D_k,$$
$$(x,y) \in C_k, 0 \leq t \leq T_{max,k} \quad (17)$$

If electrode (x,y) does not need to be activated to control any droplet in D_k, i.e., $a_1(i,x,y,t,k)$ is set to 0 for each droplet d_i in D_k, then $A_1(x,y,t,k)$ is set to 0. The constraint can be written as: for k-th subproblem ($1 \leq k \leq N$),

$$A_1(x,y,t,k) \leq \sum_{d_i \in D_k} a_1(i,x,y,t,k),$$
$$\forall\, (x,y) \in C_k, 0 \leq t \leq T_{max,k} \quad (18)$$

The constraints for $A_0(x,y,t,k)$ are similar as those for $A_1(x,y,t,k)$, as following: for k-th subproblem ($1 \leq k \leq N$),

$$A_0(x,y,t,k) \geq a_0(i,x,y,t,k), \forall\, d_i \in D_k,$$
$$(x,y) \in C_k, 0 \leq t \leq T_{max,k} \quad (19)$$
$$A_0(x,y,t,k) \leq \sum_{d_i \in D_k} a_0(i,x,y,t,k),$$
$$\forall\, (x,y) \in C_k, 0 \leq t \leq T_{max,k} \quad (20)$$

At time t, if an electrode (x,y) is not forced to be activated or deactivated, it is classified as "don't care", i.e., $A_X(x,y,t,k)$ must be set to 1. The constraint can be written as: for k-th subproblem ($1 \leq k \leq N$),

$$A_1(x,y,t,k) + A_0(x,y,t,k) + A_X(x,y,t,k) = 1,$$
$$\forall\, (x,y) \in C_k, 0 \leq t \leq T_{max,k} \quad (21)$$

Note that since "don't care" can be replaced by either '1' or '0', we use the following constraints to generate electrode-actuation status for each electrode in C_k at t: for k-th subproblem ($1 \leq k \leq N$),

$$1 \cdot A_1(x,y,t,k) + 0 \cdot A_0(x,y,t,k) + 0 \cdot A_X(x,y,t,k)$$
$$\leq as(x,y,t,k), \forall\, (x,y) \in C_k, 0 \leq t \leq T_{max,k} \quad (22)$$
$$1 \cdot A_1(x,y,t,k) + 0 \cdot A_0(x,y,t,k) + 1 \cdot A_X(x,y,t,k)$$
$$\geq as(x,y,t,k), \forall\, (x,y) \in C_k, 0 \leq t \leq T_{max,k} \quad (23)$$

9. *Broadcast constraints*: We identify whether the corresponding electrode-actuation sequences of two electrodes (e.g., (x_1,y_1) and (x_2,y_2)) are compatible by examining $as(x_1,y_1,t,k)$ and $as(x_2,y_2,t,k)$ at every time in all the subproblems ($0 \leq t \leq T_{max,k}, 1 \leq k \leq N$). At a specific time t_1 within a specific subproblem k_1, if the electrode-actuation values of (x_1,y_1) and (x_2,y_2) are different, i.e., $as(x_1,y_1,t_1,k_1)$ is not equal to $as(x_2,y_2,t_1,k_1)$, their corresponding electrode-actuation sequences are not compatible. Otherwise, if $as(x_1,y_1,t,k)$ is equal to $as(x_2,y_2,t,k)$ for every time in all the subproblems ($0 \leq t \leq T_{max,k}, 1 \leq k \leq N$), the corresponding electrode-actuation sequences of (x_1,y_1) and (x_2,y_2) are compatible. The constraint can be written as: for k-th subproblem ($1 \leq k \leq N$),

$$1 - cmp(x_1,y_1,x_2,y_2) \geq as(x_1,y_1,t,k) - as(x_2,y_2,t,k)$$
$$\forall\, (x_1,y_1),(x_2,y_2) \in C_k, 0 \leq t \leq T_{max,k} \quad (24)$$
$$1 - cmp(x_1,y_1,x_2,y_2) \geq as(x_2,y_2,t,k) - as(x_1,y_1,t,k)$$
$$\forall\, (x_1,y_1),(x_2,y_2) \in C_k, 0 \leq t \leq T_{max,k} \quad (25)$$

Note that the ILP model in [11] only checks for the compatibility of electrode-actuation sequences within each subproblem, and generates one set of control pins for each subproblem. However, the compatibility check must be performed for all the subproblems rather than for a single subproblem. This is because only one set of control pins can be used to perform droplet routing for all the subproblems. These new constraints are modeled by (24) and (25).

If the corresponding electrode-actuation sequences of two electrodes (e.g., (x_1,y_1) and (x_2,y_2)) are not compatible, (x_1,y_1) and (x_2,y_2) cannot share the same control pin. Otherwise, if the corresponding electrode-actuation sequences are compatible, (x_1,y_1) and (x_2,y_2) can (not "must") share the same control pin. The constraint can be written as:

$$cp(x_1,y_1,p) + cp(x_2,y_2,p) \leq cmp(x_1,y_1,x_2,y_2) + 1,$$
$$\forall\, (x_1,y_1),(x_2,y_2) \in B, 1 \leq p \leq P_{max} \quad (26)$$

Note that control pins are only assigned to electrodes that are used during droplet routing, those electrodes where no droplet is located during the droplet routing for all the subproblems are left without any control pins. The constraint can be written as:

$$\sum_{p=1}^{P_{max}} cp(x,y,p) = uc(x,y), \forall\, (x,y) \in B \quad (27)$$

Next we calculate the minimum number of control pins that are used. The following constraints indicate that if an electrode (x,y) is controlled by a pin p, we indicate that pin p is used; otherwise, p is not used.

$$cp(x,y,p) \leq up(p), \forall\, (x,y) \in B, 1 \leq p \leq P_{max} \quad (28)$$
$$up(p) \leq \sum_{(x,y) \in B} cp(x,y,p), 1 \leq p \leq P_{max} \quad (29)$$

Assume that for the $W_x \times W_y$ microfluidic array, the droplet routing of the bioassay is decomposed into N subproblems, the maximum number of droplets for all the subproblems is D_{max}, and the maximum completion time for droplet routing in all the subproblems is T_{max}. In the worst case, the number of variables in the ILP model is $O(NW_xW_yD_{max}T_{max})$, and the number of constraints is $O(N(W_xW_y)^2T_{max})$.

4. EXPERIMENTAL RESULTS

The proposed ILP-based optimization method for droplet-routing on a pin-constrained array was implemented on a 3.0 GHz INTEL Xeon processor, with 12 GB of memory. MOSEL [10] was used as our ILP solver. We first utilize the proposed method to design the pin-assignment for a commercial biochip from Advanced Liquid Logic, Inc. Next we evaluate the proposed method by applying to an experimental biochip for multiplexed *in-vitro* diagnostics.

4.1 Commercial Biochip Design for n-Plex Immunoassay

The n-plex immunoassay is a typical example of a multiplexed and concurrent bioassay. In this assay, a sample is analyzed for n different analytes. Sample droplets are mixed with n different reagents, and the mixed product droplets are routed to a detection site after incubation. The signal transduction for quantifying the n droplets is based on an optical signal from each droplet.

As an example of an n-plex assay, we demonstrated a 3-plex assay for the diagnosis of acute myocardial infarc-

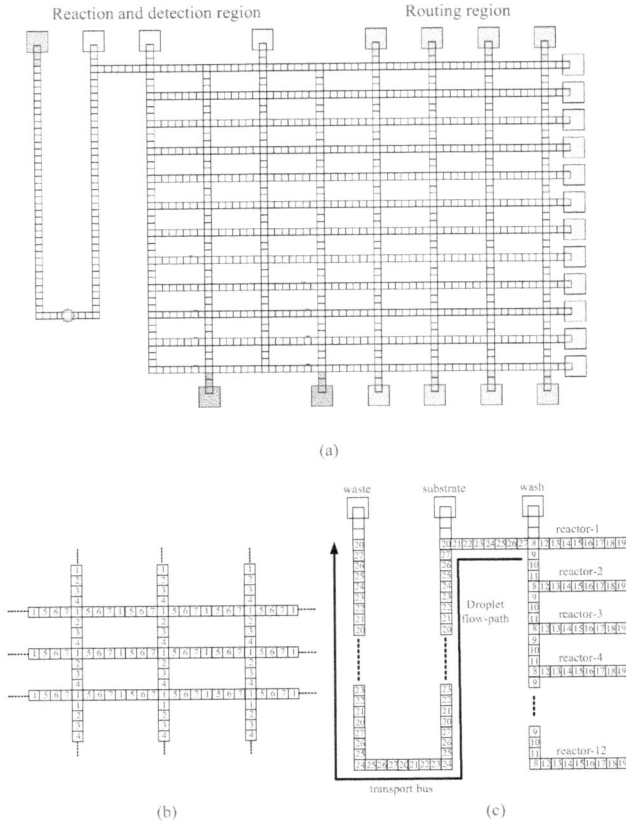

Figure 2: **Layout of fabricated biochip used for the *n*-plex assay experiment: (a) complete chip; (b) routing region; (c) reaction and detection regions.**

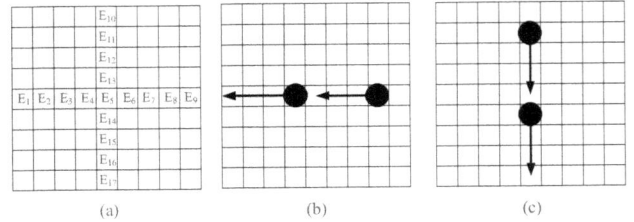

Figure 3: **Droplet-routing subproblems for the routing region of the commercial biochip: (a) the basic cell in the routing region; (b) 1st subproblem; (c) 2nd subproblem.**

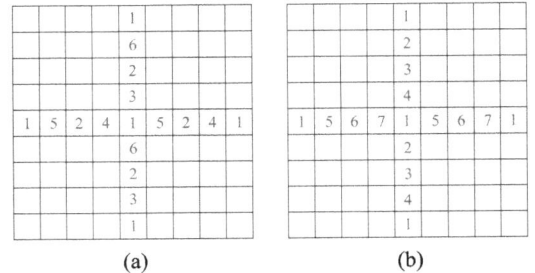

Figure 4: **Pin assignment for the basic cell of the routing region in the commercial biochip: (a) new pin assignment using ILP model (6 pins); (b) existing pin assignment in the fabricated biochip (7 pins).**

tion (AMI) [3]. The fabricated chip is a pin-constrained design that consists of over 1000 electrodes and 64 control pins [3]. Fig. 2(a) shows the layout of the complete chip. The platform consists of three regions: routing region, reaction region, and detection region. Note that the electrodes in different regions are connected to separate sets of control pins [7].

In the routing region, there are 20 reservoirs for high-throughput processing, with 8 reservoirs for sample solutions and 12 reservoirs for reagent solutions. The sample reservoirs are connected using a 4-phase vertical transport buses. Each reagent reservoir intersects with these vertical transport buses through a 4-phase horizontal transport bus. Fig. 2(b) shows the layout of the horizontal and vertical buses. The numbers in Fig. 2(b) and Fig. 2(c) refer to the pins (pin IDs) assigned to the electrodes. This pin mapping, which was done manually by the chip manufacturer, is used as a baseline here to evaluate the proposed optimization method.

Fig. 2(c) shows the layout of the reaction region and the detection region. There are 12 horizontally-placed reactors where the reagent and sample droplets are mixed. A total of 8 control pins are used for this transport bus.

For the 3-plex assay, serum (an example of human physiological fluid) is sampled and dispensed. Three assays, namely troponin-I, myoglobin, and creatine kinase-MB (CK-MB) measurements are performed on the physiological fluid. The 3-plex assay includes three stages: dispensing and rout-

ing stage, reaction stage, and detection stage, which are performed in the routing region, reaction region, and detection region, respectively.

We utilize the proposed ILP model to determine the pin assignment for the commercial biochip. First we focus on the pin assignment for the routing region. This region is an array of multiple basic cells with the same layout. The layout of a basic cell is shown in Fig. 3(a). The "dark" cells are part of the layout. Since no fluidic operation is implemented in the "light" cells, in the fabricated chip, this region is left unused and the corresponding electrodes are not fabricated. However, since the proposed ILP model is based on a regular array of electrodes, the "light" cells are used for optimization. Therefore, if we solve the pin-assignment problem for a single basic cell, the result can be directly applied to the entire routing region. The droplet routing for the basic cell can be decomposed into two subproblems, as shown in Fig. 3(b) and Fig. 3(c). For the first subproblem in Fig. 3(b), the left droplet moves from E_5 to E_1, while the right droplet moves concurrently from E_9 to E_5. For the second subproblem in Fig. 3(c), the top droplet moves from E_{10} to E_5, while the bottom droplet moves concurrently from E_5 to E_{17}.

We set the parameters $T_{max,k}$ ($k = 1, 2$) and P_{max} for the proposed ILP model. For the 1st subproblem, we set the maximum completion time for droplet routing $T_{max,1}$ to be 10 clock cycles, which is the sum of droplet-routing time for each droplet. Similarly, $T_{max,2}$ is also set to be 10 clock cycles. The maximum available control pins P_{max} is set to 81, which is equal to the number of electrodes in the basic cell. The pin-assignment results obtained by solving the corresponding ILP model is shown in Fig. 4(a); a total

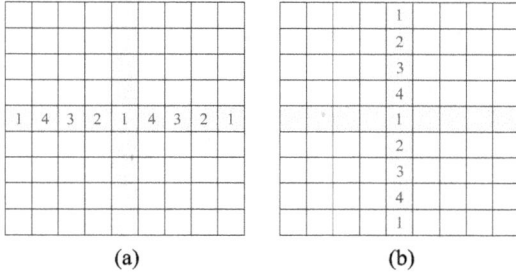

Figure 5: Pin-assignment result for the basic cell using the method in [11]: (a) 1st subproblem; (b) 2nd subproblem.

Table 2: Comparison of the number of control pins between pin assignment for the proposed ILP model and the existing design for the commercial biochip.

	No. pins (existing design)	No. pins (ILP model)
Routing region	7	6
Reaction region	19	13
Detection region	8	4
Total	34	23

of 6 pins are needed. The existing baseline pin assignment for the fabricated biochip is shown in Fig. 4(b); it requires 7 pins. The proposed ILP model saves one control pin for the routing region, while ensuring the same droplet-routing steps and routing time as in the pin-assignment implemented for the fabricated chip.

Let us now return to the ILP-based optimization model in [11], which generates the pin-assignment for each subproblem. We utilize this model to generate the pin assignment for the 1st subproblem and the 2nd subproblem separately, as shown in Fig. 5. According to [11], the pin-assignment in Fig. 5(a) should be used to execute 1st subproblem. After the 1st subproblem is finished, the pin-assignment in Fig. 5(b) must replace that in Fig. 5(a), in order to ensure that the 2nd subproblem can be executed. However, without added multiplexes and interconnects, a low-cost disposable biochip can only be fabricated according to a specific pin-assignment plan. Since the control pins and the interconnection wires are permanently etched during the fabrication, it is not practical to dynamically change the pin mapping for a fabricated biochip. Therefore, the solution provided by the method in [11] has a major fallacy and is not feasible in practice.

We also apply the proposed ILP model to optimize the pin-assignment design for the reaction region and the detection region, respectively. Table 2 shows the comparison of the number of control pins ("No. pins") between the pin-assignment design using the proposed ILP model and the existing design for the commercial biochip. The proposed ILP model can reduce the number of control pins for three regions from 34 to 23, i.e., a 32.6% reduction. The fabricated chip uses an additional 30 pins for droplet dispensing, but they are not targeted in our case study.

The CPU time needed to generate the pin assignment for all regions was over one day. However, since this is a one-time design effort, the high CPU time for this large commercial bichip is acceptable. Note that the fabricated chip has a total of 64 pins and over 1000 electrodes, and it is the most

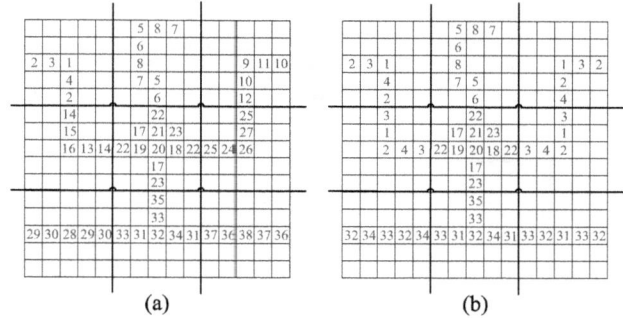

Figure 6: Pin-assignment result for the multiplexed bioassay on a 15×15 microfluidic array: (a) without time-division pin-sharing (TDPS) scheme (38 pins); (b) use TDPS scheme (20 pins).

complex chip that is currently being targeted for the clinical diagnostics market. Therefore, the proposed ILP model is adequate for handling a real design.

4.2 Experimental Biochip for Multiplexed in-vitro Diagnostics

We map a recently demonstrated multiplexed biochemical assay, which consists of a glucose assay and a lactate assay based on colorimetric enzymatic reactions, on to a 15×15 array. For each sample or reagent, two droplets are dispensed into the array. Four pairs of droplets, i.e., $\{S_1, R_1\}$, $\{S_1, R_2\}$, $\{S_2, R_1\}$, $\{S_2, R_2\}$, are routed together in sequence for the mixing operation. Mixed droplets are finally routed to the detection site for analysis.

Based on the bioassay schedule obtained using the high-level synthesis method of [16], the droplet-routing problem of the multiplexed bioassay is decomposed into five sub-problems. Two optical detectors are used to monitor the bioassay outcomes. A single mixer is used to perform the mixing operation. To make the CPU time for optimization manageable, the array is partitioned into nine sub-arrays of smaller size. We apply the proposed ILP model in each sub-array and obtain the corresponding droplet-routing paths and the pin-assignment results.

We utilize the proposed ILP model to obtain the minimum number of control pins and the corresponding droplet-routing paths for each partition. Note that different partitions have different sets of control pins. Therefore, the total number of control pins for the entire 15×15 microfluidic array is the sum of the number of control pins in all the partitions. Fig. 6(a) shows the pin-assignment results for the multiplexed bioassay on the 15×15 microfluidic array. A total of 38 control pins are used. The CPU time needed to generate pin assignment for all nine partitions was 14 hours.

Next we utilize the Time-division Pin-sharing (TDPS) scheme from [19] to further reduce the number of control pins. The basic idea here is that two partitions that have no overlapping time spans can be merged to share the same set of control pins, where a time span for a partition is defined as the period of time during which it contains a droplet. The time spans for all partitions can be easily calculated from the schedule, module placement and droplet routing results [16]; the overlaps can then be readily determined.

For example, assume that Partition 1 has a set of four control pins from Pin 1 to Pin 4, Partition 2 has a set of five control pins from Pin 5 to Pin 9. If these two partitions

Table 3: Comparison for the number of control pins for the multiplexed bioassay on a 15 × 15 microfluidic array with and without time-division pin-sharing (TDPS) scheme.

	No. pins (without TDPS)	No. pins (with TDPS)
Partition 1	4	4
Partition 2	4	4
Partition 3	4	0 (use pins in Partition 1)
Partition 4	4	0 (use pins in Partition 1)
Partition 5	7	7
Partition 6	4	0 (use pins in Partition 1)
Partition 7	3	0 (use pins in Partition 8)
Partition 8	5	5
Partition 9	3	0 (use pins in Partition 8)
Total	38	20

have no overlapping time spans, they are merged into one partition and share the same set of control pins. A subset of control pins in Partition 2, e.g., Pin 5 to Pin 8, can be used to replace Pin 1 to Pin 4 in Partition 1. Therefore, the total number of control pins is reduced from 9 to 5. This check-merge procedure continues until all partition pairs overlap in their time spans. By reducing the number of partitions, we can reduce the number of control pins needed.

According to the schedule of the multiplexed bioassay, we obtain the following 4 groups, where partitions in each group have no overlapping time spans and can be merged.

Group 1: Partition 1, Partition 3, Partition 4, and Partition 6; they share the set of control pins for Partition 1.

Group 2: Partition 2;

Group 3: Partition 5;

Group 4: Partition 7, Partition 8, and Partition 9; they share the set of control pins for Partition 8.

Therefore, the total number of control pins for the 15 × 15 array of multiplexed bioassay is reduced from 38 to 20, as shown in Fig. 6(b).

Table 3 shows the comparison for the number of control pins (No. pins) with and without time-division pin-sharing (TDPS) scheme. Note that the broadcast pin-assignment method of [20] utilizes 25 control pins, which is 5 pins more than proposed co-optimization method with TDPS scheme.

5. CONCLUSIONS

We have presented an ILP-based co-optimization method to solve the droplet routing and pin assignment problems concurrently. The proposed method simultaneously co-optimizes the droplet pathways and pin assignment, in order to minimize the number of control pins. It generates droplet-routing paths and a single pin-assignment plan that allows droplet-routing in all the subproblems without any conflict in the mapping of pins to electrodes. By targeting all the droplet-routing subproblems together, the proposed method overcomes a major limitation and fallacy associated with recent work on this problem. The effectiveness of the proposed co-optimization method has been demonstrated for a commercial biochip that is used to perform n-plex immunoassays, as well as an experimental chip used to perform multiplexed *in-vitro* diagnostics.

ACKNOWLEDGEMENT

The authors thank collaborators at Advanced Liquid Logic for providing the layout of the commercial biochip.

6. REFERENCES

[1] R. B. Fair et al., "Chemical and biological applications of digital-microfluidic devices", *IEEE Design & Test of Computers*, vol. 24, 2007.

[2] K. Chakrabarty and F. Su, *Digital Microfluidic Biochips: Synthesis, Testing, and Reconfiguration Techniques*, CRC Press, Boca Raton, FL, 2006.

[3] R. Sista et al., "Development of a digital microfluidic platform for point of care testing", *Lab on a Chip*, vol. 8, pp. 2091-2104, 2008.

[4] Z. Hua et al., "Mutiplexed real-time polymerase chain reaction on a digital microfluidic platform", *Anal. Chem.*, vol. 82, pp. 2310-2316, 2010.

[5] J. Gong et al., "Portable digital microfluidics platform with active but disposable lab-on-chip," *Proc. IEEE MEMS*, pp. 355-358, 2004.

[6] I. B.-Nad et al., "Digital microfluidics for cell-based assays", *Lab on a Chip*, vol. 8, pp. 519-526, 2008.

[7] Advanced Liquid Logic, http://www.liquid-logic.com.

[8] K. F. Bohringer, "Modeling and controlling parallel tasks in droplet-based microfluidic systems", *IEEE Trans. CAD*, vol. 25, pp. 334-344, 2006.

[9] S.-K. Fan et al., "Manipulation of multiple droplets on N × M grid by cross-reference EWOD driving scheme and pressure-contact packaging", *Proc. Int. Conf. MEMS*, pp. 694-697, 2003.

[10] Fair Isaac Corporation, http://www.fico.com

[11] T.-W. Huang and T.-Y. Ho, "A two-stage ILP-based droplet routing algorithm for pin-constrained digital microfluidic biochips", *Proc. ISPD*, 2010.

[12] M. Cho and D. Z. Pan, "A high-performance droplet-routing algorithm for digital microfluidic biochips", *IEEE Trans. CAD*, vol. 27, pp. 1714-1724, 2008.

[13] G. Medoro, et al., "A lab-on-a-chip for cell detection and manipulation", *IEEE Sensors Journal*, vol. 3, pp. 317-325, 2003.

[14] V. Srinivasan et al., "An integrated digital microfluidic lab-on-a-chip for clinical diagnostics on human physiological fluids", *Lab on a Chip*, vol. 4, pp. 310-315, 2004.

[15] R. Sista et al., "Heterogeneous immunoassays using magnetic beads on a digital microfluidic platform", *Lab on a Chip*, vol. 8, pp. 2188-2196, 2008.

[16] F. Su and K. Chakrabarty, "Unified high-level synthesis and module placement for defect-tolerant microfluidic biochips", *Proc. DAC*, 2005.

[17] F. Su et al., "Droplet routing in the synthesis of digital microfluidic biochips", *Proc. DATE*, pp. 323-328, 2006.

[18] T. Xu and K. Chakrabarty, "Integrated droplet routing in the synthesis of microfluidic biochips", *Proc. DAC*, pp. 948-953, 2007.

[19] T. Xu et al., "Automated design of pin-constrained digital microfluidic biochips under droplet-interference constraints", *ACM J. Emerging Tech. in Computing Sys.*, vol. 3, article 14, 2007.

[20] T. Xu and K. Chakrabarty, "Broadcast electrode-addressing for pin-constrained multi-functional digital microfluidic biochips", *Proc. DAC*, pp. 173-178, 2008.

[21] T. Xu, K. Chakrabarty and V. K. Pamula, "Defect-tolerant design and optimization of a digital microfluidic biochip for protein crystallization", *IEEE Trans. CAD*, vol. 29, pp. 552-565, 2010.

[22] P.-H. Yuh et al., "Placement of defect-tolerant digital microfluidic biochips using the T-tree formulation", *ACM J. Emerging Tech. Computing Sys.*, vol. 3, pp. 13.1-13.32, 2007.

[23] P.-H. Yuh et al., "BioRoute: A network flow based routing algorithm for digital microfluidic biochips", *Proc. ICCAD*, pp. 752-757, 2007.

[24] Z. Xiao and E. F. Y. Young, "CrossRouter: A droplet router for cross-referencing digital microfluidic biochips", *Proc. ASP-DAC*, pp. 269-274, 2010.

3DICs for Tera-Scale Computing
- A Case Study

Tanay Karnik, Dinesh Somasekhar, Shekhar Borkar
Intel® Labs
Hillsboro, OR, USA
tanay.karnik, dinesh.somasekhar, shekhar.y.borkar @intel.com

Abstract

TSV-based 3D chip stacking and integration technology was proposed more than twelve years ago. Since then the concept has gained tremendous traction with significant advances in the technology and planned system prototypes. The IP portfolio in 3D process, circuits, architectures, test and assembly peaked in year 2006. Today, multiple foundries offer a TSV process for 3DICs. Many design teams in academia and industry have been working towards a convincing 3D prototype that will demonstrate form factor reduction, heterogeneous integration and higher performance. However, the commercial success of true 3DICs has been limited to DRAM stacking.

This presentation will introduce technology scaling trends and new challenges. We are at an I/O inflection point due to Tera-scale computing needs. The tutorial will describe the implication to memory bandwidth and present various options. 3DICs provide an excellent alternative to address the memory bandwidth issue by providing a large near-processor memory. In this presentation we will discuss architecture, floorplanning, power routing, IO circuits, test and assembly of a large prototype processor-memory 3DIC designed for Tera-scale applications. The presentation will conclude with a discussion on test, power delivery and thermal management issues related to 3D integration.

ACM Classification Keywords: B.7 INTEGRATED CIRCUITS

General Terms: Performance, Design, Experimentation

Keywords: 3D-Stacked IC, Chip assembly, Packaging, Methodology, TSV

Bio

The speaker, Tanay Karnik, is a Principal Engineer and Program Director in Intel® Lab's Academic Research Office. He received his Ph.D. in Computer Engineering from the University of Illinois at Urbana-Champaign in 1995. His research interests are in the areas of variation tolerance, power delivery, soft errors and physical design. He has published over 40 technical papers, has 44 issued and 33 pending patents in these areas. He received an Intel Achievement Award for the pioneering work on integrated power delivery. He has presented several invited talks and tutorials, and has served on 5 PhD students' committees. He was a member of ISSCC, DAC, ICCAD, ICICDT and ISQED program committees, and JSSC, TCAD, TVLSI, TCAS review committees. Tanay was the General Chair of ISQED'08, ICICDT'08, ISQED'09 and ASQED'10. Tanay is an IEEE Senior Member, an ISQED Fellow, an Associate Editor for TVLSI and a Guest Editor for JSSC.

Copyright is held by the author/owner(s).
ISPD'11, March 27–30, 2011, Santa Barbara, California, USA.
ACM 978-1-4503-0550-1/11/03.

Advances in 3D Integrated Circuits

Robert Patti
Tezzaron Semiconductor Corporation
1415 Bond Street
Naperville, IL 60563
1 (630) 505-0404

rpatti@tezzaron.com

ABSTRACT

In this paper, we describe the developments over the past several years in field of 3D integrated circuits. 3D integration offers far greater improvements than traditional semiconductor scaling can provide today. Considered part of the "More Than Moore" category of semiconductor technology, in addition to increasing circuit and system density, 3D integrated circuits can blend a wide range of materials and technologies into a signal polylithic device acting as if these disparate items were truly fabricated together on a single wafer. 3D offers improvement in power, speed, density, cost and reliability, but to access these, new design techniques and system architectures must be used. The paper will review various 3D integration techniques, tool requirements, application of advanced test methodologies, and results.

Fig 1. SEM of 2 layer processor/memory wafer stack.

Categories and Subject Descriptors: B.7.1 [**Integrated Circuits**]: Advanced Technologies

General Terms: Design.

Keywords: 3DIC, Integrated Circuit, More than Moore, Stacking, Wafer to Wafer, Die to Wafer, Polylithic.

Copyright is held by the author/owner(s).
ISPD'11, March 27–30 2011, Santa Barbara, California, USA.
ACM 978-1-4503-0550-1/11/03.

Assembling 2D Blocks into 3D Chips

Johann Knechtel[†‡], Igor L. Markov[†] and Jens Lienig[‡]
†University of Michigan, EECS Department, Ann Arbor USA
‡Dresden University of Technology, EE Department, Dresden Germany
johann.knechtel@ifte.de, imarkov@eecs.umich.edu, jens.lienig@ifte.de

ABSTRACT

Three-dimensional ICs promise to significantly extend the scale of system integration and facilitate new-generation electronics. However, progress in commercial 3D ICs has been slow. In addition to technology-related difficulties, industry experts cite the lack of a commercial 3D EDA tool-chain and design standards, high risk associated with a new technology, and high cost of transition from 2D to 3D ICs. To streamline the transition, we explore design styles that reuse existing 2D Intellectual Property (IP) blocks in 3D ICs. Currently, these design styles severely limit the placement of Through-Silicon Vias (TSVs) and constrain the reuse of existing 2D IP blocks in 3D ICs. To overcome this problem, we develop a methodology for using TSV islands and novel techniques for clustering nets to connect 2D IP blocks through TSV islands. Our empirical validation demonstrates 3D integration of traditional 2D circuit blocks without modifying their layout for this context.

Categories and Subject Descriptors

B.7.2 [**Integrated Circuits**]: Design Aids — *Layout*

General Terms

Algorithms, Design

Keywords

3D integration, IP blocks, design styles, TSV islands

1. INTRODUCTION

Modern System-on-Chip (SoC) design faces numerous challenges, as steadily increasing demands on functionality and performance push against the limits of semiconductor manufacturing and EDA tools. Recent process-technology advances promise shorter interconnect and greater device density by means of three-dimensional integration — stacking multiple dies and implementing vertical interconnections with *Through-Silicon Vias (TSVs)*. Such 3D ICs can significantly extend the scale of system integration and facilitate new-generation electronics. While progress in commercial

3D ICs has been slow, memory-on-logic stacking has reached the market.[1] Several major design and manufacturability issues with 3D ICs currently remain unsolved [13], and a definitive commitment to 3D integration typically requires favorable cost considerations, as well as the availability of 3D EDA tools, industry standards and design methodologies. Industry experts are additionally concerned about the high risk associated with such a new technology, and potentially-prohibitive cost of transition from 2D to 3D ICs. To streamline this transition, we propose to focus on design styles that reuse existing 2D Intellectual Property (IP) blocks. As is well-known, modern chip designs are dominated by 2D IP blocks, proven in applications and considered reliable. Thus, we advocate 3D integration of legacy 2D IP blocks to circumvent many of the obstacles that currently impede wide adoption of 3D ICs.

In this paper, we make the following contributions. **First**, we describe and compare several possible design styles for 3D integration of 2D blocks, in particular the *Legacy 2D (L2D)* style which *integrates existing IP blocks* not designed for 3D integration. **Second**, we introduce a new design style for 3D integration of 2D blocks, called L2Di, where TSVs are clustered into TSV islands to reduce area overhead and provide post-silicon self-repair (similar to that in DRAM). **Third**, we propose novel algorithms and methodologies for net clustering, TSV-island insertion, deadspace alignment, and related tasks. The overall approach promises faster industry acceptance of 3D integration of legacy 2D IP blocks. **Fourth**, we empirically validate our methodology, demonstrating 3D integration of legacy 2D IP blocks without modifying their layout.

The remainder of this paper is structured as follows. In Section 2 we review important TSV characteristics and resulting integration challenges. In Section 3 we contrast possible design styles for 3D integration and discuss clustering of TSVs into TSV islands. We provide the problem formulation for the L2Di-style 3D integration in Section 4. Next, we describe our methodology in Section 5. In Section 6 we provide an empirical validation and in Section 7 we give our conclusions.

2. BACKGROUND

Since adjacent dies are connected by TSVs, TSVs are critical to 3D integration. TSVs can be manufactured in two ways: *via-first* and *via-last*. Via-first TSVs are $1\text{-}5\mu m$ in diameter and fabricated before the final metallization process; via-last TSVs are $5\text{-}20\mu m$ and fabricated after final metallization [15]. Furthermore, manufacturability demands *landing pads* and *keep-out zones* [31] which further increase TSV area footprint. At the 45nm technology node, the area

Permission to make digital or hard copies of all or part of this work for personal or classroom use is granted without fee provided that copies are not made or distributed for profit or commercial advantage and that copies bear this notice and the full citation on the first page. To copy otherwise, to republish, to post on servers or to redistribute to lists, requires prior specific permission and/or a fee.
ISPD'11, March 27–30, 2011, Santa Barbara, California, USA.
Copyright 2011 ACM 978-1-4503-0550-1/11/03 ...$10.00.

---

[1]For example, the recently released *Apple A4* SoC contains an ARM core die and two memory dies. However, the vertical interconnections are realized using wire-bonding, not TSVs.

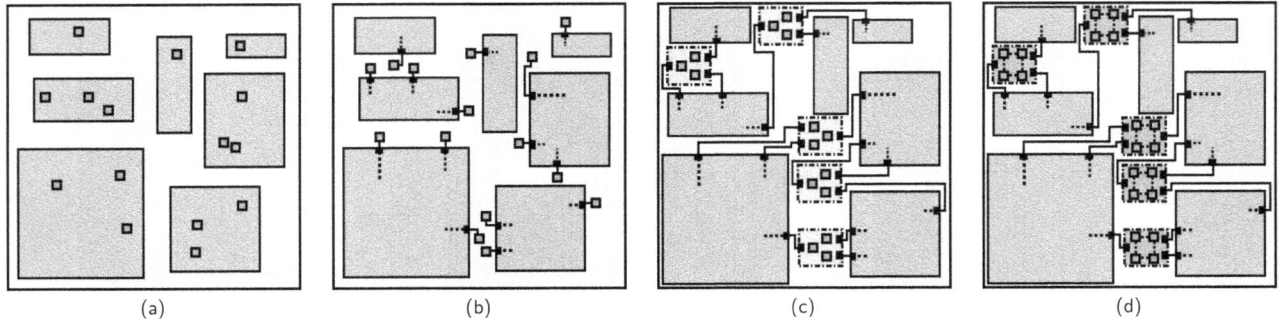

Figure 1: TSV positioning in design styles for 3D integration. (a) Gate-level and Redesigned 2D (R2D) styles place TSVs (small boxes) within the block footprint. (b) Legacy 2D (L2D) style places scattered TSVs between blocks, (c) L2D style with TSV islands (L2Di) groups TSV to blocks. (d) TSV islands can incorporate scan chains for TSV test and multiplex spare TSVs for redundancy.

footprint of a $10\mu m \times 10\mu m$ TSV is comparable to that of about 50 gates [16]. Hence, 10,000 TSVs can displace half a million gates. Previous work in physical design often neglects design constraints and overhead associated with TSVs, especially their area footprint [3, 11, 21–23, 28, 32]. Some studies explicitly consider thermal TSV insertion but not signal TSVs [20, 22, 23, 29]. Other studies incorporate signal and power TSVs in their flow, but sometimes ignore footprints of signal TSVs [18, 19].

Wirelength impact of TSVs. While the usage of TSVs is generally expected to reduce total wirelength, Kim et al. [16] observe that wirelength reduction varies depending on the number of TSVs and their characteristics. They show that TSV insertion in general and the impact of TSV footprint area in particular may increase silicon area and/or routing congestion, thereby making wires longer. Consequently, Kim et al. propose a new wirelength distribution model to estimate wirelength reduction while considering the impact of TSVs on wirelength. Case studies in [16] show that excessive usage of TSVs can undermine their potential advantages, and that this trade-off is controlled by the *granularity* of inter-die partitioning. The wirelength typically decreases for moderate (blocks with 20-100 modules) and coarse (block-level partitioning) granularities, but increases for fine (gate-level partitioning) granularities.

Another study by Kim et al. [15] suggests limiting the number of TSVs — otherwise, wirelength reduction can be undermined. Their study also reveals that using four to six dies offers most benefit for wirelength reduction. However, die stacking also raises mechanical issues, such as material stress and accurate TSV alignment [24,31], and may increase the overhead of test structures [17], exacerbate thermal problems [2], and increase the impact of intra-die variation [7]. Due to these major complications, practical 3D integration begins with only two active layers, as illustrated by a recently taped-out memory-on-processor design [9].

TSVs as layout obstacles. Via-first TSVs occupy the device layer, resulting in placement obstacles, while via-last TSVs occupy both the device and metal layers, resulting in placement and routing obstacles [16]. Hence, TSVs must be accounted for during floorplanning and/or placement. A study by Kim et al. [15] compares placing TSVs on a grid (*regular placement*) to placing scattered TSVs (*irregular placement*). The study reveals that irregular placement performs better in terms of wirelength reduction and design runtime. Since the TSVs are placed near the blocks they are connected to, there is no need for a separate TSV pin assignment process. However, regular placement helps manufacturing reliable TSVs [9, 10]. This also applies to TSV islands.

3. 3D IC DESIGN STYLES

3D integration originated with package-level integration, which connects multiple 2D chips through bonding pads, as illustrated by the quad-core variant of the *Intel Core 2* processor (two cores per die, two dies in one package). Finer granularity of 3D integration is enabled by connecting dies with TSVs, which results in 3D ICs [5]. In this section, we first discuss gate-level and block-level integration styles for 3D ICs. Gate-level integration faces multiple challenges and currently appears less practical than block-level integration. Second, we introduce a promising approach to 3D block-level integration, the Legacy 2D (L2D) style. It vertically integrates existing 2D blocks that were not originally designed for 3D integration. Third, we explain how 2D blocks are connected through TSVs.

3.1 Gate-Level Integration

One approach to 3D integration is to partition standard cells between multiple dies in a 3D assembly and use TSVs in routes that connect cells spread among active layers. This integration style promises significant wirelength reduction and great flexibility [2]. Its adverse effects include the massive number of necessary TSVs for random logic, as discussed in Section 2. The study by Kim et al. [16] reveals that partitioning gates between multiple dies may undermine wirelength reduction unless circuit modules of certain minimal size are preserved. In addition, partitioning a design block across multiple dies means it cannot be fully tested before die stacking. Moreover, after die stacking (post-bond testing), a single failed die can render several good dies unusable, thus undermining yield. Fine-grain partitioning between active layers also amplifies the impact of process variation, especially inter-die variation, on critical paths [7]. Monte Carlo SPICE simulations in [7] show that a 3D design is less likely to meet timing constraints than a comparable 2D design. Furthermore, gate-level 3D integration requires redesign of all blocks since existing IP blocks and EDA tools do not provision for 3D integration. However, 3D place-and-route tools are not yet available on the market, and IP providers have heavily invested in legacy 2D IP blocks.

3.2 Block-Level Integration

Design blocks subsume most of the netlist connectivity and are linked by a smaller number of global wires. Therefore, block-level integration reduces TSV overhead. The assignment of entire blocks to separate dies can be performed in two ways (Figure 1).

- *Redesigned 2D (R2D) style*: 2D blocks designed for 3D integration (TSVs included within the block footprints)
- *Legacy 2D (L2D) style*: 2D blocks not designed for 3D integration (TSVs placed between blocks)

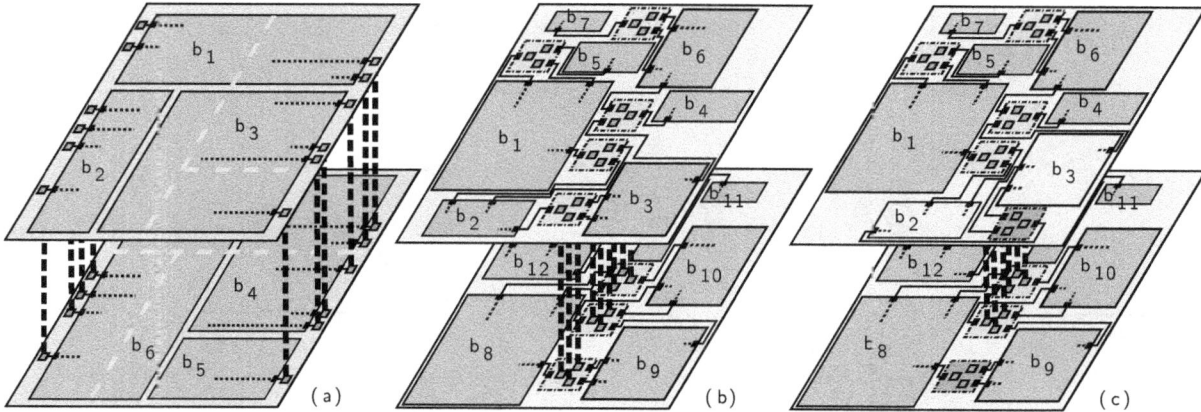

Figure 2: Deadspace alignment. Gray lines indicate (deadspace) channels and their mirror images on the adjacent die (gray dashed lines). (a) Floorplans that are small and/or regular usually contain little deadspace. Given insufficient aligned deaspace, vertical interconects (TSVs, wire-bonding) are placed at the boundary of the chip. (b) Floorplans with numerous blocks usually contain more deadspace and thus allow TSV-island insertion. Each TSV (island) requires adequate deadspace at two layers, with vertical alignment. (c) Poor alignment results in deadspace unusable for TSV insertion. A design example is shown in Figure 8.

These block-level integration styles promise the best trade-off between necessary TSV usage and wirelength reduction (Section 2). Several other important benefits of block-level integration are described next. Existing IP blocks are already equipped with Design for Testability (DFT) structures and can be tested before individual dies are stacked (pre-bond testing) [17].[2] With block-level integration, critical paths are mostly located within 2D blocks — they do not traverse multiple active layers, which limits the impact of TSV and inter-die variation on manufacturing yield. In [4], the authors propose optimal matching of *slow* and *fast* dies, based on accurate delay models with process variations considered. However, we note that this approach assumes that dies can be delay-tested before 3D stacking — a strong argument for block-level 3D integration.

Another aspect of block-level integration styles deals with design effort. The R2D style implies redesigning existing IP blocks, despite their successful track record in applications. This may require new EDA tools for physical design and verification, increasing risks of design failures and being late to market. Therefore, one hopes to avoid redesigning the broad spectrum of available functionality. It is more convenient to use legacy 2D IP blocks and to place the mandatory TSVs in the deadspace between the blocks, as provisioned by the L2D style. An extreme form of design IP reuse possible with the L2D style is *block-level mask reuse* with changes only required for global routes at high metal layers — TSVs placed in deadspace do not modify silicon layers of the blocks.

3.3 Connecting 2D Blocks by TSVs

The L2D style admits both scattered TSVs (Figure 1(b)) and TSV clusters (Figures 1(c,d)) In both cases, IP blocks are connected to TSVs with dedicated routes. TSV clusters require longer connections to block pins, but improve manufacturability by increasing exposure quality during optical lithography [10].

Clustering TSVs into TSV islands is helpful for multiple reasons. First, TSVs introduce stress in surrounding silicon which affects nearby transistors [31], but TSV islands do not need to include logic gates. The layout of TSV islands can be optimized in advance by experienced engineers. Second, clustering TSVs facilitates TSV-redundancy architectures [10, 24], where failed TSVs are shifted

within a chain structure or dynamically rerouted to spare TSVs. For example, Figure 1(d) illustrates islands with four TSVs, one of wich is spare. For the L2D style with TSV islands (L2Di), TSV islands can be placed in the deadspace between blocks.

Deadspace alignment. In some chip floorplans, the deadspace between blocks is too small to accommodate the necessary number of TSVs. In such cases, vertical interconnects can be realized by wire-bonding pads at the chip periphery or by injecting additional deadspace at the cost of a larger die footprint (Figure 2(a)). For larger designs with many diverse blocks, it is easier to find deadspace for TSV-island insertion. However, given a region of deadspace on a particular die, a TSV island also requires another region of deadspace on a neighboring die, with adequate alignment between the two regions.[3] Figure 2(b) illustrates a feasible TSV-island placement thanks to proper deadspace alignment while Figure 2(c) illustrates an infeasible TSV-island placement — blocks may need to be shifted to create sufficient aligned deadspace. In some cases, die area may need to be increased to accommodate TSV islands.

TSV placement and routing. TSV islands must be placed in aligned deadspace to facilitate short inter-layer routes that connect block pins through TSVs. A given net may be routed through one or several TSVs (islands). 2D routes may use high metal layers and/or channels between 2D blocks. The work in [30] dedicates an entire chip level to interconnect fabric.

4. PROBLEM FORMULATION

The L2Di style for 3D integration assumes the following input.

(i) **Active layers**, denoted as set $\mathcal{L}$. Each layer $l \in \mathcal{L}$ has dimensions (h_l, w_l) such that every block assigned to l can fit in the outline without incurring overlap. Note that the outlines may differ and thus allow layers to be shifted to improve deadspace alignment.

(ii) **Rectangular IP blocks**, denoted as set $\mathcal{B}$. Each block $b \in \mathcal{B}$ has dimensions (h_b, w_t) and pins, denoted as set $\mathcal{P}^b$. Each pin $p \in \mathcal{P}^b$ of block b is defined by its offset (δ_p^x, δ_p^y) with respect to the block's geometric center (origin).

---

[2]Test pins can be provisioned on each die and multiplexed/shared with other pins for pre- and post-bond testing [14].

[3]Alignment can be achieved by shifting blocks within each die, and also by displacing the dies, if the 3D chip package allows.

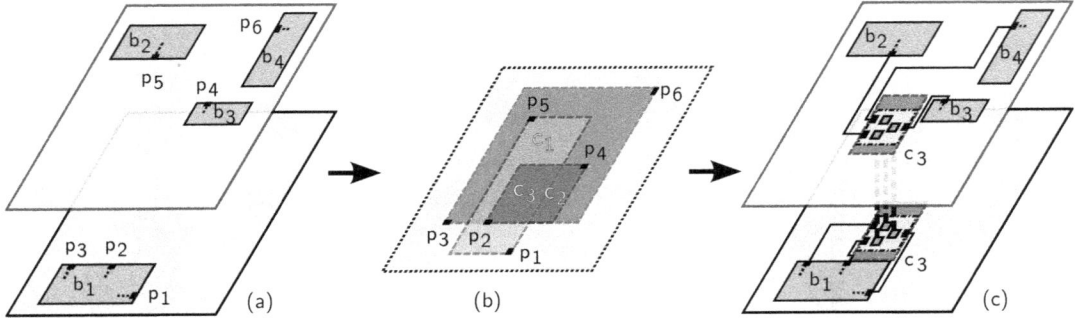

Figure 3: Net clustering and TSV-island insertion. (a) Inter-layer nets $n_1 = \{p_1, p_5\}$, $n_2 = \{p_2, p_4\}$, and $n_3 = \{p_3, p_6\}$ need to be connected through TSVs. (b) Pins are mapped to a virtual die and net bounding boxes are created. Intersections of bounding boxes mark cluster regions c_1, c_2, and c_3. The cluster region c_3 represents desired locations for a TSV island, which would facilitate shortest routes for all nets. (c) This particular cluster region is not obstructed by blocks and provides sufficient area, thus allowing TSV-island insertion without block shifting.

(*iii*) **Netlist**, denoted as set $\mathcal{N}$. A net $n \in \mathcal{N}$ describes a connection between two or more pins.

(*iv*) **TSV-island types**, denoted as set $\mathcal{T}$. Each type $t \in \mathcal{T}$ has dimensions (h_t, w_t) and capacity κ_t. Since pre-designed TSV-island types may incorporate spare TSVs, κ_t defines the number of nets that can be routed through t. While a simple formulation may deal with one type only, we believe that using pre-designed types of different shapes is essential to facilitate TSV-island insertion.

(*v*) **3D floorplan**, denoted as set $\mathcal{F}$. Each block b is assigned a location (x_b, y_b, l_b) such that no blocks overlap. (x_b, y_b) denotes the coordinate of the block's origin, l_b denotes the assigned layer.

3D integration with the L2Di style seeks to cluster inter-layer nets into TSV islands, and to insert TSV islands into aligned deadspace around floorplan blocks. If TSV-island insertion is impossible due to lack of aligned deadspace, blocks can be shifted from their initial locations, but their relative positions must be preserved. Figures 2(b,c) illustrate such block shifting. If TSV-island insertion is still impossible, additional deadspace can be inserted.

5. OUR METHODOLOGY

To connect blocks on different dies following the L2Di style, we need the locations of TSV islands. However, these locations must account for routes, so as to avoid unnecessary detours. In order to solve this chicken-and-egg problem, our techniques (*i*) clusters nets to estimate global routing demand, and (*ii*) uses these clusters to iteratively insert TSV islands. Details of our techniques are discussed in the following subsections, the overall flow is illustrated in Figure 4. For clarity of exposition, this section illustrates 3D integration in the case of only two dies. However, our techniques can be extended to more than two dies as well. In the following discussion, we refer to inter-layer nets as just nets.

Global iterations. Our clustering algorithm relies on a uniform grid. Grid-tile sizes influence per-tile net count. For example, quartering grid-tile size in Figure 5(b) would decrease the maximum per-tile net count from four to two. Having fewer nets per tile reduces the cluster size, increasing chances of TSV-island insertion. Therefore, we wrap our clustering and TSV-island insertion algorithm into an outer loop, which iteratively decreases grid-tile size from an upper bound f_{max} to a lower bound f_{min} (Table 1).

5.1 Net Clustering

The rationale for clustering nets is that placing TSV islands within net bounding boxes facilitates shortest-path connections. Also, as-

signing nets to clusters helps to select the type and capacity of each TSV island. Figure 3(c) illustrates a cluster of three nets routed through a TSV island.

To formalize the clustering process, we consider a *virtual die* — the minimum rectangle containing projections of active-layer outlines. The pins of each net are projected onto the virtual die (Figure 3(b)). Intersections of net bounding boxes in the virtual die suggest possible locations of TSV islands (Figure 3(c)).

Relevant results from graph theory. Imai and Asano [12] conducted a study on intersections of axis-aligned rectangles in the plane. First, Imai and Asano proved that n axis-aligned rectangles (e.g., bounding boxes) have a single non-empty n-way intersection *iff* each pair of these rectangles overlap. Thus, rather than check all subsets of overlapping bounding boxes, we may search

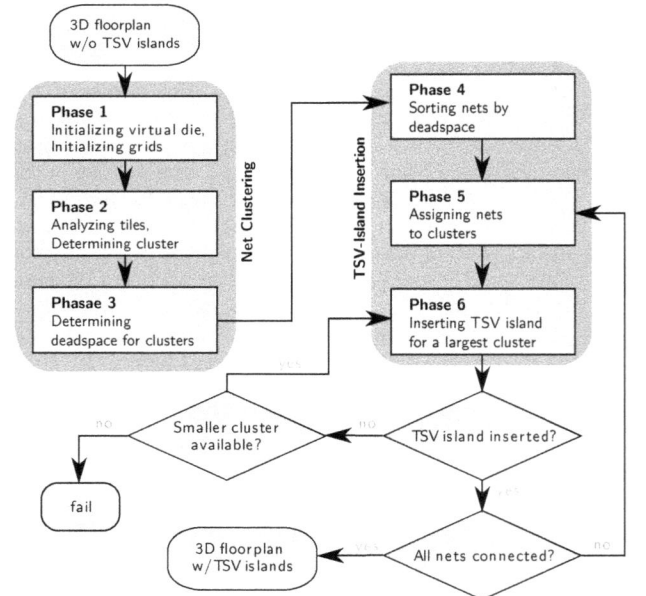

Figure 4: Flow of net clustering and TSV-island insertion. Our techniques first localize global routing demand while determining possible cluster regions, described by intersections of net bounding boxes. Second, TSV-islands insertion into cluster regions is iteratively attempted, based on dynamic scores. Depending on success of TSV-island insertion, our techniques provide a 3D floorplan with suitable placed TSV islands.

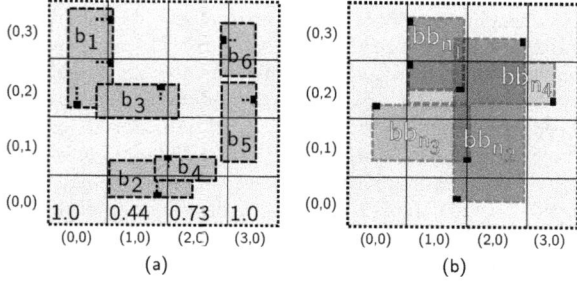

Figure 5: Uniform clustering grid $\mathcal{G}$ on virtual die. (a) Projected blocks from all active layers. We calculate the per-tile ratio of aligned deadspace, illustrated in the last row. (b) Projected net bounding boxes. According to their bounding boxes, we link nets to covered tiles. The intersection of net bounding boxes must be explicitly checked during clustering. In tile $(1,2)$, e.g., net bounding boxes bb_{n_1}, bb_{n_3} and bb_{n_3}, bb_{n_4} do not overlap pairwise, but all four nets are linked to the tile.

for cliques in a suitably defined *intersection graph*. This graph represents bounding boxes by vertices and connects overlapping boxes by edges. Second, they provided an $\mathcal{O}(n \log n)$-time algorithm for finding the maximum clique in intersection graphs with n vertices, even though this problem is NP-hard for general graphs [26]. In our context, however, a single clique is insufficient — we seek to partition the edges in the intersection graph into a small set of cliques, which is the NP-complete *clique cover* problem [6] (for *interval graphs*, this problem can be solved in polynomial time [8]).

Results by Imai and Asano imply that the largest possible net clusters correspond to maximum cliques in the intersection graph. However, in our context, large cliques may exceed the capacity of the largest available TSV island. Several TSV islands can be combined to implement such a clique, but this increases routing congestion and mechanical stress, and aggravates signal integrity problems [7,31]. Another problem with using large cliques is that corresponding (small) intersections of net bounding boxes may not include any aligned deadspace, preventing the insertion of a TSV island. On the other hand, a smaller clique would imply fewer bounding boxes and a larger intersection that is more likely to admit TSV-island insertion. Thus, we need to develop our own algorithm for identifying clique covers. Our algorithm is presented next.

Infrastructure used by our clustering algorithm. Initially, all blocks are projected onto the virtual die. In order to identify clusters (cliques) of appropriate size, a *uniform* clustering grid $\mathcal{G}$ is constructed on the virtual die (Figure 5). A clustering grid links each net n to each tile $\Xi \in \mathcal{G}$ covered by its net bounding box bb_n, and thus results in size-limited (appropriate) clusters. Cluster regions must be close to aligned deadspace, otherwise TSV islands cannot be inserted. To calculate the amount of aligned deadspace, a *non-uniform* grid is constructed. Grid lines are drawn through the four edges of each block. Grid tiles not covered by blocks define deadspace. For m blocks overlapping with a particular tile Ξ, deadspace detection runs in $\mathcal{O}(m^2)$ time [29], which is not prohibitively expensive because typically $m < 50$. In the *uniform* clustering grid, tiles with insufficient aligned deadspace ($< \Xi_{min}^d$) are marked as *obstructed*. Key parameters used in our algorithm are defined in Table 1 along with their values.

Our clustering algorithm is illustrated in Figure 6, referenced phases are also illustrated in Figure 4. **In Phase 1**, the virtual die and the grid structures are constructed. Then, each net is linked to each grid tile within the net's projected bounding box (Figure

```
CLUSTER_NETS(L, N, B, F)
 1    // Phase 1: initialize virtual die and clustering grid
 2    INITIALIZE_VIRTUAL_DIE(L)
 3    G = INITIALIZE_CLUSTERING_GRID(N, B, F)
 4    // Phase 1: link nets to grid tiles
 5    foreach net n ∈ N
 6        bb_n = DETERMINE_BOUNDING_BOX(n, B, F)
 7        foreach grid tile Ξ ∈ G where Ξ is covered by bb_n
 8            append n to Ξ.nets
 9    // Phase 2: determine possible clusters
10    foreach grid tile Ξ ∈ G where Ξ.obstructed == FALSE
11        c = DETERMINE_CLUSTER(Ξ, Ω_min, O_nets, O_link)
12        if c ∉ C
13            insert c into C
14            foreach net n ∈ c.nets
15                n.clustered = TRUE
16        elseif |c.nets| > 0
17            UPDATE_CLUSTER_REGION(c, Ξ)
18    // Phase 2: handle yet unclustered nets
19    progress = TRUE
20    while progress == TRUE
21        RESET(unclustered_nets)
22        foreach net n ∈ N where n.clustered == FALSE
23            append n to unclustered_nets
24        c = DETERMINE_CLUSTER(unclustered_nets)
25        progress = (c ∉ C)
26        if progress == TRUE
27            insert c into C
28            foreach net n ∈ c.nets
29                n.clustered = TRUE
30    // Phase 3: determine available deadspace
31    foreach cluster c ∈ C
32        foreach grid tile Ξ ∈ G where Ξ is covered by c.bb
33            c.deadspace += INTERSECTION(c.bb, Ξ) × Ξ.deadspace
```

Figure 6: Our clustering algorithm. Input data are described in Section 4. Grid tiles where aligned deadspace is below a threshold Ξ_{min}^d are pre-marked as obstructed.

5(b)). **In Phase 2**, for each unobstructed grid tile the largest cluster is determined — each linked net is considered as long as the resulting intersection of net bounding boxes is non-empty. Moreover, we impose a lower bound Ω_{min} on the overlap area between the intersection and tiles, in order to assure the intersection is covering the unobstructed tile to some minimal degree. We note that intersections in general can overlap more than one tile, depending on the net bounding boxes. An upper bound O_{nets} of nets clustered within each cluster c must not be exceeded. Also, an upper bound O_{link} for clustering each net n to clusters must not be exceeded. Next, we attempt to cluster yet-unclustered nets by relaxing the imposed bounds. Since this step allows one-net clusters, all nets are guaranteed to be clustered afterwards. **In Phase 3**, available aligned deadspace is determined for each cluster region.

5.2 TSV-Island Insertion

After running our clustering algorithm, we determined possible cluster regions (per net) where TSV islands can be inserted. However, not all clusters need to have TSV islands inserted to allow routing all nets through TSVs — according to the bound O_{link}, each net can be included in several clusters. Depending on the order of selecting clusters for TSV-island insertion, some clusters may become infeasible as TSV-island sites. Available deadspace accounted for a particular cluster may be occupied by another cluster. Furthermore, clusters containing nets which have not been clustered within unobstructed tiles need to consider nearby deadspace. This may also result in TSV islands blocking each other.

Our TSV-island insertion algorithm, illustrated in Figures 7 and 4, accounts for aligned deadspace while iteratively assigning nets to clusters and inserting TSV islands. In the following discussion, we refer to nets yet unassigned to a TSV island as *non-inserted*

```
INSERT_TSV_ISLANDS(C, N, T)
 1    // Phase 4: sort nets
 2    SORT_NETS_BY_AREA_SUPPLY(N, C)
 3    progress = TRUE
 4    while progress == TRUE
 5        // Phase 5: assign nets to clusters
 6        foreach net n ∈ N where n.inserted == FALSE
 7            c = FIND_HIGHEST_SCORED_CLUSTER(n, C, n_min^d)
 8            ASSIGN_NETS_TO_CLUSTER(c.nets, c, O_link)
 9        // Phase 6: iteratively insert TSV island for a largest cluster
10        (c, t) = INSERT_TSV_ISLAND(C, T)
11        progress = (c != NULL)
12        // Phase 6: mark & unlink handled nets from clusters
13        if progress == TRUE
14            foreach net m ∈ c.nets
15                m.inserted = TRUE
16                m.TSV_island = t
17                REMOVE_NET_FROM_CLUSTERS(m, C \ c)
18            // Phase 6: remove all assignments of nets to clusters
19            REMOVE_ASSIGNMENTS_FROM_CLUSTERS(C \ c)
20        elseif ∄ n ∈ N where n.inserted == FALSE
21            TERMINATE(success)
22        else TERMINATE(fail)
```

Figure 7: Our TSV-island insertion algorithm. Input data are described in Section 4.

nets, and to nets yet unassigned to a cluster as *unassigned nets*. **In Phase 4**, our algorithm sorts all nets by their total aligned deadspace of related clusters. Nets included in clusters with little available deadspace are considered first, since corresponding TSV islands are difficult to insert. **In Phase 5**, each unassigned net is analyzed for its associated clusters. The highest-scored cluster with respect to a dynamic *cluster score* Υ (available deadspace of cluster region divided by number of assigned nets) is chosen. Calculation and assignment of Υ for each cluster is performed dynamically within procedure FIND_HIGHEST_SCORED_CLUSTER, depending on previously assigned nets. In order to facilitate TSV-island insertion, the cluster to be chosen must provide a minimal amount of deadspace n_{min}^d for each net to be assigned to it. Then, each net of the highest-scored cluster is assigned to the cluster, subject to O_{link}. **In Phase 6**, TSV-island insertion for a largest cluster (in terms of assigned nets divided by Υ) is iteratively attempted — TSV-island insertion for clusters with many assigned nets and little available deadspace is difficult, thus these clusters are considered first. A local search (in procedure INSERT_TSV_ISLAND) over the cluster regions identifies contiguous regions with appropriate shapes to insert a TSV island with sufficient capacity.[4] Initially, aligned deadspace within the cluster regions only is considered. If no contiguous regions of deadspace can be found, a second iteration expands the cluster regions by factors c_{ext}^x, c_{ext}^y (in terms of die dimensions) to widen the search. If no contiguous regions are found again for any cluster, *block shifting* is performed to increase aligned deadspace (Section 5.3). Therefore, the cluster providing maximal amount of aligned deadspace is chosen first to minimize the total amount of shifting. After successful TSV-island insertion, all nets are unlinked from remaining clusters — according to Υ, each net may be assigned to different clusters now. Iterations continue with Phase 5 until all nets are inserted. If TSV-island insertion fails for all available clusters, our algorithm terminates with no solution.

5.3 Block Shifting using Floorplan Slacks

TSV-island insertion can fail because aligned deadspace is unavailable where it is needed. To address these failures, we propose to redistribute deadspace by shifting blocks in x- and/or y-directions

---

[4]The local search and TSV-island insertion are not described in detail due to page limitations.

without changing the floorplan outline or block ordering. We utilize the well-known concept of *floorplan slack* [1], which describes maximal displacement of a block within the floorplan. When blocks do not overlap, slacks are ≥ 0. We determine slacks for each layer separately and use standard linear-time traversals of floorplan constraint graphs, not unlike those in *Static Timing Analysis* [27]. Floorplan modifications based on constraint graphs are discussed in detail in [25]. To calculate x-slacks, we (i) pack blocks to the left boundary, and (ii) pack blocks to the right boundary. x-slack for each block is computed as the difference of the block's x-coordinates in these two packings. y-slack is calculated in the same way.

When TSV-island insertion requires redistributing deadspace by block shifting, we identify individual clusters in need of aligned deadspace. Within each cluster region, we determine the largest region R_d of aligned deadspace (if no aligned deadspace is found, we nominally consider the center of the cluster region as R_d). We then seek to consolidate additional aligned deadspace around R_d by shifting away the blocks adjacent to R_d. The distance by which each block is shifted cannot exceed its slack in the respective direction. Furthermore, the sum of such displacements in each direction cannot exceed the floorplan slack (the largest slack of any one block). Therefore, we shift blocks incrementally by small amounts so as to increase R_d until it reaches the size required to accommodate a TSV island with sufficient capacity.

Shifting a block may require shifting its abutting neighbors and other blocks. To this end, we maintain the floorplan configuration using constraint graphs and implement block shifting as follows. First, block dimension is inflated by the amount of displacement, and then standard path-tracing algorithms are applied to the constraint graph to find new block locations. To speed up this procedure, path-tracing can be performed incrementally, not unlike in incremental Static Timing Analysis.

If R_d cannot be increased sufficiently, we choose another region of aligned deadspace within the cluster region.

6. EMPIRICAL VALIDATION

We obtain 3D floorplans by running state-of-the-art software for 3D floorplanning.[5] The GSRC and MCNC benchmarks included in this infrastructure do not provide pin offsets, therefore we assume net bounding boxes to be defined by the bounding boxes of incident blocks. Two sets of floorplans are obtained, configuring the floorplanner for 10% and 15% deadspace respectively. We construct two sets of rectangular TSV islands with TSV footprints of $2\mu m^2$ and $4\mu m^2$ respectively. Each set contains TSVs islands with capacities for 2-30 nets; TSV islands are designed by packing single TSVs while considering keep-out-zone distances of $1\mu m$.

Experimental configurations. We consider two design configurations; one with guaranteed channels, one without channels. (Traditional floorplanners usually pack blocks without channels; however, many industry chips include channels between blocks to facilitate routing.) To insert channels between the blocks without modifying the floorplanner, every block was inflated before floorplanning and contracted to the original size after floorplanning. However, this increases floorplan size. An alternative is to pack blocks without channels, but carefully redistribute deadspace to facilitate TSV-island insertion. While more complex, this approach produces much more compact floorplans.

---

[5]We thank the authors of [32] and Yuankai Chen for sharing their infrastructure for 3D floorplanning.

Infrastructure for empirical validation. We implemented our algorithms using C++/STL, compiled them with g++ 4.4.4, and ran on a 64-bit Linux system with a 3.0GHz *Intel Core 2* processor and 8GB RAM. Parameters discussed in Section 5 are configured according to Table 1; Ξ_{min}^d, Ω_{min}, O_{nets}, and O_{link} control the clustering algorithm, while c_{ext}^x and c_{ext}^y control the deadspace search for TSV-island insertion, and i_{max}, f_{max}, i_{min}, and f_{min} control the global iterations. Experimental results on representative GSRC and MCNC benchmarks are reported in Table 2. In cases where our algorithm terminates with no solution, results are marked as *fail*.

TSV-island insertion and estimated wirelength. We analyze the impact of available TSV-island types, considering their capacity and dimensions. As expected, smaller TSVs increase chances of TSV-island insertion. However, wirelength overhead varies only marginally for the TSV-island dimensions that we tried. Being able to adjust the shape of TSV islands is beneficial for TSV-island insertion, while using square islands hinders insertion.

The overhead of TSV islands can be estimated by comparing actual net lengths to shortest paths (see *HPWL ratio* in Table 2), which would correspond to greedy insertion of single TSVs within net bounding boxes. Here we do not account for the increased footprint of single TSVs (due to increased keep-out-zones in comparison to packed TSV arrays) and have no way to quantify the loss of redundancy offered by TSV islands. Furthermore, this comparison would only apply to block-level interconnect, whereas most of the design's nets are subsumed within blocks. Wirelength overhead is 13.3-17.2% for the configuration without guaranteed channels, and 1.1-7.1% for the configuration with guaranteed channels. Such a moderate overhead in global interconnect, especially for the configuration with guaranteed channels, is expected and can be tolerated because 3D integration offers greater benefits [13].

Block shifting vs. guaranteed channels. Recall that inserting channels increases floorplan's deadspace. In contrast, redistributing available deadspace to facilitate TSV-island insertion supports more compact floorplans. However, we observe that floorplan slack required for block shifting is often insufficient, especially for blocks surrounding and/or covering cluster regions. Floorplans obtained with increased-deadspace configuration do not necessarily increase *aligned* deadspace because the infrastructure of [32] does not account for. For successful TSV-island insertion, our algorithms must

Deadspace & TSV area	Metrics	Design configuration					
		w/o guaranteed channels			w/ guaranteed channels		
		n100	n200	n300	n100	n200	n300
10% $2\mu m^2$	HPWL ratio	1.134	1.167	1.172	1.011	1.044	1.059
	Channel size	-	-	-	8%	13%	8%
	Runtime (s)	22.73	147.38	392.54	4.61	40.67	102.04
15% $2\mu m^2$	HPWL ratio	1.147	1.153	1.159	1.011	1.037	1.048
	Channel size	-	-	-	13%	12%	11%
	Runtime (s)	20.60	108.17	450.80	4.63	40.31	106.77
10% $4\mu m^2$	HPWL ratio	1.133	fail	fail	1.027	1.071	1.070
	Channel size	-	-	-	10%	14%	12%
	Runtime (s)	13.45	fail	fail	5.10	53.51	111.62
15% $4\mu m^2$	HPWL ratio	fail	fail	fail	1.011	1.066	1.062
	Channel size	-	-	-	10%	14%	12%
	Runtime (s)	fail	fail	fail	4.75	63.19	120.78

Table 2: L2Di integration without and with guaranteed channels. To add channels, every block was inflated before floorplanning and contracted to the original size after floorplanning, increasing floorplan outline. Binary search was used to find solutions with minimal wirelength overhead. We report channel size as the inflation factor. HPWL ratio divides wirelength of connections through TSVs by shortest-path wirelengths.

use aligned deadspace further from their cluster regions, thus increasing wirelength overhead.

Inserting deadspace channels not only reduces wirelength overhead and runtime, but also helps L2Di integration succeed where it otherwise fails. We performed a binary search for channel insertion with block inflations ranging from 6% to 14%. Considering the trade-off between floorplan increase and wirelength overhead, our results represent lowest wirelength overheads. Figures 8(a,b) illustrate L2Di integration of *n300* without and with guaranteed channels respectively. Some TSV islands are placed outside cluster regions due to deadspace limitations within cluster regions, introducing wirelength overhead. Cluster regions differ for the two design configurations. Without guaranteed channels, less aligned deadspace is available in general. Since our clustering algorithm accounts for available deadspace, cluster regions are more likely to intersect in this configuration.

7. CONCLUSION

Our work seeks to streamline the transition from existing practice in chip design to 3D integration. In addition to manufacturing and cost considerations, this transition is hampered by the lack of relevant standards and commercial EDA tools. A key insight in our work is that many of the benefits of 3D integration can be obtained while reusing existing 2D blocks. Therefore, we analyzed different design styles for 3D integration of 2D blocks. We conclude that, in the near future, the most promising and least risky design style for 3D integration is the L2Di style.

To enable the L2Di style, we contribute novel techniques for clustering of nets and inserting TSV islands. We provide an empirical validation of our techniques, demonstrating the possibility for 3D integration of 2D blocks without modifying their layout. We observe that the use of TSV islands may increase wirelengths beyond shortest paths, depending on the given floorplan. However, this effect can be mitigated by using a larger number of (smaller) TSV islands and by inserting deadspace channels between blocks.

Acknowledgements. We are thankful to Jin Hu for proofreading drafts of this paper.

Metric	Meaning	Value
Ξ_{min}^d	Min aligned deadspace per clustering-grid tile (*tile size*)	0.9
Ω_{min}	Min overlap area between cluster region and grid tile (*tile size*)	0.25
O_{nets}	Max nets per cluster	30
O_{link}	Max clusters per net	5
n_{min}^d	Min aligned deadspace per net in a cluster (*TSV footprint*)	1.05
c_{ext}^x, c_{ext}^y	Factors for extending cluster region to search for nearby deadspace (*die dimensions*)	0.05
bb_{min}	Area of the smallest net bounding box	floorplan specific
i_{max}	Global iterations for decreasing tile size from f_{max} to bb_{min}	5
f_{max}	Max tile size (bb_{min})	1.05
i_{min}	Global iterations for decreasing tile size from bb_{min} to f_{min}	35
f_{min}	Min tile size (bb_{min})	0.6

Table 1: Parameters for net clustering and TSV-island insertion algorithms, along with their values.

References

[1] S. N. Adya, I. L. Markov. Consistent placement of macro-blocks using floorplanning and standard-cell placement. *ISPD '02*, pp. 12–17.

[2] J. Cong, Y. Ma. Thermal-aware 3D floorplan. *Three Dimensional Integrated Circuit Design*, pp. 63–102. Springer US, 2010.

[3] J. Cong, J. Wei, Y. Zhang. A thermal-driven floorplanning algorithm for 3D ICs. *ICCAD '04*, pp. 306–313.

[4] C. Ferri, S. Reda. R. I. Bahar. Strategies for improving the parametric yield and profits of 3D ICs. *ICCAD '07*, pp. 220–226.

[5] R. Fischbach, J. Lienig, T. Meister. From 3D circuit technologies and data structures to interconnect prediction. *SLIP '09*, pp. 77–84.

[6] R. J. Fowler, M. S. Paterson, S. L. Tanimoto. Optimal packing and covering in the plane are NP-complete. *IPL*, 12(3):133–137, 1981.

[7] S. Garg, D. Marculescu. 3D-GCP: An analytical model for the impact of process variations on the critical path delay distribution of 3D ICs. *ISQED '09*, pp. 147–155.

[8] M. C. Golumbic. *Algorithmic graph theory and perfect graphs*. Elsevier, 2004.

[9] M. B. Healy et al. Design and analysis of 3D-MAPS: A many-core 3D processor with stacked memory. *CICC '10*.

[10] A.-C. Hsieh et al. TSV redundancy: Architecture and design issues in 3D IC. *DATE '10*, pp. 166–171.

[11] W.-L. Hung et al. Interconnect and thermal-aware floorplanning for 3D microprocessors. *ISQED '06*, pp. 98–104.

[12] H. Imai, T. Asano. Finding the connected components and a maximum clique of an intersection graph of rectangles in the plane. *J. Algorithms*, 4(4):310–323, 1983.

[13] International technology roadmap for semiconductors.
http://www.itrs.net/Links/2009ITRS/Home2009.htm.

[14] L. Jiang, Q. Xu, K. Chakrabarty, T. M. Mak. Layout-driven test-architecture design and optimization for 3D SoCs under pre-bond test-pin-count constraint. *ICCAD '09*, pp. 191–196.

[15] D. H. Kim, K. Athikulwongse, S. K. Lim. A study of through-silicon-via impact on the 3D stacked IC layout. *ICCAD '09*, pp. 674–680.

[16] D. H. Kim, S. Mukhopadhyay, S. K. Lim. Through-silicon-via aware interconnect prediction and optimization for 3D stacked ICs. *SLIP '09*, pp. 85–92.

[17] H.-H. S. Lee, K. Chakrabarty. Test challenges for 3D integrated circuits. *IEEE Design & Test of Computers*, 26(5):26–35, 2009.

[18] Y.-J. Lee, R. Goel, S. K. Lim. Multi-functional interconnect co-optimization for fast and reliable 3D stacked ICs. *ICCAD '09*, pp. 645–651.

[19] Y.-J. Lee, M. Healy, S. K. Lim. Co-design of reliable signal and power interconnects in 3D stacked ICs. *IITC '09*, pp. 56–58.

[20] X. Li et al. LP based white space redistribution for thermal via planning and performance optimization in 3D ICs. *ASP-DAC '08*, pp. 209–212.

[21] X. Li, Y. Ma, X. Hong. A novel thermal optimization flow using incremental floorplanning for 3D ICs. *ASP-DAC '09*, pp. 347–352.

[22] Z. Li et al. Integrating dynamic thermal via planning with 3D floorplanning algorithm. *ISPD '06*, pp. 178–185.

[23] Z. Li et al. Efficient thermal-oriented 3D floorplanning and thermal via planning for two-stacked-die integration. *TODAES*, 11(2):325–345, 2006.

[24] I. Loi et al. A low-overhead fault tolerance scheme for TSV-based 3D network on chip links. *ICCAD '08*, pp. 598–602.

[25] M. D. Moffitt, J. A. Roy, I. L. Markov, M. E. Pollack. Constraint-driven floorplan repair. *TODAES*, 13(4):1–13, 2008.

[26] C. H. Papadimitriou, K. Steiglitz. *Combinatorial Optimization : Algorithms and Complexity*. Dover Publications, 1998.

[27] S. S. Sapatnekar. *Timing*. Kluwer, 2004.

[28] S. Sridharan et al. A criticality-driven microarchitectural three dimensional (3D) floorplanner. *ASP-DAC '09*, pp. 763–768.

[29] E. Wong, S. K. Lim. Whitespace redistribution for thermal via insertion in 3D stacked ICs. *ICCD '07*, pp. 267–272.

[30] X. Wu et al. Cost-driven 3D integration with interconnect layers. *DAC '10*, pp. 150–155.

[31] J.-S. Yang et al. TSV stress aware timing analysis with applications to 3D-IC layout optimization. *DAC '10*, pp. 803–806.

[32] P. Zhou et al. 3D-STAF: scalable temperature and leakage aware floorplanning for three-dimensional integrated circuits. *ICCAD '07*, pp. 590–597.

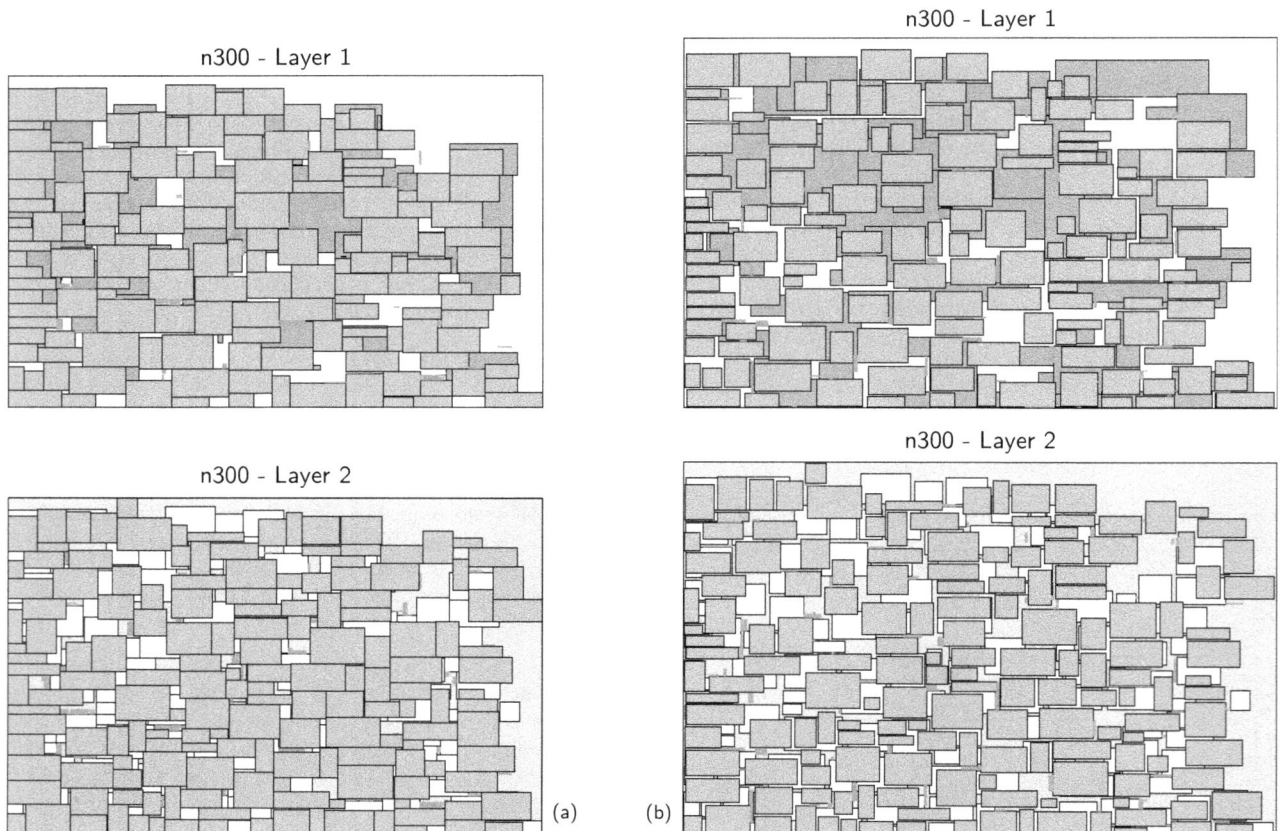

Figure 8: L2Di-style integration of *n300*, (a) without guaranteed channels and (b) with guaranteed channels. Grey rectangles represent blocks, red (black) rectangles TSV islands (up to 30 TSVs per island). Cluster regions are shown in brown (dark grey) on Layer 1, while aligned deadspace is shown in beige (light grey) on Layer 2. Channel insertion in (b) increased die size by 21%.

Invited Talk

Automated Placement for Custom Digital Designs

Tung-Chieh Chen
SpringSoft Inc.
Hsinchu, Taiwan
donnie_chen@springsoft.com

Abstract

Custom layouts of digital blocks are often used in mixed-signal designs in order to meet the critical performance requirements. Unlike traditional standard-cell based digital placement, custom-cell based digital placement may need to consider special routing patterns and pre-wires, very high design utilization, topology placement constraints, etc.

This talk will address several unique placement problems for custom digital designs. First, spine style routing is often used in custom digital designs because of limited layer routing; it is necessary for placement to consider the spine topology to improve the routability. Second, the cell-level oxide diffusion (OD) sharing can help to reduce placement area, if custom cells are designed with common power and ground portions. Third, placement also needs to consider the electrostatic discharge (ESD) issue by enlarging the spacing between some cells to avoid direct power ground path. Last but not least, placement migration is another important and challenging issue; the resulting placement should follow not only the physical hierarchy of the design but also all additional topology placement constraints so that the resulting layout can meet the performance requirements.

Categories & Subject Descriptors: B.7.2 [Hardware, Integrated Circuits, Design Aids]: Placement and routing; G.4 [Mathematics of Computing, Mathematical Software]: Algorithm Designs and Analysis

General Terms: Algorithms, Design, Performance

Bio

The speaker is a principal engineer in the Physical Design Group of SpringSoft Inc. He received his Ph.D. from National Taiwan University (NTU) in 2008. He was a visiting scholar at the University of Texas at Austin in 2007. His primary interests in research and product development cover VLSI physical design, floorplanning, and placement for digital and analog designs. He has published four book chapters, six IEEE Trans. on CAD journal papers, and 11 ACM/IEEE conference papers, all on floorplanning and placement. He is the third place winner in 2006 ACM ISPD Placement Contest and the first place winner in 2007 ACM SIGDA CADathlon Contest. He received the Best Dissertation Award from the Graduate Institute of Electronics Engineering (GIEE) at NTU in 2008 and the Outstanding Research Award from GIEE at NTU in 2007 and 2008.

Copyright is held by the author/owner(s).
ISPD'11, March 27–30, 2011, Santa Barbara, California, USA.
ACM   978-1-4503-0550-1/11/03..

Quantifying Academic Placer Performance on Custom Designs

Samuel Ward
IBM STG
11400 Burnet Rd.
Austin, TX 78758
siward {@us.ibm.com}

David A. Papa
IBM Austin Research
11501 Burnet Rd.
Austin, TX 78758
iamyou {@eecs.umich.edu}

Zhuo Li
IBM Austin Research
11501 Burnet Rd.
Austin, TX 78758
lizhuo{@us.ibm.com}

Cliff Sze
IBM Austin Research
11501 Burnet Rd.
Austin, TX 78758
csze {@us.ibm.com}

Charles Alpert
IBM Austin Research
11501 Burnet Rd.
Austin, TX 78758
alpert {@us.ibm.com}

Earl Swartzlander
The University of Texas at Austin
Electrical and Computer Engineering
Austin, TX 78712 USA
eswartzla {@aol.com}

ABSTRACT

There have been significant prior efforts to quantify performance of academic placement algorithms, primarily by creating artificial test cases that attempt to mimic real designs, such as the PEKO benchmark containing known optimas [5]. The idea was to create benchmarks with a known optimal solution and then measure how far existing placers were from the known optimal. Since the benchmarks do not necessarily correspond to properties of real VLSI netlists, the conclusions were met with some skepticism. This work presents two custom constructed datapath designs that perform common logic functions with hand-designed layouts for each. The new generation of academic placers is then compared against them to see how the placers performed for these design styles. Experiments show that all academic placers have wirelengths significantly greater then the manual solution; solutions range from 1.75 to 4.88 times greater wirelengths. These testcases will be released publically to stimulate research into automatically solving structured datapath placement problems.

Categories and Subject Descriptors

B.7.2 [Integrated Circuits]: Design Aids – Placement and Routing

General Terms: Algorithms, Design, Experimentation

Keywords: Standard Cell Placement, Datapath Placement, Placement Benchmarks

1 INTRODUCTION

Automatic VLSI placement algorithms have improved significantly since the ISPD placement contests in 2005 and 2006 [14] [15]. These contests released 16 new placement benchmarks derived from industrial designs.

The benchmarks contained a number of important features that were not present in the previous set of benchmarks: (i) they ranged

Permission to make digital or hard copies of all or part of this work for personal or classroom use is granted without fee provided that copies are not made or distributed for profit or commercial advantage and that copies bear this notice and the full citation on the first page. To copy otherwise, or republish, to post on servers or to redistribute to lists, requires prior specific permission and/or a fee.
ISPD'11, March 27–30, 2011, Santa Barbara, California, USA.
Copyright 2011 ACM 973-1-4503-0550-1/11/03...$10.00.

in size from 211k cells to 2.18M cells, much larger than previous benchmarks (ii) they contained large fixed obstacles not seen in previous benchmarks (iii) they contained large movable objects which cover more than a single circuit row in height, a feature that was added to existing benchmarks [1].

In addition, the structure of the contest forced placement algorithms to optimize half-perimeter wirelength (HPWL), runtime and a target density, which is used in practice to improve both timing and routability of circuits in physical synthesis. Prior to the contests, few academic placers could solve these realistic problem instances, though that is certainly not true today [3] [4] [10] [16]. It is easy to observe that benchmarks are important to guide the development of practical placement algorithms.

Prior to these contests, there were attempts to quantify the suboptimality of placement heuristics. Hagen, *et al.* [7] had the idea of taking copies of small circuits and replicating them, then loosely connecting their ports together, in order to create a much larger benchmark. For example, by connecting four copies of a well-placed circuit together in 2 x 2 grid, they obtained a placement wirelength that was no more than four times that of the original circuit. While interesting, this experiment is arguably unrealistic since these defined connections between the copies do not correspond to real logic functions. Furthermore, no pin locations are defined for the circuit (nor were there any for the original). This work overcomes both prior objections.

More recently, Chang, *et al.* [5] created the placement examples with known optima (PEKO) and placement examples with known upperbounds (PEKU) algorithms and released two sets of benchmarks with solutions that are known to be optimal or close to optimal. Optimality was achieved by adding nets to cells in configurations that cannot be shortened. In other words, they created a design where every net was a super-short net, though the pin distributions of cells matched that of a typical VLSI circuit. Reported results show wirelengths in the range of 1.43 to 2.40 times the optimal value. Again while interesting, these netlists did not correspond to any logic function at all. It could be argued that the PEKO and PEKU testcases are artificially hard and that no placer would ever need to solve them.

Given the renaissance in automated placement technology that has occurred, it seems like a good time to revisit this issue of quantifying placement algorithms. Perhaps this new generation is close to optimal, especially given that placer improvements are worthy of publication when they manage to obtain a 1-2% improvement in wirelength. However, unlike previous efforts, this work quantifies placement algorithms on useful logic function. It is accepted folklore that current placement tools do not perform particularly well on custom or structured designs. Due to their regular structure, datapath designs enable a designer to construct highly compact custom layouts. To shed light on this issue, the solutions of academic placement tools are compared on two manual designs created for this purpose.

The initial design for each circuit was developed using standard, custom design practices. Logic gates from automated design standard cell libraries were hierarchically built within a custom schematic design framework. Each design used a reduced library of basic 2-, 3-, and 4-input NAND, NOR, INVERT, MUX and XOR gates and latches. These gates were manually placed and fixed for both benchmarks and design inputs and outputs placed directly on the driving or receiving gate.

The combinational logic gates for both benchmarks were allowed to move during the course of automatic placement by several academic tools [19]. We observed that every placer[1] produced a solution having wirelength at least 1.75 times that of the custom solution while having no density constraints.

Many have speculated that poor performance of placers on datapath designs is due to very tight density constraints. Perhaps placers could find the right structures but simply had trouble with the legalization. Consequently, eight variants of each design were created where additional whitespace was inserted to provide more opportunity for the placers. While wirelength was improved, all placers still generated solutions with wirelengths at least 1.44 times that of the custom solution. The empirical results confirm there remains significant room for improvement in modern academic placement algorithms.

The paper is organized as follows. Section 2 presents a rotate circuit, and shows a common manual layout solution. Section 3 does the same for a compare logic circuit. Experiments results comparing six placers are presented in Section 4. Conclusions are presented in Section 5.

2   DESIGN 1: ROTATE LOGIC

Rotate circuits, also known as cyclic shifters [8] [11] [12], are a simple and common bit operation generally found throughout microprocessors, cryptography, imaging, and biometrics [2] [13]. Traditionally, rotators are custom designed because of their highly regular structure and significant routing complexity [6] [18] though some work on automated placement has been explored [9].

2.1 Overview

A standard rotate function consists of cascaded 2-input MUXes, as shown in Figure 1. A rotator circuit receives a set of inputs $d[0:n-1]$ and $r[0:m-1]$ and produces an output $s[0:n-1]$, where $d[0:n-1]$ has been rotated by some amount encoded by $r[0:m-1]$. In the following notation, & indicates a logical AND, + indicates a logical OR, and ! indicates a logical NOT. To mathematically define the rotate functions, let $k[i,j]$ denote the internal point at ith

row and jth column in Figure 1, where $i = (0:m-1)$ for $r[0]$ to $r[m-1]$ and $j = (0:n-1)$ for $d[0]$ to $d[n-1]$. Then, $k[0,j] = !\ (\ r[0]\ )\ \&\ d[i]\ +\ r[0]\ \&\ d[i+1]$, where j = 0, ..., n-1 and note that $n = 2^m$. Thus the general equations are:

$$k[i,j] = !\ (\ r[i]\ )\ \&\ k[i-1\ ,j] + r[i]\ \&\ k[\ i-1\ ,j+2^i]$$
where i = 0, ..., m-1, j = 1, ..., n-1

$$k[i, j] = k[i, j + z * n],\quad \text{where } z \text{ is } 0, 1, 2, ...,$$

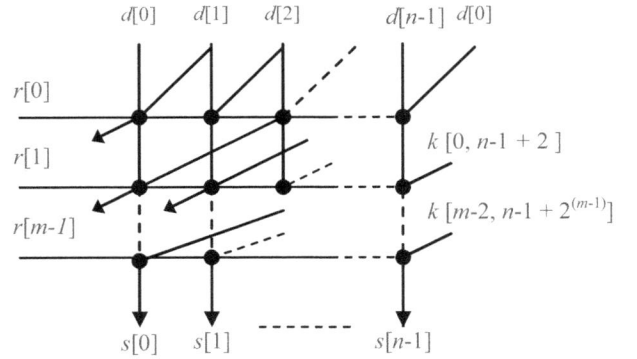

Figure 1. Rotate Block Diagram

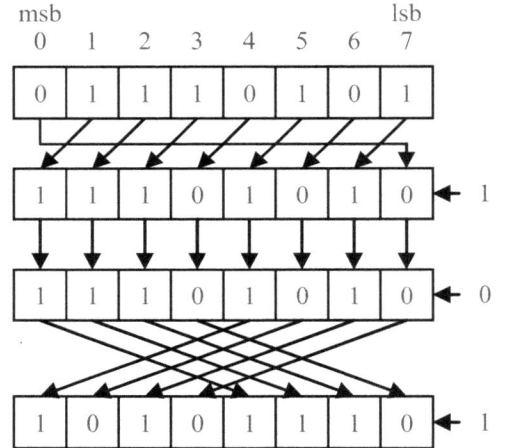

Figure 2. Rotate Example

Figure 2 shows an example of an eight-way rotate function. The initial input vector $d[0:7] = \{\ 01110101\ \}$ and $r[0:2] = \{\ 101\ \}$, indicating a rotation of five. In the first stage, $r[0]$ rotates the input vector one bit position, $r[1]$ in stage two does not rotate the vector, and in the third stage, $r[2]$ rotates the vector four more bit positions for a result of $s[0:7] = \{\ 10101110\ \}$.

Rotator designs present automated design tools with the challenge of producing a densely-packed placement solution while minimizing routing congestion. There are two parts to the routing challenge, local routing and global routing. Local routes between each MUX must be lined up very carefully to leave space for the global select lines $r[0:m-1]$. At each stage, the route from the previous stage shifts one more column over, creating a congested routing network. Design placement that minimizes jogging global

---

[1] APlace failed to produce a legal solution.

routes is critical to achieve a routable design that meets area and timing constraints. In addition, careful attention to the design of global routes is necessary for optimal delay.

2.2 Benchmark Details

The first benchmark in this paper, Design 1, derives from the manual placement of an actual high-speed microprocessor rotate function. The logic implementation also includes two enable signals at each rotate circuit. Certain portions of the design are modified, such as the intermediate output pins and the latch points, without modifying overall functionality. Figure 3 displays the basic MUX and the enable building block for the design, which is referred to as a complex subcell. Each complex subcell is comprised of a two-to-one MUX with a corresponding select signal $r[i]$ and enable signals $e_h[i]$ and $e_v[j]$. Enable signal $e_h[i]$ runs horizontally to each bit stack in the ith row, and $e_v[j]$ runs vertically to each complex subcell within a bit stack at jth column. Exact circuit implementation can vary, depending on the specific technology; however, in this design, a single two-to-one MUX and a three-input NAND gate are used for the implementation. Each following stage is inverted to maintain polarity without impacting TWL calculations.

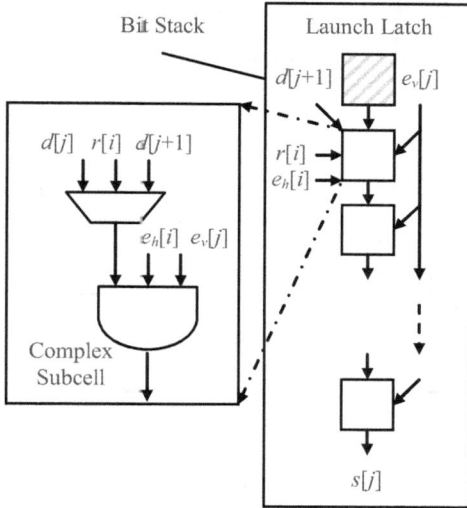

Figure 4. 8-Bit Rotate Physical Layout with 3-bit Encoding

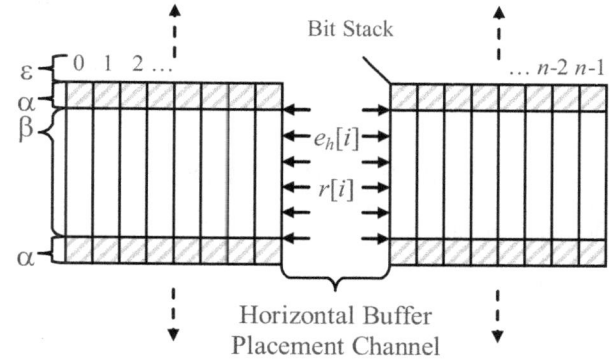

Figure 3. Rotate Sub-block

Using the notation from Figure 1, Design 1 contains n=512 and m=9, which means it is a 512 bit rotate circuit with 9 encoding bits. Each $d[0:n-1]$ is stored in a latch with a fixed location and drives the stacked MUX structure, which is nine complex subcells high. Primary input (PIs) and output pins (POs) were placed directly on top of their respective connections minimizing PI/PO routing distance.

2.3 Placement Details

Figure 4 shows a representative layout for an 8-bit rotator as in Figure 2. The data bus $d[0:7]$ initially resides in the latches denoted *lat* with each complex subcell stacked directly on top. The *mux* is the 2:1 MUX in the complex subcell and *and* is the AND-3 gate in the complex subcell. The rotate result, $s[0:7]$ leaves the top of each bit stack driven from the last complex subcell.

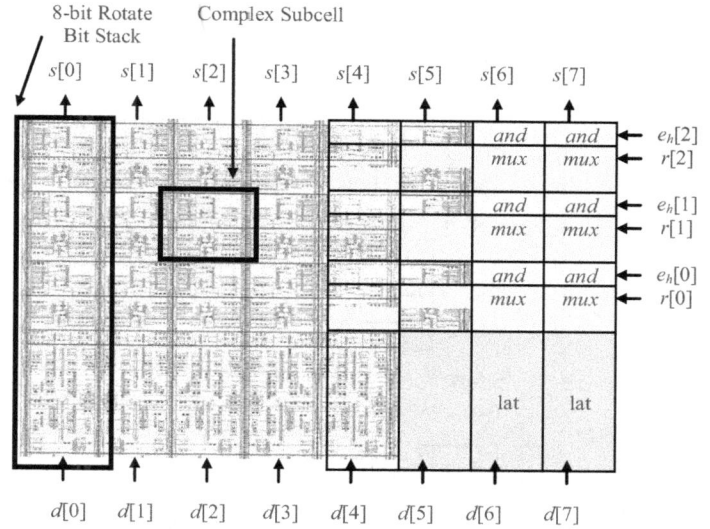

Figure 5. Rotate Row

Figure 5 displays the next level of hierarchy in which bit slices are placed next to each other to form a rotate row, n-bits wide. Let α denote the total latch height, β denote the total logic height of the stacked complex subcells (nine in this example, corresponding to r[0:8]), and ε denote the any added whitespace in the bit slice. For the base design, ε equals 7% of the total height of the bit slice. Each bit stack is ordered, as in Figure 3, to line up the MUX rotate signals $r[i]$ and enable signals $e_h[i]$ and $e_v[j]$ with their corresponding complex subcell. This is critical for both routability and minimizing TWL, since the fanout on $r[i]$, $e_h[i]$ and $e_v[j]$ is very large. Between bit stacks $(n/2 - 1)$ and $(n/2)$, space for buffer placement is added where the rotate line bus $r[0:m-1]$ and enable signal bus $e_h[0:m-1]$ are lined up to drive horizontally to each bit slice.

The top level of hierarchy is shown in Figure 6 where each row from 0 to p-1 is an independent copy of the rotate row shown in Figure 5. In the middle of the block, space for buffer placement is added where the enable signal bus $e_v[j]$ is lined up to drive vertically to each row.

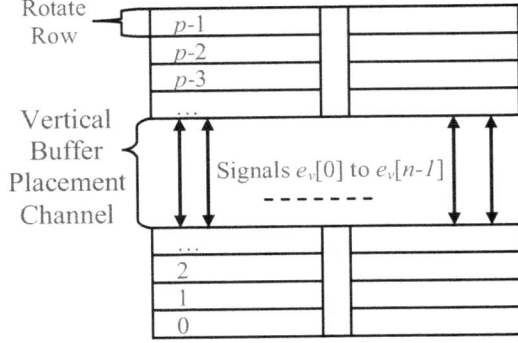

Figure 6. Fully Placed Rotate Block

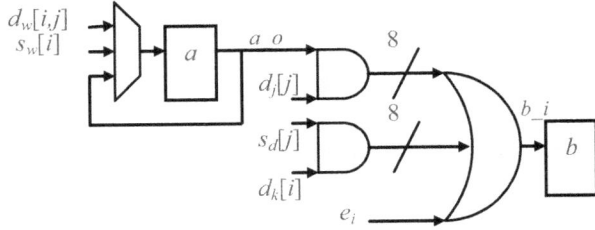

Figure 7. Design 2 Bit Stack Structure

out for illustrative purposes of the manual layout solution. In the full implementation, each bit stack consists of 16 AND gates driving a 16 way OR gate configuration. The logic gates from Figure 7 are interleaved into a single circuit row, and pins between rows for each select line are lined up evenly to reduce branch routing. Latch a and the MUX that drives it are placed at the bottom of the stack, the data flows through the AND/OR reduce logic.

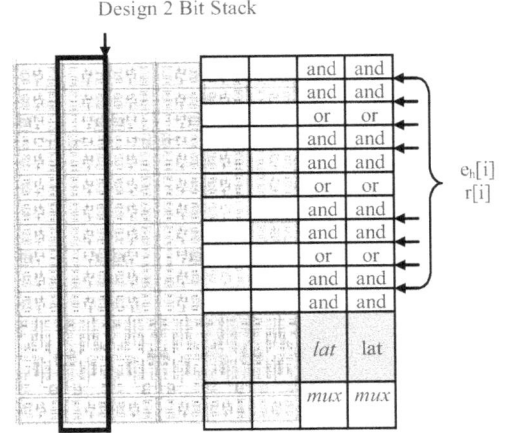

Figure 8. Design 2 Eight Bit Stack Physical Layout

3   DESIGN 2: AND/OR LOGIC

Design 2 is a standard AND/OR logic tree [8] [11] common throughout datapath design[2] with the bit stack logic structure shown in Figure 7. This structure is used in many applications, such as translation buffers and structured content addressable-memory circuits. Two signals, a_o and $s_d[j]$, are driven into an AND gate with the data inputs and then into an OR tree. The output of the OR tree is then ORed with a set signal e_i, and the result is latched.

3.1 Benchmark Details

Design 2 is a simplified version of a custom placed industrial design. Careful packing of the repeated logic enables optimization of both timing and area while reducing congestion. The bit stack in Figure 7 is repeated $n=257$ times in one row and there are $m=12$ rows for a total of 3084 bit stack instances placed within Design 2. Signal $d_w[i,j]$, $d_j[i]$, and $d_k[i]$, where $i = 0, 1, ..., n$-1 columns and $j = 0, 1, ..., m$ rows, are primary data input signals; e_i and $s_d[i]$ are high fanout select lines running through the ith row where i = 0, 1, ..., n-1, and j = 0, 1, ..., m-1 for the bit stack with m rows and n columns. Select line $s_w[i]$ runs within row i and is a write enable select signal to latch new data into latch a. If $s_w[i]$ is not enabled, the prior value in a is selected and stored. Enable signal e_i is an override signal that will set latch b.

3.2 Placement Details

Custom placement of Design 2 leads to regularly placed rows with tightly packed cells, shown in Figure 8, where eight total bit stack cells have been placed. Figure 8 represents a partial 8 bit stack lay

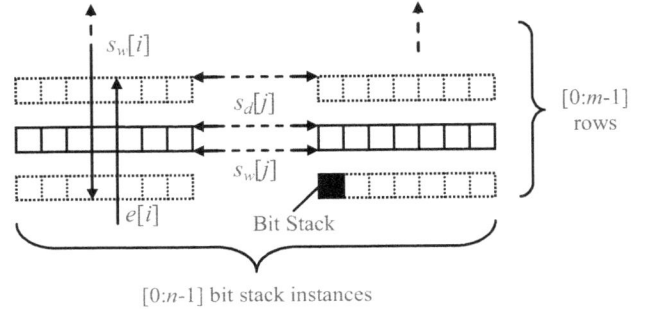

Figure 9. Placement for Design 2

Figure 9 shows the overall layout of the entire design with one bit stack shaded. Each bit stack is placed side-by-side $n=257$ times in one row, where there are 12 rows in total and placement space is added in the middle, both vertically and horizontally for the global wire drivers.

Table 1. Design Characteristics

Designs	Aspect Ratio	Num Nets	Num Pins	Num Terminals	# Movable Cells
DESIGN 1	1.1	33977	113968	3496	29312
DESIGN 2	2.6	40811	191926	6886	37008

4   EXPERIMENTAL RESULTS

Table 1 outlines the design details for each circuit that was constructed. Design 1 contains 29,312 movable cells, and Design 2 contains 37,008 movable cells. These are both reasonably small, custom-placement designs built using common structured placement tools, including schematic capture and layout. Once built, the netlist was exported to Bookshelf [14][15] format and

---

[2] Standard cell design practice allows for the interchange between logic gates of the same size. Thus, this implementation can be modified to many representative circuits, such as magnitude comparators, standard equality circuits or parity circuits.

the wirelength measured. The custom layout solution is compared against the following placers for these two designs:

- mPL6 v6 [3]
- CAPO v10.2 [16]
- FastPlace v2.0 [20]
- NTUPlace3 v7.10.19 [4]
- APlace v1.0 [10]
- Dragon v3.01 [21]

The authors of Timberwolf [19] were contacted, but they were unable to provide a version compatible with the Bookshelf [14][15] format at this time. This simulated annealing approach may perform well on these moderately sized test cases. For all placers, a target density constraint[3] was not imposed to give maximum freedom to pack cells.

Table 2. TWL Results

Placer	Design 1			Design 2		
	TWL	TWL Ratio	Run Time (s)	TWL	TWL Ratio	Run Time (s)
Custom	1041383	1.000	n/a	1684059	1.000	n/a
Capo	1971020	1.893	34	4588620	2.725	112
mPL6	1826466	1.754	188	2975343	1.767	366
ntuPlace3	2191076	2.104	59	0	0.000	ABORT
APlace*	n/a	n/a	n/a	n/a	n/a	n/a
Dragon	3421607	3.286	413	8233024	4.889	474
FastPlace	1932727	1.856	21	3625048	2.153	38

*APlace totals are after global placement only

4.1 Initial Placement Results

Table 2 displays the results of the custom placement versus the academic placer. Column one shows all different placement algorithms, where "custom" corresponds to the manual-placed designs. Column two displays the measured TWL compared to the custom solution for each placement method on Design 1, and column five displays the TWL for Design 2. For both designs, APlace failed to find a legal placement solution. Columns three and six correspond to the percentage increases in TWL compared to the custom-placed solution. Columns four and seven display the placement runtime in seconds for each design.

The custom-placement method resulted in a TWL of 1,041,383 for Design 1 and of 1,684,059 for Design 2. For Design 1, the mPL6 algorithm again produced the best automated placement result with 1,826,466, a 75% increase in TWL. The run time for all placers is less than seven minutes because of the small design size. For Design 2, the mPL6 algorithm resulted in the best automated placement result with 2975343, a 1.76 TWL ratio. NTUPlace3 resulted in an assertion error for the base case and the run time for all other placers is less than eight minutes for Design 2.

4.2 Adding Additional Whitespace

As mentioned earlier, it is important to understand whether placers could not find the right structure, or could not legalize. Seven

---

[3] This was achieved by supplying each placer with a target density requirement of 100% density as defined as in ISPD placement contests [14] [15]

additional variations of each benchmark were generated by increasing white space using the following scheme. As shown in Figure 10, let η denote the total height of default bit slack, α denote the height of latch logic, β denote the height of placeable logic, and ϵ denote the white space added to the bit stack.

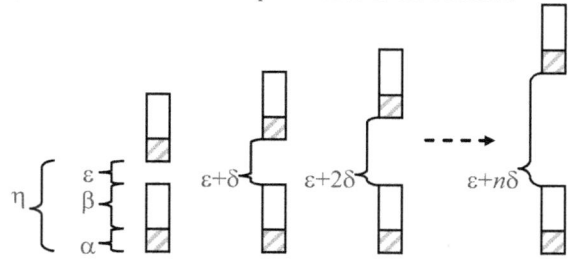

Total Cell Height: $\eta = \alpha + \beta + n\epsilon$

Figure 10. Structured Placement Experiments

The original testcase experiment for both designs was set up with ϵ=7% of the design height, i.e., there was about 7% free space added to each bit stack. Then the white space was increased using the scheme in Figure 10 and manually replaced the custom solution so that we could compare to the automated placement algorithms.

4.2.1 Whitespace Placement Results

Eight experiments were run on both designs, incrementally increasing the available whitespace to allow more room for automated placement tools while not applying any constraints. However, no significant improvements were achieved even when overall utilization dropped to 50%.

Table 3. TWL Results of Design 1

Placer	1.07	1.13	1.20	1.27	1.34	1.40	1.47	1.54
CAPO	1.89	1.78	1.76	1.88	1.75	1.74	1.72	1.74
mPL6	1.75	1.68	1.61	1.63	1.64	1.66	1.65	1.64
NTUPlace3	2.10	1.98	1.93	1.88	1.88	1.85	1.87	1.86
APlace*	1.49	1.51	1.42	1.42	1.39	1.22	1.23	1.22
Dragon	3.29	3.66	3.94	4.29	4.34	4.34	4.63	5.41
FastPlace	1.86	1.80	1.85	1.72	1.73	1.73	1.87	1.85

Tables 3 and 4 display the TWL placement results for each experiment compared to the custom design solution at the same whitespace percentage. The actual wirelength results are presented in graphical form in Figures 11 and 12.

Table 4. TWL Results for Design 2

Placer	7%	13%	20%	27%	34%	40%	47%	54%
CAPO	2.72	2.01	1.80	1.78	1.75	1.72	1.65	1.63
mPL6	1.77	1.69	1.62	1.57	1.52	1.49	1.47	1.44
NTUPlace3	0.00	0.00	0.00	0.00	0.00	1.51	1.48	1.46
APlace*	1.38	1.29	1.29	1.29	1.29	1.23	1.24	1.23
Dragon	4.89	4.99	5.17	5.15	4.88	4.84	5.05	4.92
FastPlace	2.15	1.80	2.00	1.99	1.91	1.94	1.85	1.86

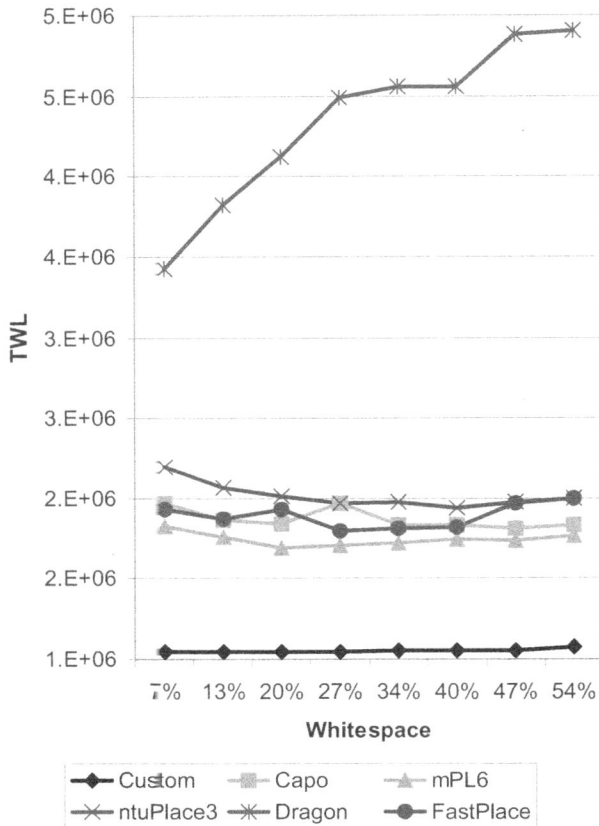

Figure 11. Design 1 TWL Results

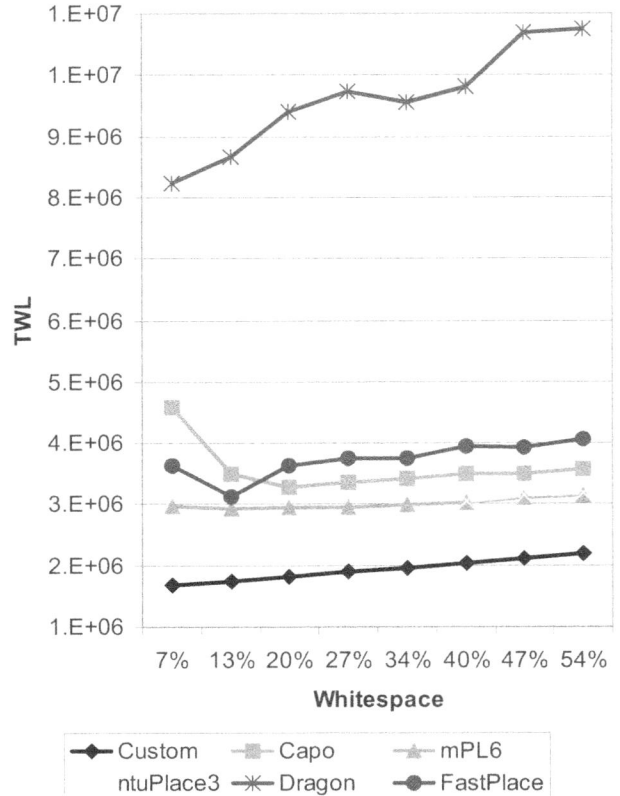

Figure 12. Design 2 TWL Placement Results

Figure 11 displays the change in TWL for Design 1 over each of the eight experiments. All placers except Dragon show wirelength improvement as more whitespace is added. When more than 27% of whitespace is added however, the improvement of TWL and its ratio starts to saturate. After that point, adding additional white space did not significantly improve the overall TWL ratio compared to the custom placed design. TWL for the custom solution does increase at a small percentage as whitespace increases, however, not enough to be visible on this graph. Results for APlace are not shown because it failed to find a legal placement solution. One interesting result is the significant TWL increase in Dragon placer as available whitespace is added. In general, the overall *TWL ratios* are improving as whitespace increases however; this is primarily due to the TWL increase of the custom solution.

Figure 12 displays the TWL results for each placement solution throughout the eight experiments for Design 2. TWL at only 7% whitespace is significantly higher for all placers but quickly drops to within a 50% increase for mPL6. The custom TWL for Design 2 increases significantly as whitespace increases for the design helping the overall TWL ratios for the placers. This again does not point to an improved placement solution with increased whitespace, but is instead a result of the logic spreading from the manual solution. Dragon placement results also exhibit the significant TWL increase trend seen in Design 1 as whitespace increases. Only results for NTUPlace3 are reported for experiments with 40%, 47%, and 54% whitespace because runs before that point aborted.

Minimum overall TWL for the placers occurs within the range for 13% to 27% whitespace after which TWL begins to increase at a similar slope to the custom solution.

The following observations were made:

- Generally, placers did not improve much with additional white space, though CAPO improved from 1.89 down to 1.74 on Design 1 and from 2.72 to 1.63 on Design 2. Part of the improvement comes from the TWL increase from the manual solution.

- For Design 1, the best overall result for each experiment came from mPL6 with a 61% increase in TWL at 20% whitespace. The algorithm NTUPlace3 was unable to run[4] until the whitespace increased by 40%, at which point it generated comparable results to mPL6. The TWL increase of the manual solution is not obvious, but it also increases with added area.

- There was a gradual improvement in the TWL ratio for Design 2 as area increased for the design, with mPL6 decreasing from 77% to 44%.

- By industry standards, both designs are small relative to state-of-the-art work yet all placers presented significantly suboptimal TWL results.

5   ANALYSIS

Obviously, it is disappointing that academic placers perform poorly on these real designs, so further examination is necessary.

---

[4] Working with the NTUPlace3 authors to provide results for Design 2.

Figure 13 displays a zoomed in snapshot of three rows of the custom placed layout for Design 1 with the placeable logic between the latches from one row highlighted in light blue. The design was placed again using one of the better performing placement algorithms to see what happens to the blue cells that are densely packed in the custom layout. This is shown in Figure 14.

Observe the irregularity present in the blue highlighted logic. Most of the blue logic is placed within the bounds of the correct rows of latches, but a significant portion is left outside, despite adequate whitespace. This may occur because:

- Multilevel placement algorithms employ clustering to abstract a netlist into larger components that will be placed together. This manifests itself in loosely connected "blobs" of logic rather than densely packed structures.
- Analytical placement algorithms commonly employ net models such as cliques or stars of two-pin edges to represent hyperedges. Such models derate the weights of edges representing high-fanout nets to compensate for the increased number of edges needed to represent them. As a result, the impact of a 512-bit select line is very low, yet these nets could provide clues to the structure within the design.
- Clustering algorithms do not typically account for logic functions, and make decisions purely on local connectivity. This often leads to merging of gates across bit slices rather than merging the slice into a large cluster.

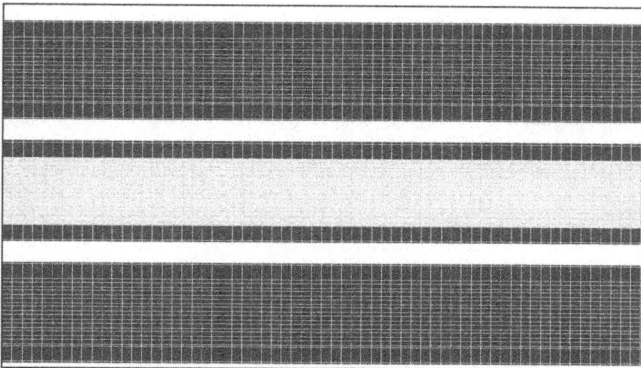

Figure 13. Custom Design 1 Placement Solution

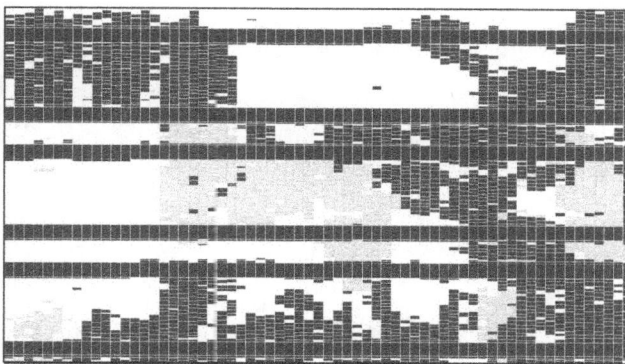

Figure 14. Automatic Design 1 Placement Solution

6   CONCLUSIONS

Recent years have seen truly significant improvements in runtimes, quality, and scalability in standard-cell placement algorithms. This work measures their performance on real datapath placement examples and compares them to hand-designed layouts. Academic placers still have a long way to go in order to match the quality of custom design solutions. An important contribution of this work is to release these benchmarks publically.

In order to keep pace with technology innovation, design automation must strive to improve productivity. This work highlights poor performance of modern placement tools on a key design style that is lacking in automation --- structured datapaths. One of the challenges posed by this problem is identifying regular datapaths within a larger design. Hence, new layouts will be constructed that combine both datapath and control logic in the same benchmark, providing even more challenging placement testcases. Future work will seek ways to improve automatic placement algorithms in terms of wirelength, routability, and timing closure of datapath placement problems.

7   REFERENCES

[1] S. N. Adya and I. L. Markov, *"Consistent Placement of Macro-blocks Using Floorplanning and Standard-Cell Placement"* ISPD 2002, pp. 12-17.

[2] P.W. Bosshart and Q.D. An. *Shifter circuit for an arithmetic logic unit in a microprocessor.* US. patent 5896305, 1999.

[3] T. F. Chan, J. Cong, J. R. Shinnerl, K. Sze, and M. Xie. *mPL6: Enhanced multilevel mixed-size placement.* In Proc. ISPD, pp 212–214, 2006.

[4] T.-C. Chen, Z.-W. Jiang, T.-C. Hsu, H.-C. Chen, and Y.-W. Chang. *A high-quality mixed-size analytical placer considering preplaced blocks and density constraints,* In Proc. ICCAD, 2006.

[5] Chin-Chih Chang, Jason Cong, Michail Romesis, and Min Xie; *Optimality and Scalability Study of Existing Placement Algorithms;* IEEE Transactions on Computer-aided Design Vol..23, pp. 537-549 2004

[6] R.L. Davis. *Uniform shift networks.* Computer, 7:327-334, September 1974.

[7] Lars W. Hagen, Dennis J.-H. Huang, Andrew B. Kahng; *Quantified Suboptimality of VLSI Layout Heuristics Design Automation,* DAC pp 216-221, 1995.

[8] J.L. Hennessy and D.A. Patterson. *Computer Architecture: A Quantitative Approach.* San Mateo, CA: Morgan Kaufmann Publishers, Inc., 2nd edition, 1996.

[9] Hillebrand, M.A.; Schurger, T.; Seidel, P.-M.; *How to half wire lengths in the layout of cyclic shifters,* Fourteenth International Conference on VLSI Design, 2001, pp. 339 – 344.

[10] A. B. Kahng, S. Reda, and Q. Wang. *Architecture and details of a high quality, large-scale analytical placer.* In Proc. ICCAD, pp 890–897, 2005.

[11] Koren. *Computer Arithmetic Algorithms.* Englewood Cliffs, NJ: Prentice-Hall Inc., 1993.

[12] T. Machida. *Bidirectional barrel shift circuit.* US. Patent 4665538, 1987.

[13] S.M. Mueller and W.J. Paul. *Computer Architecture: Complexity and Correctness.* Springer Verlat, 2000.

[14] G.-J. Nam, *"ISPD 2006 Placement Contest: Benchmark Suite and Results,"* ISPD 2006, p. 167.

[15] G.-J. Nam, C. J. Alpert, P. G. Villarrubia, B. B. Winter, M. C. Yildiz *"The ISPD2005 Placement Contest and Benchmark Suite,"* ISPD 2005, pp. 216-220.

[16] J. A. Roy, S. N. Adya, D. A. Papa, and I. L. Markov. *Min-cut floorplacement.* IEEE Transactions on Computer-Aided Design, Vol. 25, pp. 1313–1326, July 2006.

[17] P.-M. Seidel. *On the Design of IEEIT Compliant Floating-point Units and Their Quantitative Analysis.* PhD thesis, University of Saarland, December 1999.

[18] L. Sigal *et al. Circuit design techniques for the high-performance CMOS IBM S/390 Parallel Enterprise Server G4 microprocessor.* IBM Journal of Research and Development, Vol. 41, pp. 489-503, 1997.

[19] Wern-Jieh Sun and Carl Sechen, *Efficient and Effective Placement for Very Large Circuits*, IEEE Transactions on Computer-Aided Design, Vol 14, pp. 349-359, 1995

[20] Natarajan Viswanathan and Chris Chong-Nuen Chu, *FastPlace: Efficient Analytical Placement Using Cell Shifting, Iterative Local Refinement, and a Hybrid Net Model* IEEE Transactions on Computer-Aided Design, VOL. 24, pp. 722-733 2005

[21] M. Wang, X. Yang, and M. Sarrafzadeh, *Dragon2000: Standard-cell placement tool for large industry circuits,* in Proc. IEEE/ACM Int. Conf. Computer-Aided Design, 2000, pp. 260–263.

Regularity-Constrained Floorplanning
for Multi-Core Processors

Xi Chen
Department of ECE
Texas A&M University
College Station, TX 77843
xchen19@tamu.edu

Jiang Hu
Department of ECE
Texas A&M University
College Station, TX 77843
jianghu@ece.tamu.edu

Ning Xu
College of CST
Wuhan University of Technology
Wuhan 430070, China
xuning@whut.edu.cn

Abstract

Multi-core technology becomes a new engine that drives performance growth for both microprocessors and embedded computing. This trend asks chip floorplanners to consider regularity constraint since identical processing/memory cores are preferred to form an array in layout. As chip core count keeps growing, manual floorplanning will be inefficient on the solution space exploration while conventional floorplanning algorithms do not address the regularity constraint. In this work, we investigate how to enforce regularity constraint in a simulated-annealing based floorplanner. We propose a simple and effective technique for encoding the regularity constraint in sequence-pairs. To the best of our knowledge, this is the first work on regularity-constrained floorplanning in the context of multi-core processor designs. Experimental comparison with a semi-automatic method shows that our approach yields an average of 22% less wirelength and mostly smaller area.

Categories and Subject Descriptors

B.7.2 [**Integrated Circuits**]: Design Aids - Layout

General Terms

Algorithms, Design

Keywords

VLSI, Physical design, Floorplanning, Multi-core processors

1. Introduction

When the Moore's law is near its end, continuing chip performance growth will inevitably rely on the improvement of system level integration. This is evidenced by the popularity of multi-core technology for both microprocessors and embedded processors. In a foreseeable future, current multi-core processors will advance to many-core processors, which allow hundreds of cores on a chip. This trend presents new challenges to the design and design automation technologies. This paper will discuss floorplanning problem for multi-core and many-core processors and propose an algorithmic solution to this problem.

Permission to make digital or hard copies of all or part of this work for personal or classroom use is granted without fee provided that copies are not made or distributed for profit or commercial advantage and that copies bear this notice and the full citation on the first page. To copy otherwise, or republish, to post on servers or to redistribute to lists, requires prior specific permission and/or a fee.
ISPD'11, March 27–30 2011, Santa Barbara, California, USA.
Copyright 2011 ACM 978-1-4503-0550-1/11/03...$10.00.

Floorplanning is the first primary physical design step that decisively affects chip layout area and on-chip communication. When the number of cores on a chip is small, the floorplanning can be managed by manual designs, especially for CMP (Chip Multiprocessors). For instance, a 4-core processor can be manually placed in a 2×2 array. When the core count exceeds one hundred, the options of floorplans increase dramatically. Then, it would be very difficult for manual design to quickly and thoroughly explore these options. Besides processing cores, a processor chip usually contains cache, I/O blocks and communication fabrics. Further, CMP technology will move from homogeneous to heterogeneous cores [1] like IBM Cell processor. These facts imply heterogeneous entities which make manual floorplanning even more difficult. Therefore, there is a strong need for automatic floorplanning techniques for many-core CMP designs.

Multi-core technology is also widely adopted in SoC (System-on-Chip) designs and leads to the so-called MPSoC (Multi-Processor SoC). SoC designs are often targeted to embedded computing and require much shorter design turn-around time than microprocessors. Although conventional floorplanning techniques are applicable to current MPSoC designs, there is a new problem as the system grows from multi-core to many-core. That is, if multiple identical cores are adopted, they are preferred to be placed in a regular array. If ever possible, regularity is desired in chip layout for the sake of design simplicity, modularity and easy management of physical resources.

The regularity issue is rarely considered in conventional floorplanning. One similar case is analog circuit layout [2, 3] where components are often placed in a symmetric fashion. One may want to fulfill the regularity constraint by enforcing the symmetry constraints. Even though symmetry and regularity are related, regularity is actually more complex than symmetry and often more difficult to achieve. A chip with m cores can be placed in a $p \times q$ array and there are often multiple ways for the factorization of $m = p \times q$, e.g., $m = 30 = 1 \times 30 = 2 \times 15 = 3 \times 10 = 5 \times 6 = 6 \times 5 \ldots 30 \times 1$. Even for a specific factorization, symmetries about different axes need to be maintained to obtain a regular array. The work of [4] addressed the array-type constraint for analog placement. However, there is a key difference between the array constraint in analog placement and regularity constraint in multi-core processor floorplanning. In analog placement, array blocks of the same type of device are compacted together in order to reduce the effect of spatially-dependent variations. In multi-core processor floorplanning, in contrast, non-array blocks can be placed between array blocks and one group of array blocks can be placed inside of another group of array blocks.

In this paper, we present our work on floorplanning with regularity constraint, which is oriented toward multi-core processor designs. Our floorplanner is a simulated annealing algorithm using sequence-pair representation. Our key contributions are on how to encode the regularity constraint in sequence-pair and how to achieve the regularity in packing procedure. To the best of our knowledge, this is the first work studying regularity-constrained floorplanning for multi-core processors. We compared our approach with a naïve method that first places array blocks manually and then performs conventional floorplanning for non-array blocks while the array blocks are fixed. The experimental results indicate that our approach can achieve an average of 22% less wirelength than the semi-automatic method. At the same time, our approach usually leads to smaller area.

2. Previous Work with Symmetry Constraints

Among previous works on floorplanning, those for analog integrated circuits have the closest problem formulation as our regularity-constrained floorplanning work. Since an analog circuit typically has a small number of elements, its placement is often equivalent to the floorplanning of a digital integrated circuit. In analog circuit designs, one important requirement is to place blocks or devices symmetrically with respect to one common axis so as to improve the tolerance to common-mode noise [5]. There is a large body of analog circuit layout works focusing on the symmetry constraints. In [2], a symmetry-constrained analog block placement method is proposed. This work is based on a typical floorplanning approach – simulated annealing with sequence-pair representation. The symmetry constraint is described through sequence-pairs. For a sequence-pair (α, β), α_A^{-1} denotes the position of block A in sequence α. Consider a group of blocks G that must be placed symmetrically around a vertical axis. A sequence-pair (α, β) is symmetric-feasible for G if for any blocks A and B in G

$$\alpha_A^{-1} < \alpha_B^{-1} \Leftrightarrow \beta_{\sigma(B)}^{-1} < \beta_{\sigma(A)}^{-1} \qquad (1)$$

where $\sigma(A)$ is the block symmetric to A. It is pointed out in [6] that condition (1) is sufficient but not necessary. More recently, the work of [3] presented another sequence-pair based approach for simultaneously satisfying symmetry and centroid constraints. A B*-tree based method is proposed in [7] for handling both 1-D and 2-D symmetry constraints. Another symmetry-constrained analog placement work [8] uses O-tree representation. In [9], a symmetry-aware placement work is proposed based on Transitive Closure Graphs (TCG) data structure.

To certain extent, regularity constraint can be treated as an extension to the symmetry constraints. However, the extension is not trivial as the number of implicit symmetry constraints embedded in a regularity constraint can be quite large. A $p\times q$ array requires horizontal (vertical) symmetry for all pairs of p rows (q columns). For instance, a 4×5 array implies $\binom{4}{2}=6$ horizontal symmetry constraints and $\binom{5}{2}=10$ vertical symmetry constraints. These numbers are much greater than the case of analog placement which usually has at most 1 symmetry constraint horizontally or vertically.

In [4], the array-type constraint is considered for analog placement. In order to mitigate the effect of PVT (Process, Voltage and Temperature) variations, which are usually spatially correlated, the work of [4] packs array blocks of the same type right next to each other, with no other blocks in between. By

contrast, multi-core processors often place non-array blocks between array blocks or place one group of array blocks inside another group of array blocks. In Figure 1, non-array blocks SIU and CCX are placed in between the 2×4 array of L2T blocks and the L2T blocks are placed in between the 2×8 L2D blocks.

Figure 1: Floorplan of SUN Niagara-3 processor.

3. Floorplanning with Regularity constraint
3.1 Problem Formulation

The input to floorplanning includes a set of n blocks, each with area A_i where $i = 1, 2, ...n$, a set of l nets $N_1, N_2, ..., N_l$ among the n blocks, a set of k array groups $G_1, G_2, ..., G_k$. Each *array group* is a subset of the blocks that must be placed in a regular array. If a block is in an array group, it is called an *array block*; otherwise, it is called a *non-array block*. The problem is to construct a floorplan F that satisfies non-overlapping and the regularity constraint, and minimizes the cost function $cost(F) = (1 - \lambda) \times area(F) + \lambda \times wirelength(F)$, where λ is a weighting factor, $area(F)$ is the total area of F and $wirelength(F)$ is the total wire length of F. In this work, we use the half-perimeter model for wirelength estimation. The main difference from conventional floorplanning is the regularity constraint. An array group is composed by blocks of identical size and shape, which are usually processor cores or memory cores. Although the cores are required to be placed in an array, the shape of the array, which is decided by the number of rows and columns, is flexible. For example, for an array group of m blocks, any array of $p\times q=m$ is allowed and considered in the floorplanning as long as aspect ratio constraint is satisfied. Moreover, the blocks in an array group do not have to be placed next to each other. The floorplan of SUN Niagara-3 processor in Figure 1 demonstrates such example. It contains 16 SPARC processor cores, which are placed in a 2×8 array. However, the two rows are not adjacent to each other and allow other blocks to be placed in between.

3.2 Regularity in Relative Order

Floorplan representation is critical to the efficiency of a floorplanning method, especially for the popular simulated-annealing approach. There are two main categories of floorplan representations: absolute representation and topological representation. In an absolute representation, every block is specified in terms of its absolute coordinates. The solution search in the absolute representations tends to be complex and difficult. Therefore, topological representations become more popular and lead to many research results including sequence-pair [10], O-tree [11], B*-tree [12], Corner Block List (CBL) [13] and Transitive Closure Graphs (TCG) [14].

In this work, we focus on sequence-pair based floorplanning as sequence-pair is one of the most known floorplan representations and has been successfully applied in handling symmetry constraints [2, 3]. In section 3.5, we will briefly discuss the regularity constraint in other floorplan representations. For a set of blocks, a sequence-pair consists of two sequences of block IDs corresponding to the orders along two diagonal directions. A sequence-pair like $(<... i ... j ...>, <... i ... j ...>)$ implies that block $i(j)$ is to the left (right) of block $j(i)$. Similarly, $(<... i ... j ...>, <... j ... i ...>)$ means that block $i(j)$ is above (below) block $j(i)$. A sequence-pair for 6 blocks $(<1 2 4 5 3 6>, <3 6 2 1 4 5>)$ is demonstrated in Figure 2.

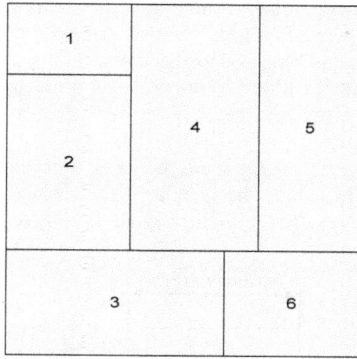

Figure 2: Floorplan for the sequence-pair $(<1 2 4 5 3 6>, <3 6 2 1 4 5>)$.

The regularity constraint to an array group of m blocks implies that these blocks must be placed in a $p \times q$ array, where $p \times q$ is a factorization of m. The q blocks in one row must appear as a common subsequence [15] in the sequence-pair.

Definition 1: Common subsequence. *A set of q blocks $b_1, b_2... b_q$ form a common subsequence in a sequence-pair (α, β) if $\alpha_1^{-1} < \alpha_2^{-1} < \cdots < \alpha_q^{-1}$ and $\beta_1^{-1} < \beta_2^{-1} < \cdots < \beta_q^{-1}$ where α_i^{-1} (β_i^{-1}) indicates the position of block b_i in sequence α (β).*

For example, the floorplan of Figure 3 can be specified by sequence-pair $(<0 1 2 3 4 5>, <2 1 0 5 4 3>)$, which contains 3 common subsequences $(0, 3)$, $(1, 4)$ and $(2, 5)$ for the 3 rows. This concept of common subsequence is similar to H-alignment in [16].

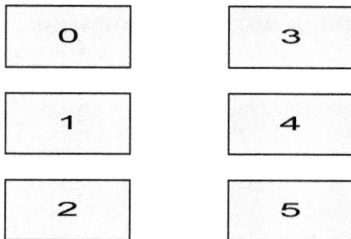

Figure 3: An array group placed in a 3×2 array.

Likewise, the blocks in a column must appear as a reversely common subsequence in the sequence-pair.

Definition 2: Reversely common subsequence. *A set of q blocks $b_1, b_2... b_q$ form a reversely common subsequence in a sequence-pair (α, β) if $\alpha_1^{-1} < \alpha_2^{-1} < \cdots < \alpha_q^{-1}$ and $\beta_1^{-1} > \beta_2^{-1} > \cdots >$*

β_q^{-1} *where α_i^{-1} (β_i^{-1}) indicates the position of block b_i in sequence α (β).*

For the example in Figure 3, block 0, 1 and 2 form a reversely common subsequence.

Lemma 1 *The necessary condition that m blocks lead to a $p \times q$ array floorplan: the m blocks constitute p common subsequences of length q and q reversely common subsequences of length p in the sequence-pair.*

The above condition is necessary but not sufficient because a sequence-pair specifies only a relative order. Generation of an array-type floorplan also depends on the packing procedure, which will be discussed in Section 3.3.

Definition 3: Regularity subsequence-pair (RSP). *A <u>contiguous</u> subsequence of length m that satisfies Lemma 1 in a sequence-pair is called regularity subsequence-pair.*

In Figure 3, there are no other blocks/cores in the middle of 3×2 array and the corresponding sequence-pair is a RSP. In fact, the mapping from an array floorplan to sequence-pair is not unique. For instance, the floorplan in Figure 3 can be alternatively specified by sequence-pair $(<0 3 1 4 2 5>, <2 5 1 4 0 3>)$, where each row is in a contiguous subsequence. This is in contrast to $(<0 1 2 3 4 5>, <2 1 0 5 4 3>)$ where each column is a contiguous subsequence.

Definition 4: Row (column) based regularity subsequence-pair *is a regularity subsequence-pair where each (inversely) common subsequence corresponding a row (column) is contiguous.*

If non-array blocks are allowed in the middle of an array, they can be embedded within the corresponding RSP. One such example is given in Figure 4 where block 8 is inside of both subsequences of the RSP.

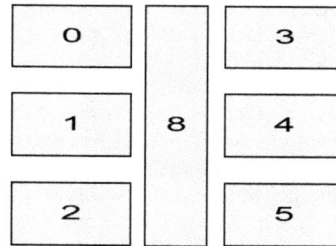

Figure 4: Block 8 is placed in the middle of the regularity subsequence $(<0 1 2 \underline{8} 3 4 5>, <2 1 0 \underline{8} 5 4 3>)$.

Rule 1: *A non-array block can be inside both or neither of α and β sequences of a RSP. A non-array block cannot be inside one of α and β sequences but outside of the other for a RSP.*

For the example of Figure 5, for sequence-pair (α, β), we do not allow $(<0 1 2 \underline{8} 3 4 5>, <\underline{8} 2 1 0 5 4 3>)$. Here block 8 is outside the α part of RSP $(0 1 2 3 4 5)$, but inside the β part. This is a violation of Rule 1 as it may lead to misalignment in later packing procedure. A packing result of this sequence-pair is depicted in Figure 5.

Rule 2: *A non-array block can be inside both or neither of α and β part of a contiguous (reversely) common subsequence in a row (column) based RSP. A non-array block cannot be inside one of α and β part but outside of the other for a contiguous (reversely) common subsequence in a row (column) based RSP.*

For example, $(<0 1 2 \underline{8} 3 4 5>, <2 \underline{8} 1 0 5 4 3>)$ is not allowed as non-array block 8 is outside the α part of the reversely common subsequence (0 1 2), but inside its β part. The reason of rule 2 is the same as rule 1. We enforce rule 1 and rule 2 in the floorplanning algorithm in order to avoid ambiguity for subsequent packing that may lead to misalignment.

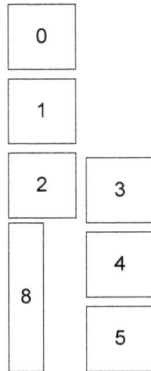

Figure 5: Misalignment due to violation of Rule 1.

The enforcement of rule 2 implies that a non-array block cannot be placed between two columns (rows) for a row (column) based RSP. Consequently, row-based and column-based RSPs may lead to different floorplans if non-array blocks are allowed between array blocks. In order to explore the complete space, both row-based and column-based RSP need to be separately examined.

3.3 Regularity in Packing

A sequence-pair only specifies a relative order for the blocks and the absolute locations for the blocks need to be further decided through a packing procedure. In [10], horizontal and vertical constraint graphs are constructed for a sequence-pair. Then, the packing is obtained by performing the longest path algorithm on the graphs. Later, a faster packing algorithm based on longest common sequence (LCS) is proposed in [15]. In this work, we adopt the packing approach of [15].

Regularity in the packing implies the alignment and spacing constraints. Array blocks of each row (column) must be horizontally (vertically) aligned. Sometimes, one may additionally prefer identical spacing between rows (columns). This is illustrated in Figure 6. The alignment and spacing constraints can be expressed according to the block locations. For example,

$$X_{i,j} - X_{i,j-1} = X_{i,j+1} - X_{i,j} \qquad (2)$$
$$Y_{i,j} - Y_{i-1,j} = Y_{i+1,j} - Y_{i,j} \qquad (3)$$

where X, Y are x coordinate and y coordinate of the lower-left corner of an array block, and $i(j)$ represents row (column) index.

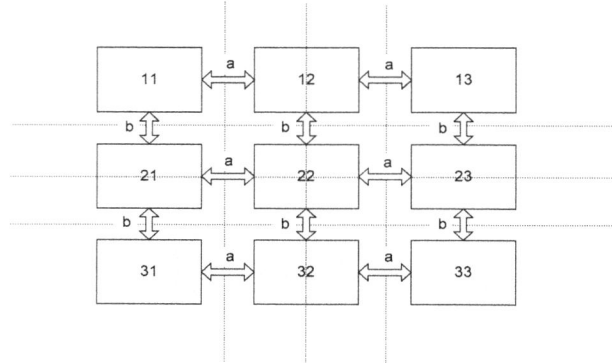

Figure 6: Alignment and uniform spacing.

If there is no non-array block inside an array, i.e., no non-array block is inside a RSP, then the array can be pre-packed into a single object, with or without spacing. Then, the LCS (Longest Common Subsequence) packing algorithm [15] can be applied directly.

If there is any non-array block inside an array, then the minimum uniform spacing is decided by the largest dimension among the non-array blocks. We perform a pre-processing to the non-array blocks before calling the LCS engine. The pre-processing is to temporarily expand the dimensions of the non-array blocks. Consider the example in Figure 7 where there are 5 non-array blocks: 6, 7, 8, 9, and 10. Block 10 is between two regular array columns, it will not affect the regularity of two array columns, so we only need to consider other 4 non-array blocks. If block 6 has the maximum height and width among 6, 7, 8 and 9, we expand the height of 7, 8 and 9 to be the same as block 6. This height is called *virtual height*. We also expand the width of block 7 to be the same as block 6, to reach a *virtual width*. We do not expand the width of block 8 and 9 as they are in the rightmost column of the array and the expansion will only affect blocks outside of this array.

After the expansion is finished, the LCS algorithm is performed to do the packing. Once the packing result is obtained, the non-array blocks are restored to their original dimensions.

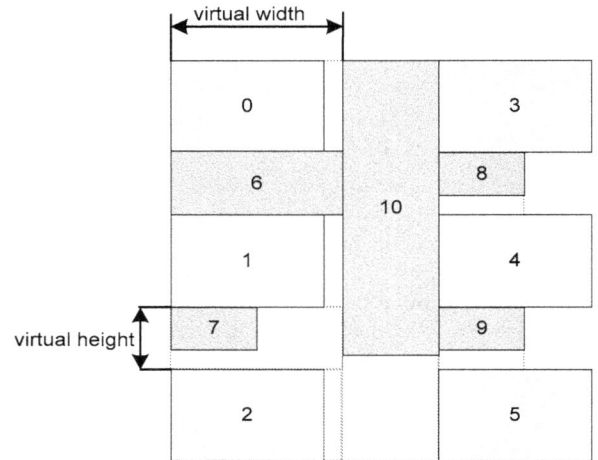

Figure 7: Non-array blocks within an array.

3.4 The Floorplanning Algorithm

Once the regularity constraint is encoded in sequence-pairs, the floorplanning algorithm is a straightforward extension of conventional simulated annealing based approach. Initially, we generate a sequence-pair that satisfies Lemma 1 for each array group. For each array group with m blocks, we do a random factorization $m = p \times q$. Then, the m cores are randomly allocated to p subgroups. Each subgroup, which corresponds to a row, is arranged as a common subsequence in the initial sequence-pair. Next, the p common subsequences are arranged in a reversely common order so that a regularity subsequence-pair is obtained for the array group. After the initial sequence-pair is obtained, a packing procedure (as described in Section 3.3) is performed. The result is evaluated according to the cost function $cost(F) = (1 - \lambda) \times area(F) + \lambda \times wirelength(F)$, which is defined in Section 3.1.

The initial solution is iteratively revised through moves like in a typical simulated annealing procedure. Our algorithm includes the following moves:

- Changing the factorization of an array group. For example, a 2×6 array can be changed to a 3×4 array or other factorizations.
- Changing the RSP for an array group between row-based and column-based.
- Moving a non-array block into (or outside) a RSP (Regularity Subsequence-pair).
- Swapping two non-array blocks.
- Rotating a non-array block.
- Rotating all blocks in an array group.
- Swapping two blocks in the same array group.

Most of the moves are first performed on the sequence-pair and then the packing is performed. The cost function for the packing result is evaluated. The last type of move, which swaps two blocks in the same array group, needs further discussion. Since any two blocks in the same array group have the same size and shape, swapping them does not affect area and non-overlapping constraint, and sometimes may reduce wirelength. This swap affects only wirelength and therefore can be skipped if $\lambda = 0$, i.e., the floorplanning is only for area minimization. Since the interconnect within an array group is typically symmetric, this swap affects more on wires between the array group and blocks outside this group.

3.5 Other Floorplan Representations

Besides sequence-pair, there are numerous other floorplan representations: B*-Tree, O-Tree, Corner Block List (CBL) and Transitive Closure Graph (TCG), to name a few. Many of them, including B*-Tree, O-Tree and CBL, are for compacted floorplans. The compacting nature of these representations often conflicts with the alignment requirement for multi-core processor designs. On the left of Figure 8, B*-Tree forces block 3 right next to block 1. Consequently, block 3 is not aligned with block 4 and 5 like the sequence-pair result on the right. Of course, one can redo the compaction but this nullifies a main advantage of these representations. TCG is not a compacted floorplan representation and is similar to sequence-pair in nature. We will study how to use TCG for multi-core processor floorplanning in our future research.

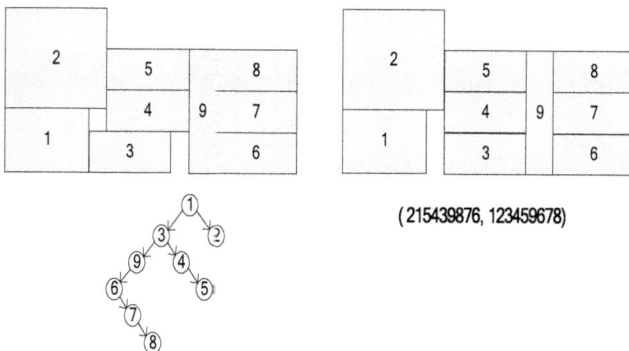

(215439876, 123459678)

Figure 8: Floorplan compaction from B*-Tree and sequence-pair.

4. Experimental Results

Since there is no previous work on regularity-constrained floorplanning for multi-processors, we compared our approach with a manual prefix method. That is, the blocks in each array group are manually placed in an array and then the other (non-array) blocks are placed using conventional floorplanning [10] while the array blocks are fixed. The input to both our method and the manual prefix method includes a set of blocks containing pre-identified array groups. The blocks in the same array group should have the same size and shape. In the manual prefix approach, different factorization options of an array group are explored. For instance, if a chip has an array group of 16 blocks (like figure 1), the manual prefix approach is performed 5 times with arrays of 1×16, 2×8, 4×4, 8×2 and 16×1, respectively. Then the each array group can be treated as a single big block during the floorplanning. The best result among these options is selected to be compared with our approach.

Existing public domain benchmarks are mostly from old designs and do not have cases for multi-core processors. Therefore, we made modifications to the MCNC and GSRC floorplan benchmark by converting a subset of blocks into array groups. The floorplanning algorithms are implemented in C++ and the experiments are performed on a Windows machine with a 2.5GHz Intel Core 2 Duo processor and 2GB memory.

The main results on the MCNC benchmark circuits are reported in Table 1. Here, the value of λ is 0.5, i.e., the weighting factors for the area and wirelength are the same. Each circuit has one array group. The second column lists the array factorization that leads to the minimum cost for manual prefix method. Our approach can reduce wirelength by 22% on average. At the same time, our approach achieves the same or less area and mostly faster runtime.

A more detailed diagnosis of the results in Table 1 is presented in Figure 9, where one chart is generated for each circuit. Each chart displays our solution (in small triangle) and the solutions from manual prefix (in small diamonds) in term of area and wirelength. Since we examined multiple array options for the manual prefix methods, solutions from all options are plotted in the charts. One can see that our solutions always dominate the manual prefix solutions, i.e., our solutions are superior to or at least the same on both area and wirelength.

We also compared the two approaches for area-driven only formulation. The results are summarized in Table 2. The third column lists the number of blocks in each array group. The GSRC cases are relatively large and each has two array groups. Except Apte, our approach always results in better area usage (or less dead-space). The area advantage from our approach is more obvious for the larger cases from GSRC benchmark. In Figure 10, we provide visualization on the floorplanning result on GSRC circuit n100. Figure 10 (a) indicates that our floorplanning approach allows non-array blocks inside an array. Further, our approach allows one array to be embedded in another array.

5. Conclusions and Future Research

In this work, we propose a floorplanning approach for multi-core processors where processor and memory cores are required to be placed under regularity constraint. This is the first work on applying regularity constraint floorplanning to multi-processor. Experimental results indicate that our approach significantly outperforms a naïve semi-automatic method. In future research, we will further study the multi-core processor floorplanning using other representations like TCG. We will also perform a comparative study with regularity-constrained floorplanning for analog circuit such as [4] and how it performs under fix-outline constraint.

Table 1. Experimental results of manual prefix and our approach on MCNC circuits, with λ = 0.5.

MCNC Circuit	Manual Prefix (MP)				Our Approach				
	Min cost array	Area(mm²)	Wirelength(mm)	CPU(s)	Area(mm²)	Area reduction vs. MP	Wirelength (mm)	Wirelength reduction vs. MP	CPU(s)
Apte	4*1	48.21	628.5	19.6	48.21	0%	472.3	24.8%	22.0
Hp	1*4	10.65	344.8	30.5	9.67	9.2%	279.4	18.9%	27.2
Xerox	1*4	25.74	1016.1	144.6	25.45	1.1%	687.5	32.3%	102.0
Ami33	4*2	1.22	83.9	525.8	1.19	2.5%	77.9	7%	474.3
Ami49	4*4	50.85	2095.3	1931.5	49.53	2.6%	1559.5	25.5%	1354.6

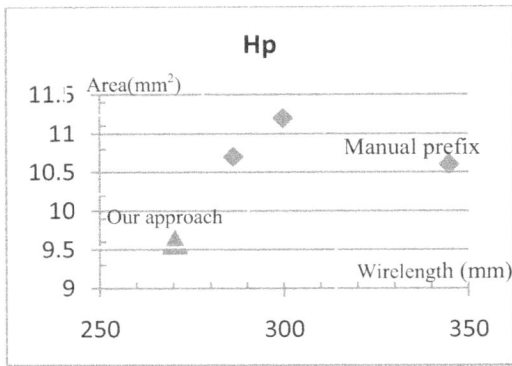

(a)

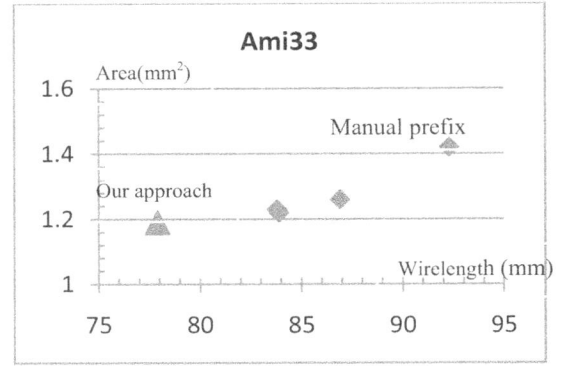

(c)

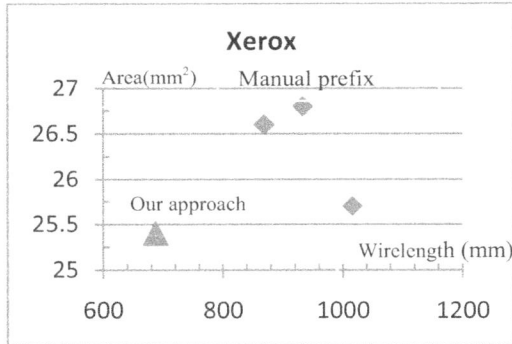

(b)

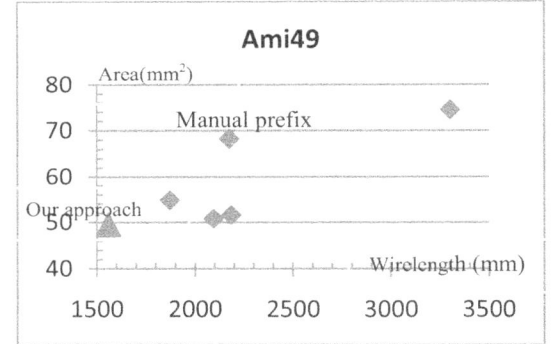

(d)

Figure 9: (a-d). Area and wirelength results from our approach (triangle) and multiple options of prefix approach (diamond), λ = 0.5.

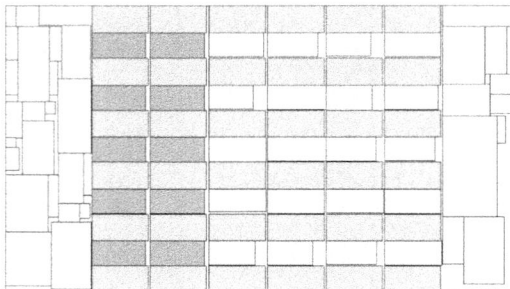

(a)

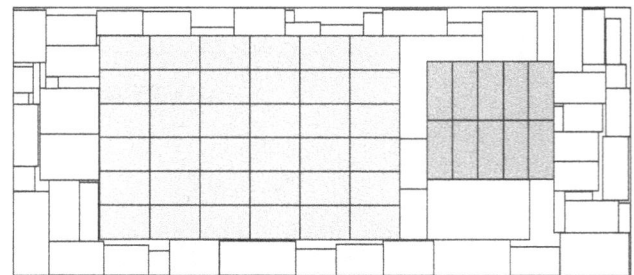

(b)

Figure 10: Floorplan of n100 generated by (a) our approach and (b) the manual prefix method.

Table 2. Area-driven floorplanning results on MCNC and GSRC circuits.

Circuit	Total No. of blocks	No. of array blocks	Manual Prefix			Our Approach	
			Min area arrays	Area Usage (%)	CPU (s)	Area Usage (%)	CPU (s)
Apte	9	4	4*1	95.56	32.52	96.56	3.20
Hp	10	4	2*2	90.63	22.59	90.64	16.41
Xerox	11	4	1*4	96.71	14.07	97.13	29.87
Ami33	33	8	2*4	94.63	379.74	95.42	331.30
Ami49	49	16	8*2	93.69	713.93	93.80	231.3
n50	50	16,12	4*4, 4*3	88.06	71.367	93.05	42.89
n70	70	24,9	4*6, 3*3	87.02	149.45	90.53	465.1
n100	100	36,10	6*6, 2*5	90.16	461.33	92.20	259.3
n200	200	56,21	7*8,7*3	84.11	3016.45	92.89	5007.4
n300	300	81,40	9*9,10*4	86.25	5429.79	89.82	6370.9

6. References

[1] R. Kumar, D. M. Tullsen, N. P. Jouppi and P. Ranganathan, "Heterogeneous Chip Multiprocessors," *IEEE Computer*, Vol. 38, No. 11, pp. 32-38, 2005.

[2] F. Balasa and K. Lampaert, "Symmetry within the Sequence-Pair Representation in the Context of Placement for Analog Design," *IEEE Transactions on CAD*, Vol. 19, No. 7, pp. 721-731, July 2000.

[3] L. Xiao and E. F. Y. Young, "Analog Placement with Common Centroid and 1-D Symmetry Constraints," *ACM/IEEE Asia and South Pacific Design Automation Conference*, pp.353-360, 2009.

[4] S. Nakatake, "Structured Placement with Topological Regularity Evaluation", *ACM/IEEE Asia and South Pacific Design Automation Conference*, pp. 215-220, 2007.

[5] J. M. Cohn, D. J. Garrod, R. A. Rutenbar and L. R. Carley, *Analog Device-Level Layout Automation*, Kluwer Academic Pub., 1994.

[6] S. Kouda, C. Kodama and K. Fujiyoshi, "Improved Method of Cell Placement with Symmetry Constraints for Analog IC Layout Design," *ACM International Symposium on Physical Design*, pp.192-199, 2006.

[7] P. H. Lin and S. C. Lin, "Analog Placement Based on Novel Symmetry-Island Formulation," *ACM/IEEE Design Automation Conference*, pp.464-467, 2007.

[8] Y. Pang, F. Balasa, K. Lampaert and C. K. Cheng, "Block Placement with Symmetry Constraints Based on the O-tree Non-slicing Representation", *ACM/IEEE Design Automation Conference*, pp. 454-467, 2000.

[9] L. Zhang, C. J. Shi and Y. Jiang, "Symmetry-Aware Placement with Transitive Closure Graphs for Analog Layout Design," *ACM/IEEE Asia and South Pacific Design Automation Conference*, pp. 180-185, 2008.

[10] H. Murata, K. Fujiyoshi, S. Nakatake and Y. Kajitani, "VLSI Block Placement Based on Rectangle-Packing by the Sequence-Pair," *IEEE Transactions on CAD*, Vol. 15, No. 12, pp. 1518-1524, December 1996.

[11] P.-N. Guo, C.-K. Cheng and T. Yoshimura, "An O-Tree Representation of Non-Slicing Floorplan and Its Applications," *ACM/IEEE Design Automation Conference*, pp. 268-273, 1999.

[12] Y. C. Chang, Y. W. Chang, G. M. Wu, and S. W. Wu, "B*-Trees: A New Representation for Non-Slicing Floorplans," *ACM/IEEE Design Automation Conference*, pp. 458-463, 2000.

[13] X. Hong, G. Huang, Y. Cai, J. Gu, S. Dong, C.-K. Cheng, and J. Gu, "Corner Block List: An Effective and Efficient Topological Representation of Non-Slicing Floorplan," *IEEE/ACM International Conference of Computer-Aided Design*, pp. 8-12, 2000.

[14] J. M. Lin and Y.W. Chang, "TCG: a Transitive Closure Graph Based Representation for General Floorplans," *IEEE Transactions on VLSI Systems*, vol. 13, No. 2, pp.288-292, February 2005.

[15] X. Tang, R. Tian and D. F. Wong, "Fast Evaluation of Sequence-pair in Block Placement by Longest Common Subsequence Computation," *ACM/IEEE Design Automation and Test in Europe*, pp.106-111, 2000.

[16] X. Tang and D. F. Wong, "Floorplanning with Alignment and Performance Constraints," *ACM/IEEE Design Automation Conference*, pp. 848-853, 2002.

Power-Driven Flip-Flop Merging and Relocation*

Shao-Huan Wang
National Tsing Hua University
shwang@cs.nthu.edu.tw

Yu-Yi Liang
National Tsing Hua University
yuyiliang@cs.nthu.edu.tw

Tien-Yu Kuo
National Tsing Hua University
u9562105@oz.nthu.edu.tw

Wai-Kei Mak
National Tsing Hua University
wkmak@cs.nthu.edu.tw

ABSTRACT

We propose a power-driven flip-flop merging and relocation approach that can be applied after conventional timing-
driven placement and before clock network synthesis. It targets to reduce the clock network size and thus the clock
power consumption as well as the switching power of the
nets connected to the flip-flops by selectively merging flip-
flops into multi-bit flip-flops and relocating them under timing and placement density constraints. The experimental
results are very encouraging. For a set of benchmarks, our
approach reduced the clock wirelength by 30 to 50%. Meanwhile, the switching power of signal nets connected to the
flip-flops were reduced by 2 to 43%.

Categories and Subject Descriptors

J.6 [**COMPUTER-AIDED ENGINEERING**]: Computer-
aided design (CAD)

General Terms

Algorithms, Design

Keywords

Clock Network, Low Power, Multi-bit Flip-Flop, Post Placement

1. INTRODUCTION

Clock network plays an important role in power consumption as it accounts for up to 50% [1] of dynamic power in
some real circuits for its highest switching rate. Many kinds
of power reduction technique have been proposed. [2, 3]
worked on buffer sizing for clock power minimization. [4, 5,
6] designed some new low-power flip-flop structures. [7, 8, 9]
discussed clock gating. [10, 11, 12] minimized the power of
clock network by considering the location of registers in the

*This work was partially supported by NSC under grant
NSC 99-2220-E-007-007.

Permission to make digital or hard copies of all or part of this work for
personal or classroom use is granted without fee provided that copies are
not made or distributed for profit or commercial advantage and that copies
bear this notice and the full citation on the first page. To copy otherwise, to
republish, to post on servers or to redistribute to lists, requires prior specific
permission and/or a fee.
ISPD'11, March 27–30, 2011, Santa Barbara, California, USA.
Copyright 2011 ACM 978-1-4503-0550-1/11/03 ...$10.00.

placement stage. They try to group registers into clusters
and place registers in a cluster closer to reduce the wirelength of the leaf level of a clock tree. In this paper, we
make use multi-bit flip-flop to reduce the clock power and
switching power of signal nets.

The use of multi-bit flip-flop (MBFF) was first proposed
in [13] for reducing clock delay, controlling clock skew, and
improving routing resource utilization. [14] introduced a design methodology for MBFF inference during logic synthesis
for area and power reduction. However, to form MBFFs in
early design stage, it is hard to consider its effect on timing. Recently, [15] introduced a new algorithm to reduce
the power consumption of flip-flops by incrementally forming more multi-bit flip-flops at the post-placement stage.

Figure 1 shows an example of replacing traditional flip-
flops by a multi-bit flip-flop. Because of manufacturing
ground rules in advanced process technology, inverters tend
to be oversized so that an inverter can drive more than one
traditional flip-flop. By merging multiple 1-bit flip-flops into
one MBFF, it is possible to eliminate some inverters. For
example, we can eliminate two inverters after merging in
Figure 1. It will reduce the total area and power consumption of the flip-flops. This is the motivation for [15].

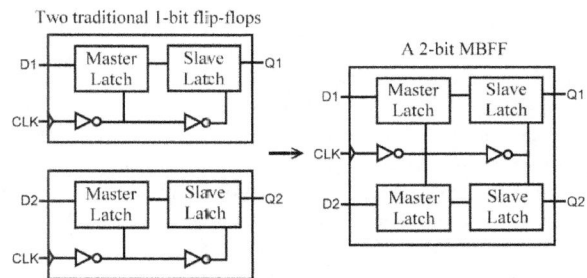

Figure 1: Replacing two traditional flip-flops by a
2-bit MBFF.

More importantly, using multi-bit flip-flop can also reduce
the number of clock sinks. This will reduce the wirelength of
the clock network and the buffers required in the clock network to maintain the slew and balance the skew. Therefore,
power consumed by the clock network will be reduced.

In this paper, we propose a power-driven flip-flop merging and relocation approach that can be applied after conventional timing-driven placement and before clock network
synthesis. To apply flip-flop merging after initial placement,

the following issues need to be considered: *Decide which flip-flops should be merged into MBFFs and decide their locations considering placement density, timing constraints and switching power of signal nets to obtain the best possible power saving.* There are a lot of flip-flops on a chip at different locations but we cannot merge every flip-flop due to timing constraint. Moreover, there are many possible combinations of flip-flops and each will result in different power consumption. Since a library offers many kinds of MBFF, in terms of number of bits, area and capacitance, we have to find the target MBFFs we should use. Besides, without careful planning, MBFFs that are generated may be put into new places that may cause area overflow and routing congestion.

The rest of this paper is organized as follows. We define the power-driven flip-flop merging and relocation problem in details in section 2. We define some useful terms and give the proof of NP-hardness of the problem in section 3. Section 4 contains the detailed descriptions of the proposed algorithm stage by stage. The experimental results and conclusions are in sections 5 and 6, respectively.

2. PROBLEM FORMULATION

We want to reduce the net switching power of the clock and the signal nets by selectively relocating and merging flip-flops into multi-bit flip-flops at the post-placement stage. The net switching power of a net i is given by $k\alpha_i C_i V^2$ where k is a constant, α_i is the switching rate, C_i is the capacitance to be charged and discharged, and V is the supply voltage.

Before clock network synthesis, the wirelength and hence the capacitance of the clock network are not known. However, our experiment shows the reduction percentage of the number of sinks is roughly proportional to the reduction percentage of the wirelength of clock network. In addition, [10] observed that most of the wire capacitance of a clock tree is at the leaf level. In some industrial designs, the percentage can reach 40%. Hence, reducing the number of clock sinks can result in a fine level of clock power saving. However, merging a group of flip-flops into a single MBFF will affect the wirelength and hence the net switching power of the signals connected to these flip-flops. The clock power reduction by flip-flop merging may be canceled out by the change in switching power of signal nets. So, we have to consider them simultaneously. We define the post-placement *Power-Driven Flip-Flop Merging and Relocation* problem as follows:

Input: We are given a pre-placed design with a set of flip-flops[1] and their corresponding fanin and fanout locations, and a MBFF library. We are also given some placement density constraint and timing constraint.

Objective: Selectively relocate and merge flip-flops into multi-bit flip-flops to minimize the number of clock sinks and signal net switching power subject to the given placement density and timing constraints.

We use the normalized cost below in order to strike a balance between minimization of clock network and switching power of the nets connected to the flip-flops.

$$w \times \frac{Final\ \#sinks}{Original\ \#sinks} + (1-w) \times \frac{\sum_{net_i \in F} (\alpha_i \times WL_i)}{\sum_{net_i \in F} (\alpha_i \times OWL_i)}$$

(1)

---

[1]They are not necessary to be 1-bit. Our algorithm can also handle original flip-flops with multiple bits.

where $0 \le w \le 1$ is a user defined weight, F is the set of all signal nets connected to the flip-flops. And α_i is the switching rate of signal net_i, OWL_i and WL_i are the wirelength of net_i before and after merging.

We must ensure that the design meets certain constraints after flip-flop merging. The first one is timing constraint. When multiple 1-bit flip-flops are merged into a multi-bit flip-flop, the new multi-bit flip-flop may incur additional routing length due to the change of the location, and thus changes the timing. Figure 2 shows an example.

But we can take advantage of the original timing slacks between a flip-flop and its fanin and fanout pins. The timing slack between a flip-flop and a pin can be converted into a maximum distance constraint from the pin using a desired delay model. Thus the *feasible region* that a flip-flop can be placed in would be a 45-degree tilted rectangular region, as shown in Figure 3.

We also consider the placement density of each local area to avoid routing congestion. Therefore, we partition a layout into $b_N \times b_M$ bins, and each bin has a maximum placement density constraint. We assume every multi-bit flip-flop can only be placed on an unoccupied grid point within a bin without violating the placement density constraint.

3. PRELIMINARIES

The feasible region of each original FF is a 45-degree tilted rectangular region, and these feasible regions may intersect

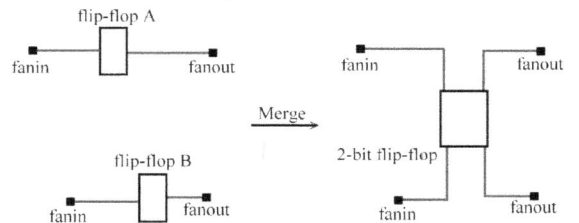

Figure 2: Wirelength and timing will be changed by merging.

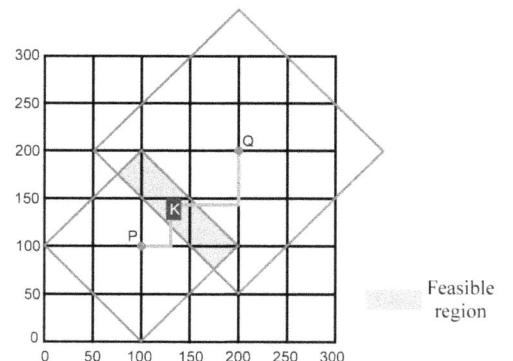

Figure 3: Feasible region of flip-flop K which is connected to pins P and Q. The maximum allowable distances from P and Q to K are assumed to be 100 and 150, respectively.

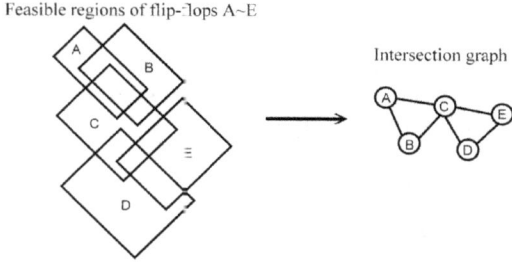

Figure 4: The intersection graph.

each other. Every intersected region represents the feasible region of a MBFF for the corresponding original FFs. As shown in Figure 4, each intersection of feasible regions of original FFs is a clique in the corresponding rectangle intersection graphs[2]. Therefore, if we consider the case to minimize the number of clock sinks with timing constraint only, the problem is exactly the same as *minimum clique partition of a rectangle intersection graph (MCPRIG)*. The MCPRIG problem is defined as follows:

Input: Given a rectangle intersection graph $G = (V, E)$. V is the vertex set representing all rectangles and E is the edge set representing all intersection relations.

Objective: Find k disjoint cliques to cover all the vertices in V where k is minimum.

MCPRIG has been proved to be NP-hard in [16], and it is a special case of our power-driven flip-flop merging and relocation problem by setting w to 1 in equation (1) and removing the constraints of placement density and library. By the definition of MCPRIG, it can be reduced to our problem easily. Hence, we get the following theorem:

THEOREM 1. *The power-driven flip-flop merging and relocation problem is NP-Hard.*

Although MCPRIG is NP-Hard, it still has some property which can help us to solve the power-driven flip-flop merging and relocation problem. We define some terms and introduce the property as follows.

Definition 1. A maximal clique is a clique that cannot be extended by including one more adjacent vertex.

Definition 2. For a vertex $v_i \in V$, participating degree of v_i is the number of maximal cliques which cover v_i.

Definition 3. A vertex v_i is a critical vertex if and only if the participating degree of v_i is 1.

Definition 4. A maximal clique c_i is called an essential maximal clique if and only if c_i contains at least one critical vertex.

LEMMA 1. *Let C_e be the set of all essential maximal cliques. $|C_e|$ is a lower bound of k. In other words, $k \geq |C_e|$.*

PROOF. By the definition of essential maximal clique, we know there exist $|C_e|$ critical vertices covered by different maximal cliques. Hence, in order to cover these vertices we need at least $|C_e|$ cliques. □

---

[2]In a rectangle intersection graph, each vertex corresponds to a rectangle, and if two rectangles intersect with each other, there is an edge between the corresponding vertices.

Lemma 1 gives a lower bound of the number of sinks of our merging problem. Therefore, if we want to minimize the number of sinks, it is good to start from finding all maximal cliques. In the next section, we will use these definitions and lemma to construct an efficient algorithm to solve the power-driven flip-flop merging and relocation problem.

4. POWER-DRIVEN FLIP-FLOP MERGING AND RELOCATON ALGORITHM

The proposed design flow is shown in Figure 5. First, we compute the feasible region of each original flip-flop according to the timing slacks after placement. Second, we form the corresponding rectangle intersection graph and compute all the maximal cliques in order to get all groups of flip-flops that can be merged without violating any timing constraint (section 4.1). The third step is to find a set of non-conflicting[3] MBFFs by dividing maximal cliques into MBFFs (section 4.2). The last step is to determine the location of each MBFF (section 4.3). Noticeably, there might exist 1-bit MBFFs in the set of non-conflicting MBFFs. They are also relocated to further optimize the switching power of signal nets.

Because our algorithm can take care of original FFs which are more than 1-bit, we can apply our algorithm iteratively until there is no further improvement. We note that, although we assign MBFFs to unoccupied grid points subject to placement density constraints, there may still exist some overlappings between the MBFFs and pre-placed cells. Placement legalization can be done using techniques like[17, 18].

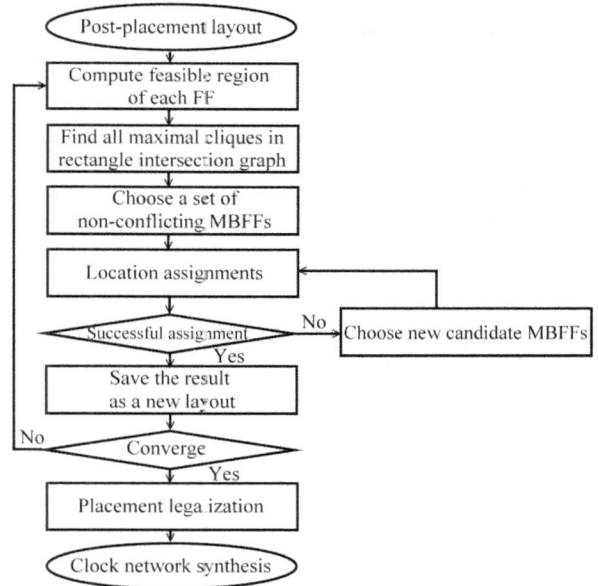

Figure 5: Design flow with power-driven flip-flop merging and relocation.

---

[3]Non-conflicting means if one FF has been merged into some MBFF, then this FF cannot be merged into other MBFF.

4.1 Mergeable Flip-Flop Groups Determination

Each group of flip-flops that can be merged without violating any timing constraint corresponds to a clique in the rectangle intersection graph. There can be a large number of cliques in the rectangle intersection graph. But we do not have to explicitly generate and store all of them as they are all contained in the maximal cliques. So we will just find all maximal cliques.

Finding a maximum clique in a general graph is NP-Hard, but to find maximum clique/all maximal cliques in a rectangle intersection graph is in P. We utilize the *sweep line method*[19] to find all maximal cliques. Sweep line method is a kind of scanning to determine the relation between rectangles on a plane. It keeps track of *in-edge and out-edge* to determine whether we are now inside a particular rectangle or not.

To perform the method, we rotate the entire chip by 45-degree. Then, we extend all vertical boundaries of feasible regions. Every two adjacent extended boundaries form a *region* as in Figure 6. By the scanning from the left-most region to the right-most one and each region from top to bottom, we can get all maximal cliques. More details can be found in [19]. This algorithm runs in $O(n^2 \log n)$, where n is the number of rectangles.

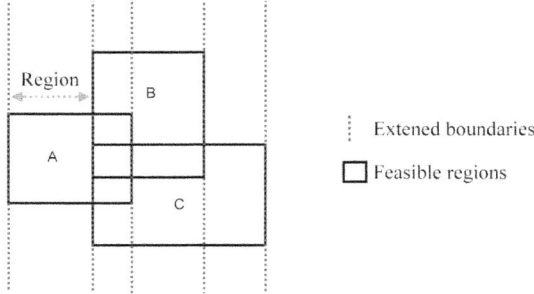

Figure 6: A, B, C are three original FFs. We can find the maximum clique ABC in the second region from the left.

4.2 MBFF Extraction Considering Participating Degree

In this stage, we want to extract MBFFs from the maximal cliques according to the given library. The main idea of MBFF extraction is to create MBFFs with the smallest cost greedily. First, for each maximal clique, we generate a set of possible MBFFs. Then, for each set, we take the one with smallest cost as the *candidate MBFF* of the corresponding maximal clique. Second, we put the one with the smallest cost from these candidate MBFFs into the set of non-conflicting MBFFs, said $\mathcal{N}$, repeatly. Each time we put a MBFF β into $\mathcal{N}$, we must make sure that β does not conflict with any existing MBFF belonging to $\mathcal{N}$. After that, we create a new candidate MBFF from the clique c which generated β. This new candidate MBFF also should have no conflict with any existing MBFF belonging to $\mathcal{N}$. After the set of candidate MBFFs becomes empty, $\mathcal{N}$ is the target set of non-conflicting MBFFs to do location assignment.

Since, we want to avoid creating a MBFF whose fanin(s)

Algorithm 1 MBFF extraction

Input: The set of all maximal cliques $\mathcal{C}$.
Output: A set of non-conflicting MBFFs $\mathcal{N}$.
1: $H \leftarrow \phi$ //A heap of candidate MBFFs. The key value is the the cost of each MBFF.
2: $\mathcal{N} \leftarrow \phi$
3: $\mathcal{P} \leftarrow \phi$ //$\mathcal{P}$ is the set of FFs which are allocated into $\mathcal{N}$
4: **for all** $c \in \mathcal{C}$ **do**
5:     H.Push(sample(c)); //sample(c) returns a new feasible candidate MBFF with the smallest cost among $|c|$ MBFFs generated from clique c.
6: **end for**
7: **while** H is non-empty **do**
8:     $\beta \leftarrow H$.ExtractMin();
9:     **if** $M \cap \mathcal{P} = \phi$ **then** //β does not conflict with $\mathcal{N}$
10:         $\mathcal{N} \leftarrow \{\beta\} \cup \mathcal{N}$;
11:         $\mathcal{P} \leftarrow \mathcal{P} \cup \beta$;
12:     **end if**
13:     $c(\beta) \leftarrow c(\beta) - \mathcal{P}$; //$c(\beta)$ is the clique which generated β.
14:     H.Push(sample($c(\beta)$));
15: **end while**
16: **return** $\mathcal{N}$;

---

and fanout(s) are too far away to reduce the total wire length of signal nets. The cost of a MBFF β is defined as follows:

$$cost = \lambda \times \frac{\mathcal{D}(\beta)}{\mathcal{B}(\beta)} + (1-\lambda) \times \frac{\sum_{net_i \in Net(\beta)} (\alpha_i \times EstWL_i)}{\sum_{net_i \in Net(\beta)} (\alpha_i \times OWL_i)} \quad (2)$$

$\mathcal{D}(\beta)$ is the *total participating degree*[4] of MBFF β, and $\mathcal{B}(\beta)$ is the total number of bits of β. $Net(\beta)$ is the set of all the signal nets connected to β, and $EstWL_i$ is the estimated wire length of net_i.

An original flip-flop with smaller participating degree has less merging choices, and so should be merged first to encourage more merging overall. If the number of bits of β is larger, it means we can reduce more sinks by using β. Hence, we use $\frac{\mathcal{D}(\beta)}{\mathcal{B}(\beta)}$ as a measure of the effect on total number of sinks by using β.

However, finding all the possible MBFFs contained in a clique takes exponential time and memory space. For example, a clique c is formed by eight 1-bit FFs, and the library only supports 2-bit MBFF, so C_2^8 different MBFFs can be generated by c. Thus, we generate at most n MBFFs randomly according to the library from a clique, where n is the clique size. Then, we calculate the cost of each one and take the one with the smallest cost as a candidate MBFF of the clique. We call this operation as *sample(c)*. Note that, the original FFs might not be 1-bit, so there are many ways to form a k-bit MBFF. We can get all kinds of combination by solving the *changes-making problem*[20] using dynamic programming. For instance, if there are 1/2/3/4-bit MBFFs in the library, we can generate a 4-bit MBFF with four 1-bit original FFs, or using one 1-bit original FF and one 3-bit original FF, etc. We store all the combinations in *changes table* for MBFF sampling.

The MBFF extraction algorithm is shown in Algorithm 1.

---

[4]The total participating degree is defined as the sum of participating degree of each original FF in β.

In line 5, we generate one candidate from each maximal clique and push it into a heap. From lines 8 to 12, we pick the best candidate β from the heap and put it into $\mathcal{N}$ if it does not conflict with any MBFF belonging to $\mathcal{N}$. And in line 14, we generate another candidate by sampling the maximal clique that generated β again. Each time we do the MBFF sampling of a clique, we will select a new target feasible candidate MBFF which does not conflict with any MBFF belonging to $\mathcal{N}$. We repeat these steps until all the FFs are allocated into some MBFFs of $\mathcal{N}$.

We use an example to illustrate the procedure of MBFF extraction. We are given a MBFF library that contains 1/2/4-bit MBFFs. There are 7 FFs $v_1 \sim v_7$ and all of them are 1-bit FF. After applying sweep line method, we get two maximal cliques, $c_1 = \{v_1, v_2, v_3, v_6, v_7\}$ and $c_2 = \{v_4, v_5, v_6\}$. Suppose after the initial sampling, we get MBFF $\{v_1, v_2, v_3, v_6\}$ from c_1 and $\{v_4, v_5\}$ from c_2 where $Cost(\{v_1, v_2, v_3, v_6\}) < Cost(\{v_4, v_5\})$. In the first iteration, we extract $\{v_1, v_2, v_3, v_6\}$ and put it into $\mathcal{N}$, then generate a new MBFF, said $\{v_7\}$, from c_1 after updating c_1. The value of each set after this iteration would be: $H = \{\{v_4, v_5\}, \{v_7\}\}$, $\mathcal{N} = \{\{v_1, v_2, v_3, v_6\}\}$, $\mathcal{P} = \{v_1, v_2, v_3, v_6\}$, $c_1 = \{v_7\}$ and $c_2 = \{v_4, v_5, v_6\}$. Again, we extract a new MBFF, $\{v_4, v_5\}$, with the smallest cost from H. After the second iteration, the sets would be updated as follows: $H = \{\{v_7\}\}$, $\mathcal{N} = \{\{v_1, v_2, v_3, v_6\}, \{v_4, v_5\}\}$, $\mathcal{P} = \{v_1, v_2, v_3, v_4, v_5, v_6\}$, $c_1 = \{v_7\}$ and $c_2 = \{\}$. In the last iteration, we extract $\{v_7\}$ from H, and get $\mathcal{N} = \{\{v_1, v_2, v_3, v_6\}, \{v_4, v_5\}, \{v_7\}\}$ which is the set of MBFFs ready for location assignment.

4.3 Decide MBFF Locations

For a non-conflicting MBFF β, we first calculate the *weighted median interval* of the x-(y-)coordinate of its fanin(s) and fanout(s), where the weight of a pin is the switching rate of the signal net between β and the pin.

We call the region formed by the intersection of these two weighted median intervals the *preferred region* of β, as shown in Figure 7. We call the bins covered by the preferred region the *preferred bins*. If we can put MBFF β in its preferred region, then the switching power of the signal nets connected to β, $\sum_{net_i \in Net(\beta)} (\alpha_i \times WL_i)$, will be minimized. We set the rank of preferred bins to 0 and assign increasing ranks to other bins inside β's feasible region as in Figure 8. Therefore, each selected MBFF has a ranked-list of bins, and we perform bin assignments according to these ranked-lists, and

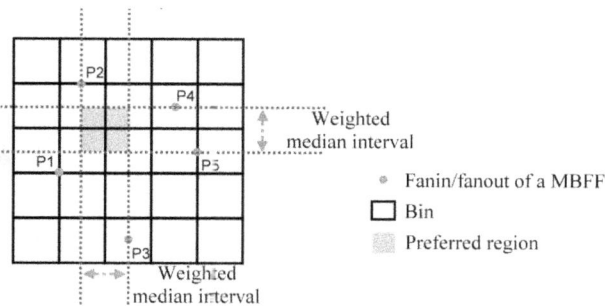

Figure 7: Example of preferred region. The switching rate ratio of the nets connected to a certain MBFF from P1 ~ P5 are 2:1:1:3:1.

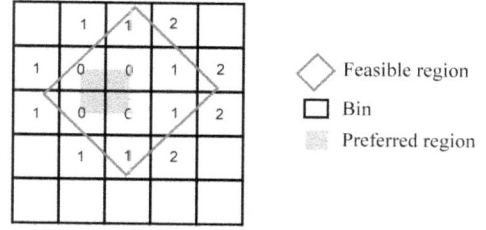

Figure 8: Ranking of bins.

then do the grid assignments. Nevertheless, we might fail to assign some non-conflicting MBFFs due to placement density constraint. So, if a non-conflicting MBFF generated cannot be successfully assigned, we will generate a smaller MBFF from its originating clique until we can assign it successfully.

The schema of grid point assignment is like bin assignment. For each MBFF, if the bin assignment will not cause bin density violation, then we will find all the preferred grid points. And we run a breath-first search to find an unoccupied grid point to place the MBFF starting from preferred grid points. With this assignment method, we can optimize the switching power of the signal nets connected to the MBFFs.

5. EXPERIMENTAL RESULTS

We implemented the algorithm in C++ and conducted our experiments on an Ubuntu workstation with 8GB memory and Intel(R) Xeon(R) E5506 @ 2.13GHz CPU. Cases t0~t3 are from [21] and r1~r5 are from [22]. For r1~r5, we assume all the original sinks are 1-bit flip-flops, original timing slacks and placement densities are generated randomly. For these 9 cases, we assume the unit resistance and capacitance of wire are 0.0001(Ohm/nm) and 0.0002(fF/nm), and use Bounded-Skew Clock Tree Routing(Version 1.0) [23] to perform zero-skew clock tree synthesis. Switching rate of signal nets connected to flip-flops of these cases are randomly generated in the range from 0.05 to 0.15.

We tested the 9 test cases with a MBFF library containing 1/2/4-bit MBFFs. The experimental results with the given library are shown in Table 1 and Table 2. We note that the ratio between the wirelength of signal nets connected to FFs and the wirelength of clock network varies from case to case, the weight w in equation (1) is adjusted accordingly.

Table 1 shows that our algorithm can achieve very good reduction in terms of number of clock sinks and wirelength of clock network. For the given library, we can reduce the number of sinks by 63.90% in average. The same table also shows, due to the reduction of sinks, the wirelength of clock tree can be reduced significantly by 40.51%. We can observe from Figure 9 that the reduction of the number of clock sinks is roughly proportional to the reduction of the wirelength of clock network. Figure 10 shows the clock tree synthesis results of case r1 using [23] before and after merging.

Table 2 shows the overall improvement of $\alpha_i \times WL_i$ of signal nets connected to the FFs. This product is proportional to the switching power consumption of the signals. We can make a rough estimation of the total switching power reduc-

Table 1: Reduction of clock sinks and clock tree wire length.

Test cases	#FF			FF area reduction	Clock tree WL(nm)			Run time(s)
	original	final	reduction		original	final	reduction	
r1	267	101	62.17%	3.30%	1325183	930661	29.77%	1.53
r2	598	223	62.70%	3.38%	2621623	1781824	32.03%	5.91
r3	862	298	65.42%	3.47%	3357327	2184565	34.93%	6.98
r4	1903	592	68.89%	3.56%	6839628	4185940	38.79%	28.20
r5	3101	921	70.29%	3.63%	10145960	6002024	40.84%	51.62
t0	120	37	69.16%	3.58%	39637	22545	43.12%	0.03
t1	60000	15040	74.93%	3.75%	3981765	1955086	50.89%	1053.12
t2	5524	1525	72.39%	3.68%	985348	543020	44.89%	1.98
t3	953	246	74.18%	3.73%	201755	102271	49.39%	1.14
avg.			68.90%	3.56%			40.51%	

Table 2: Reduction of wire length and estimated switching power of the nets connected to the flip-flops.

Test cases	$\sum_{net_i \in F} WL_i$			$\sum_{net_i \in F} \alpha_i \times WL_i$		
	original	final	reduction	original	final	reduction
r1	1743703	1756925	-0.75%	179802	176664	1.74%
r2	3930879	3732569	5.04%	400928	370500	7.58%
r3	5672241	5191913	8.46%	574642	511426	11.00%
r4	12616681	12066921	4.35%	1266302	1187960	6.18%
r5	20528314	19012768	7.38%	2061324	1856472	9.93%
t0	83285	74365	10.71%	8755	7482	14.53%
t1	53624875	33077705	38.31%	5356145	3157927	41.04%
t2	3562985	2099595	41.07%	357151	204907	42.62%
t3	576710	448090	22.30%	58576	43931	25.00%
avg.			15.21%			17.74%

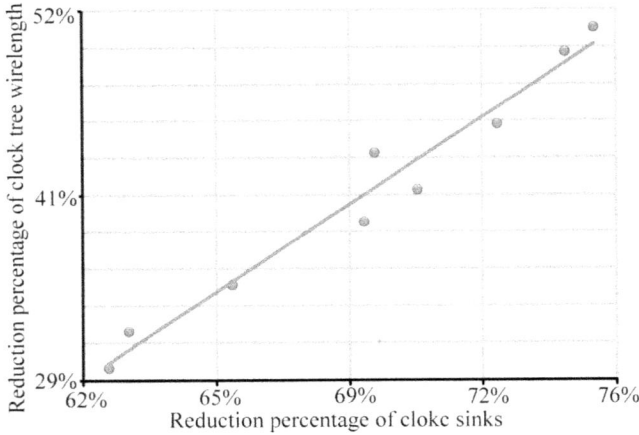

Figure 9: The reduction percentage of clock tree wirelength versus the reduction percentage of clock sinks.

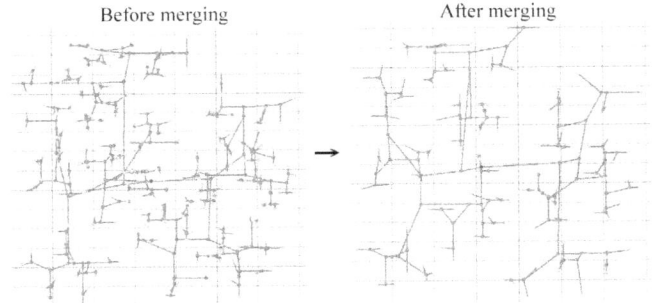

Figure 10: CTS result of r1 before and after merging.

Table 3: MBFF library from [15].

Bit Number	Normalized Power consumption per bit	Normalized Area per bit
1	1.00	1.00
2	0.86	0.96
4	0.78	0.71

tion. Assume the switching power due to the clock network and the signal nets are 40% and 60%, respectively. Furthermore 80% of the clock network switching power is due to wire capacitance and one fourth of the signal nets are connected to flip-flops. Then, the experiment results show that the total switching power consumption of the entire chip can be reduced by about 15.62% in average after applying our power-driven flip-flops merging and relocation approach.

5.1 Experiments Using the Objective of [15]

We note that in a related work [15], a different objective function is targeted. [15] addressed the problem of merging flip-flops into multi-bit flip-flops to optimize the power consumption of flip-flops only while considering the wirelength of nets connected to flip-flops. Our algorithm can be modified to target their objective function.

Table 4: Comparison with [15].

Test cases	#FFs	[15]			Ours		
		FF Power Red.	HPWL Red.	Run time(s)	FF Power Red.	HPWL Red.	Run time(s)
c1	98	14.8%	8.7%	0.01	15.64%	8.2%	0.01
c2	423	16.9%	5.3%	0.04	17.52%	11.1%	0.05
c3	1692	17.1%	5.2%	0.10	17.41%	11.5%	0.22
c4	5129	16.8%	5.5%	0.28	17.07%	11.5%	0.72
c5	10575	17.1%	5.1%	0.60	17.29%	13.4%	1.89
c6	169200	17.2%	5.1%	78.92	17.52%	11.8%	36.12
avg.		16.65%	5.82%		17.03%	11.25%	

Since a larger MBFF would lead to more flip-flop power reduction as shown in Table 3, we modify our algorithm as follow. In the stage of MBFF extraction, for each clique, we find the maximum MBFF size, k, that can be generated, and we randomly sample ten k-bit MBFFs from each clique. Then, the one resulting in the largest wirelength reduction of the nets connected to FFs is chosen as the candidate MBFF. Finally, we filter out some conflicting candidate MBFFs as before, and do the location assignment.

We compared the 6 test cases from [15] as shown in Table 4. "FF Power Red." is the reduction of power consumed by FFs, and "HPWL Red." is the wirelength reduction of the nets connected to FFs. The experimental results for [15] were taken from their paper which were run on a 2.66GHz Intel i7 PC. We get more flip-flop power and wirelength reduction in these cases, and our run time is better than [15] in the largest test case. If we compare the runtime in Table 1 and Table 4, we can see that it is more time-consuming to minimize the clock network and switching power of the nets connected to the flip-flops simultaneously. The reason is that forming a larger MBFF would not necessarily be better because it may greatly increase the switching power of the nets connected to it. So we can not greedily pick the largest MBFF each time, as a result sampling will take more time and the while loop from line 7 to line 15 in Algorithm 1 will also take more iterations to terminate.

6. CONCLUSIONS

In this paper, we presented a power-driven flip-flop merging and relocation approach to reduce the switching power consumption of the entire circuit. Our algorithm does a non-conflicting candidate searching and determines the location of a multi-bit flip-flop using weighted median interval and ranking. Experimental results indicated that the number of clock sinks can be reduced by about 70%. Moreover, decreasing the number of clock sinks leads to shorter wirelength of clock tree and hence less power consumption. It is shown that our works can reduce clock tree wirelength by 30~50% and thus the power consumption. Additionally, our work can also save the power taken by those signal nets connecting to FFs and the experimental results indicate a reduction of about 17% in average.

7. ACKNOWLEDGMENTS

We would like to thank Prof. Mark Lin of NCCU for providing us the benchmarks used in [15].

8. REFERENCES

[1] D. Liu and C. Svensson. Power consumption estimation in cmos vlsi circuits. *IEEE Solid-State Circuits*, 29:663–670, 1994.

[2] K. Wang and M. Marek-Sadowska. Buffer sizing for clock power minimization subject to general skew constraints. In *Design Automation Conference*, pages 153–156, 2004.

[3] G. Wilke and R. Reis. A new clock mesh buffer sizing methodology for skew and power reduction. In *IEEE Computer Society Annual Symposium on VLSI*, pages 227–232, 2008.

[4] A.S. Seyedi, S.H. Rasouli, A. Amirabadi, and A. Afzali-Kusha. Low power low leakage clock gated static pulsed flip-flop. In *IEEE International Symposium on Circuits and Systems*, pages 3658–3611, 2006.

[5] S. Naik and R. Chandel. Design of a low power flip-flop using cmos deep sub micron technology. In *International Conference on Recent Trends in Information, Telecommunication and Computing*, pages 253–256, 2010.

[6] C.C. Yu. Design of low-power double edge-triggered flip-flop circuit. In *IEEE Conference on Industrial Electronics and Applications*, pages 2054–2057, 2007.

[7] S.K. Teng and N. Soin. Low power clock gates optimization for clock tree distribution. In *International Symposium on Quality Electronic Design*, pages 488–492, 2010.

[8] M. Donno, A. Ivaldi, L. Benini, and E. Macii. Clock-tree power optimization based on RTL clock-gating. In *Design Automation Conference*, pages 622–627, 2003.

[9] Q. Wu, M. Pedram, and X. Wu. Clock-gating and its application to low power design of sequential circuits. *IEEE Transactions on Circuits Systems I*, 47(3):415–420, 2000.

[10] Y. Cheon, P.H. Ho, A.B. Kahng, S. Reda, and Q. Wang. Power-aware placement. In *Design Automation Conference*, pages 227–232, 2008.

[11] Y. Lua, C.N. Sze, X. Hong, Q. Zhou, Y. Cai, L. Huang, and J. Hu. Navigating registers in placement for clock network minimization. In *Design Automation Conference*, pages 176–181, 2005.

[12] W. Hou, D. Liu, and P.H. Ho. Automatic register banking for low-power clock trees. In *International Symposium on Quality Electronic Design*, pages 647–652, 2009.

[13] R.P. Pokala, R.A. Feretich, and R.W. McGuffin.

Physical synthesis for performance optimization. In *IEEE International ASIC Conference and Exhibit*, 1992.

[14] Y. Kretchmer. Using multi-bit register inference to save area and power: the good, the bad, and the ugly. In *EE Times Asia*, 2001.

[15] Y.T. Chang, C.C. Hsu, P.H. Lin, Y.W. Tsai, and S.F. Chen. Post-placement power optimization with multi-bit flip-flops. In *IEEE/ACM International Conference on Computer-Aided Design*, 2010.

[16] G. Chabert, L. Jaulin, and X. Lorca. A constraint on the number of distinct vectors with application to localization. Technical report, Small workshop on Interval Methods, 2009.

[17] U. Brenner and J. Vygen. Legalizing a placement with minimum total movement. *IEEE Transactions on Computer-Aided Design of Integrated Circuits and Systems*, pages 1597–1613, 2004.

[18] S. Chou and T.Y. Ho. OAL: An obstacle-aware legalization in standard cell placement with displacement minimization. In *IEEE International SOC Conference*, pages 329–332, 2009.

[19] M.I. Shamos and D. Hoey. Geometric intersection problems. In *IEEE Symposium on Foundations of Computer Science*, 1991.

[20] H. Abelson, J. Sussman, and J. Sussman. *Structure and Interpretation of Computer Programs*. MIT, 1984.

[21] CAD contest of taiwan. `http://cad\_contest.ee.ntu.edu.tw/cad10/Problems/B1\_Faraday\_091223\_MultiBitFF.pdf`.

[22] R.S. Tsay. Exact zero skew. In *IEEE International Conference on Computer-Aided Design*, pages 336–339, 1991.

[23] A.B. Kahng and C.W. Tsao. Bounded-skew clock tree routing - version 1.0. `http://vlsicad.ucsd.edu/GSRC/bookshelf/Slots/BST/`.

INTEGRA: Fast Multi-Bit Flip-Flop Clustering for Clock Power Saving Based on Interval Graphs

Iris H.-R. Jiang
Dept. of Electronics Eng. &
Inst. of Electronics
National Chiao Tung University
huiru.jiang@gmail.com

Chih-Long Chang
Dept. of Electronics Eng. &
Inst. of Electronics
National Chiao Tung University
paralost.ee96@nctu.edu.tw

Yu-Ming Yang
Dept. of Electronics Eng. &
Inst. of Electronics
National Chiao Tung University
yuming.yyang@gmail.com

Evan Y.-W. Tsai
Faraday Technology Corp.
Hsinchu, Taiwan
ywtsay@faraday-
tech.com

Lancer S.-F. Chen
Faraday Technology Corp.
Hsinchu, Taiwan
lancer_c@faraday-
tech.com

ABSTRACT

Clock power is the major contributor to dynamic power for modern IC design. A conventional single-bit flip-flop cell uses an inverter chain with a high drive strength to drive the clock signal. Clustering such cells and forming a multi-bit flip-flop can share the drive strength, dynamic power, and area of the inverter chain, even can save the clock network power and facilitate the skew control. Hence, in this paper, we focus on multi-bit flip-flop clustering at post-placement to gain these benefits. Utilizing the properties of Manhattan distance and coordinate transformation, we model the problem instance by two interval graphs and use a pair of linear-size sequences as our representation. Without enumerating all compatible combinations, we extract only partial sequences that are necessary to cluster flip-flops at a time, thus leading to an efficient clustering scheme. Moreover, our coordinate transformation brings fast operations to execute our algorithm. Experimental results show the superior efficiency and effectiveness of our algorithm.

Categories and Subject Descriptors

B.7.2 [**Integrated Circuits**]: Design Aids—*Placement and Routing*

General Terms

Algorithms, Design, Theory

Keywords

Clock power, multi-bit flip-flops, post-placement optimization, interval graph, coordinate transformation

Permission to make digital or hard copies of all or part of this work for personal or classroom use is granted without fee provided that copies are not made or distributed for profit or commercial advantage and that copies bear this notice and the full citation on the first page. To copy otherwise, to republish, to post on servers or to redistribute to lists, requires prior specific permission and/or a fee.
ISPD'11, March 27–30, 2011, Santa Barbara, California, USA.
Copyright 2011 ACM 978-1-4503-0550-1/11/03 ...$10.00.

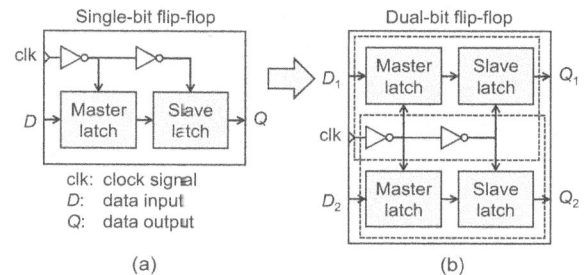

Figure 1: (a) A (single-bit) flip-flop. (b) A dual-bit flip-flop with two sets of data input and output pins.

1. INTRODUCTION

Clock power is the major dynamic power source since the clock signal toggles in each cycle [10]:

$$P_{clk} = C_{clk} V_{dd}^2 f_{clk}, \qquad (1)$$

where P_{clk} is clock power, f_{clk} is the clock frequency, V_{dd} is the supply voltage, and C_{clk} is the switching capacitance including the gate capacitance of flip-flops (sequential elements) controlled by the clock signal and the capacitance associated with the interconnect and buffers/inverters in the clock network. To minimize the switching capacitance C_{clk}, a new type flip-flop cell–multi-bit flip-flop (MBFF)–is designed [3, 7]. A single-bit flip-flop is composed of two chained inverters and two cascaded latches (see Figure 1(a)); the inverter chain has a high drive strength to shorten the delay from the clock edge to data output. Hence, clustering several single-bit flip-flops together can share the drive strength of the inverter chain (see Figure 1(b)). The benefits of MBFFs are twofold: 1) For the first term of C_{clk}, clustered flip-flops consume less dynamic power and area. 2) For the second term of C_{clk}, the clock network can have a more regular topology, easier skew control [7], and lower power due to fewer clock sinks, a shallower depth, and fewer clock buffers.

Hence, MBFF clustering is proposed to minimize clock power [3, 5, 11, 2]. Chen *et al.* in [3] and Hou *et al.* in [5] leverage on register banking at logic synthesis and at early physical synthesis, respectively. However, the subsequent timing and routing cost of the clustered result may some-what deviate from what is expected at such early stages.

On the other hand, Yan and Chen in [11] and Chang *et al.* in [2] postpone this task to post-placement to further consider the timing and even routing issues. Yan and Chen in [11] analyze the timing-safe region for each flip-flop and then construct an intersection graph to record the pairwise compatibility. They reduce MBFF clustering to minimum clique partitioning and solve it by greedily merging flip-flops with fewest compatible flip-flops. However, they assume the bit numbers of the MBFF library are contiguous and unlimited. Considering a discrete and finite MBFF library, Chang *et al.* in [2] present a window-based clustering method. They enumerate maximal cliques and reduce MBFF clustering to maximum independent set; they greedily cluster flip-flops inside the processing window and place MBFFs considering the routing cost and the placement density. However, the clustered flip-flops cannot cross the window boundary, and window sliding repeatedly visits layout bins. Moreover, minimum clique partitioning and maximum independent set are both NP-hard, so [11] and [2] resort them to greedy heuristics. Even so, they may still incur a large storage requirement and a long runtime to solve a large-scale design.

In this paper, we tackle the MBFF clustering problem at post-placement: Given an MBFF library and the timing slack of each flip-flop in a design, cluster flip-flops to minimize clock power (as well as the routing cost) subject to timing and placement density constraints. Although clustering flip-flops may change the wire power of data input and output pins, the amount of change is quite small compared with the significant reduction on clock power. Since MBFF clustering is NP-hard, our goal is to solve it effectively and efficiently instead of optimally. Utilizing the properties of Manhattan distance and coordinate transformation, we successfully encode the timing-safe regions of flip-flops into two interval graphs [6, 8]. Particularly, to remedy the time and space complexities, we neither convert two interval graphs into one intersection graph nor enumerate all possible combinations of compatible flip-flops. Instead, we adopt a pair of linear-size sequences as the representation. We identify "decision points" in the sequences, at which there exist some essential flip-flops to be clustered. The essential flip-flops and their related flip-flops extracted at each decision point maintain the global view of compatibility. We then place MBFFs as close to the optimum location for the routing cost as possible. Our key features are as follows:

- From the time and space complexities' viewpoint, the representation—a pair of linear-size sequences—implies an efficient data structure, and our coordinate transformation brings fast operations to execute our algorithm.

- The number of decision points is significantly smaller than the number of flip-flops. We cluster flip-flops at only decision points thus leading to an efficient clustering scheme.

- Without enumerating all compatible combinations, we sweep and interleave two sequences to extract the compatibility on the fly and maintain the global view.

- Our data structure and algorithm are both independent of the number of layout grids and/or bins.

Table 1: MBFF Library: Power vs. Area.

Bit number	Power	Area	Normalized power per bit	Normalized area per bit
1	100	100	1.00	1.00
2	172	192	0.86	0.96
4	312	285	0.78	0.71

The experiments are conducted on seven industrial circuits. Experimental results show that the concise representation delivers superior efficiency and effectiveness. Compared with the very recent work [11] and [2], we can deliver the best power saving (only 0.17% away from the power lower bound) and the shortest run times (359X and 17X speedup). Our results also show that the number of decision points is significantly smaller than the number of flip-flops indeed (only 12% of flip-flop count). Moreover, it can be seen that the impact on combinational dynamic power (wire power of data input and output pins) is relatively small with respect to the clock power saving. Compared with MBFF clustering at logic synthesis [3], flip-flop replacement rate and the post-clock-tree-synthesis clock power can considerably be improved by post-placement MBFF clustering, even under timing and placement density constraints.

The remainder of this paper is organized as follows. Section 2 gives the problem formulation; Section 3 derives the properties of Manhattan distance and coordinate transformation and gives our representation; Section 4 presents our algorithm–INTEGRA; Section 5 shows our experimental results; finally, Section 6 concludes this paper.

2. PROBLEM FORMULATION

In this section, we describe the problem formulation and detail the design information and constraints.

The Multi-Bit Flip-Flop Clustering Problem: Given a (contiguous or discrete) MBFF library and the timing slacks of flip-flops in a design, cluster flip-flops to minimize flip-flop power (as well as the routing cost) subject to timing slack and placement density constraints.

Flip-flop power is the primary objective, while the routing cost is secondary. The input design may contain pre-clustered MBFFs. An MBFF library cell is associated with its bit number, consumed power, and occupied area (see Table 1). In the sequel, a flip-flop means a single-bit flip-flop, while an MBFF means a multi-bit one.

Assume the chip area contains $W \times H$ grids. The chip area is divided into $W_c \times H_c$ bins, and a bin is further divided into $W_b \times H_b$ grids (see Figure 2). Each gate should be placed on some grid point, and each grid can be occupied by at most one gate. The placement density constraint restricts the area utilization of each bin:

$$A_{fb} \leq T_b(W_b H_b A_g - A_{pb}) - A_{cb}, \qquad (2)$$

where A_{fb} is the flip-flop area, T_b is the target density, A_g is the grid area, A_{pb} is the macro area, A_{cb} is the area occupied by combinational elements within each bin.

The routing cost $L(i)$ of flip-flop i is the sum of its data input and output wirelength:

$$L(i) = L_{fi}(i) + L_{fo}(i), \forall i, \qquad (3)$$

where $L_{fi}(i)$ (respectively $L_{fo}(i)$) is the Manhattan distance

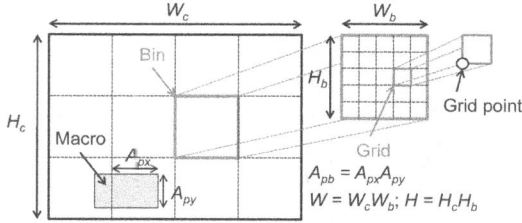

Figure 2: Layout bins and grids.

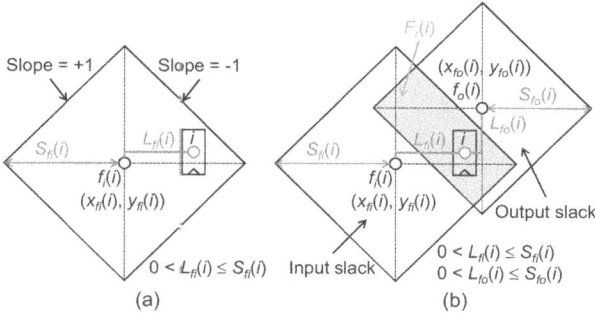

Figure 3: (a) Flip-flop i's input slack. (b) Flip-flop i's feasible region.

to its fanin $f_i(i)$ (respectively fanout $f_o(i)$) gate. The routing cost also reflects the change on the wire power of data input and output pins. The amount of change is quite small compared with the clock power saving.

On the other hand, the input (respectively output) timing slack of a flip-flop is the maximum allowable wire delay between the flip-flop and its fanin (respectively fanout) gate without timing violations; the wire delay bound can be expressed as an effective wirelength[1] bound as follows:

$$0 < L_{fi}(i) \le S_{fi}(i), 0 < L_{fo}(i) \le S_{fo}(i), \forall i, \quad (4)$$

where $S_{fi}(i)$ (respectively $S_{fo}(i)$) is the input (respectively output) slack. If flip-flop i has multiple fanouts, each fanout $f_o(i)$ corresponds to a routing cost and an output slack.

3. PROPERTIES AND REPRESENTATION

In this section, we derive the properties and introduce the representation used in our algorithm.

3.1 The Feasible Region

According to Equation (4), each flip-flop i's input (respectively output) slack constraint defines a diamond whose center is located at i's fanin (respectively fanout) gate and whose half diagonal length is the slack value (see Figure 3(a)). For timing-safety, flip-flop i's feasible region $F_r(i)$ is the overlap of these diamonds (see Figure 3(b)).

---

[1]For the Synopsys Liberty library, the delay of a gate, lumped with its output wire delay, is dominated by its output loading [4]. Since the placement of combinational elements is unchanged during post-placement MBFF clustering, the timing slack between a flip-flop and its fanin/fanout gate depends on only the wire loading, i.e., the Manhattan distance between them. The timing model passes the static timing check.

3.2 Coordinate Transformation

Overlapping diamonds and checking if a grid point is located within the feasible region is computationally intensive. Hence, we accelerate the operations by coordinate transformation. After being rotated by 45° clockwise, the diamonds become squares; the feasible region can then be retrieved by overlapping squares, more computationally efficient than overlapping diamonds.

Let all gates and flip-flops be placed at grids, integer coordinates in a Cartesian coordinate system C. Consider a new coordinate system C' with origin at $(0,0)$ in C. The coordinate transformation between grid point (x, y) in C and its counterpart (x', y') in C' is defined as follows.

$$\begin{pmatrix} x' \\ y' \end{pmatrix} = \begin{pmatrix} 1 & 1 \\ -1 & 1 \end{pmatrix}\begin{pmatrix} x \\ y \end{pmatrix}; \begin{pmatrix} x \\ y \end{pmatrix} = \begin{pmatrix} \frac{1}{2} & \frac{-1}{2} \\ \frac{1}{2} & \frac{1}{2} \end{pmatrix}\begin{pmatrix} x' \\ y' \end{pmatrix}. \quad (5)$$

The unit length in C' equals $\sqrt{2}$ unit length in C. Because of the scaling factor, the transformation from C to C' can be done by simple anf fast integer addition/subtraction. Figure 4(a) shows the transformed chip area: $0 \le x \le W$, $0 \le y \le H$; $0 \le x' \le H + W$, $-W \le y' \le H$. By Equation (5), the x' and y' coordinates of every grid point must be both even or odd:

$$x' \mod 2 = y' \mod 2. \quad (6)$$

Hence, for each grid, its left-bottom corner is defined as the grid point, while its center is a non-grid point. As shown in Figure 4(b), the feasible region of a flip-flop is a rectangle defined as the overlap of its input and output squares. Let $s_{x'}(i)$, $e_{x'}(i)$, $s_{y'}(i)$, and $e_{y'}(i)$ denote the left, right, bottom, and top boundaries of flip-flop i's feasible region. We have

$$\begin{aligned} s_{x'}(i) &= \max(y_{fi}(i) + x_{fi}(i) - S_{fi}(i), \\ & \quad y_{fo}(i) + x_{fo}(i) - S_{fo}(i)); \\ e_{x'}(i) &= \min(y_{fi}(i) + x_{fi}(i) + S_{fi}(i), \\ & \quad y_{fo}(i) + x_{fo}(i) + S_{fo}(i)); \\ s_{y'}(i) &= \max(y_{fi}(i) - x_{fi}(i) - S_{fi}(i), \\ & \quad y_{fo}(i) - x_{fo}(i) - S_{fo}(i)); \\ e_{y'}(i) &= \min(y_{fi}(i) - x_{fi}(i) + S_{fi}(i), \\ & \quad y_{fo}(i) + x_{fo}(i) + S_{fo}(i)). \end{aligned} \quad (7)$$

3.3 Representation

After coordinate transformation, flip-flop i's feasible region is a rectangle in the new coordinate system C'. By projecting this rectangle on x' and on y' axes, we have two intervals, $I_{x'}(i)$ and $I_{y'}(i)$:

$$\begin{aligned} I_{x'}(i) &= [s_{x'}(i), e_{x'}(i)], \\ I_{y'}(i) &= [s_{y'}(i), e_{y'}(i)], \forall i. \end{aligned} \quad (8)$$

Their starting and end points are defined by Equation (7). A b-bit MBFF j including flip-flops $j_1, \ldots, j_b$ implies j_1's, $\ldots$, and j_b's feasible regions overlap thus forming a clique not only on x' interval graph but also on y' interval graph.

To avoid $O(n^2)$ space complexity, we adopt a pair of linear-size sequences, $(X'.Y')$, instead of one intersection graph, as the representation. The two sequences store the starting and end points of x' and y' intervals in ascending order; if several points have the same coordinate, starting points are ordered before end points. We have sequence X' for x' intervals and sequence Y' for y' intervals, $|X'| = |Y'| = 2n$ for

117

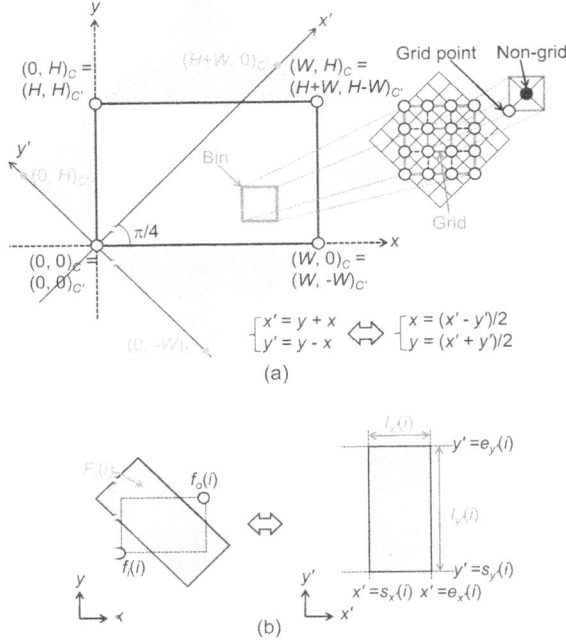

Figure 4: Coordinate transformation: (a) The chip area. (b) 2 intervals for each feasible region.

n flip-flops.

$$X' = <x'_1, x'_2, \ldots, x'_{2n}>, x'_1 \leq x'_2 \leq \cdots \leq x'_{2n},$$
$$x'_j \in \{s_{x'}(i),\ e_{x'}(i) : i = 1..n\}, j = 1..2n;$$
$$Y' = <y'_1, y'_2, \ldots, y'_{2n}>, y'_1 \leq y'_2 \leq \cdots \leq y'_{2n},$$
$$y'_j \in \{s_{y'}(i),\ e_{y'}(i) : i = 1..n\}, j = 1..2n. \qquad (9)$$

Please note that in our algorithm, X' is constructed at the beginning while Y' is dynamically and partially generated on the fly (see Section 4). Figure 5(a) gives a subdesign with flip-flops $0, 1, \ldots, 7$, where the tilted boxes are feasible regions. Figure 5(b) shows the transformed feasible regions. Figure 5(c)(d) illustrates the corresponding x' and y' intervals. Figure 5(e) lists the corresponding sequence X'.

4. MBFF CLUSTERING–INTEGRA

In this section, we detail our MBFF clustering algorithm–INTEGRA. We introduce the concept of decision points to reduce the times that flip-flop clustering is applied, and thus INTEGRA elaborates an efficient clustering scheme. INTEGRA contains 3 phases: initialization, flip-flop clustering, flip-flop placement. Flip-flop clustering and placement are compounded. INTEGRA is briefly summarized as follows.

1. INTEGRA analyzes the design intent (see Figure 5(e)).

2. INTEGRA finds a decision point in X' and extracts the essential flip-flops and their related flip-flops (see Figure 5(e)(f)).

3. INTEGRA finds the maximal clique in the partial Y' for each essential flip-flop (see Figure 5(g)).

4. INTEGRA clusters each essential flip-flop (see Figure 5(g)).

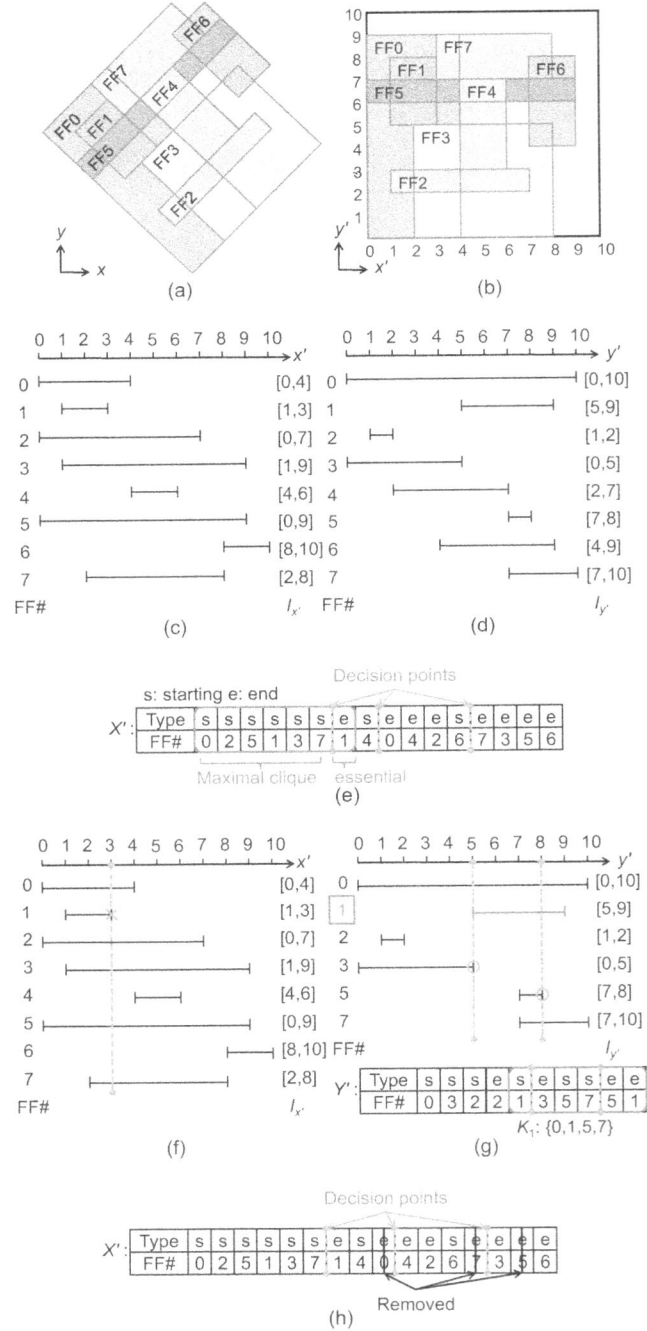

Figure 5: MBFF clustering: (a) The feasible regions of a subdesign with 8 flip-flops. (b) The transformed feasible regions, aligned with x' and y' axes. (c) x' intervals. (d) y' intervals. (e) Sequence X' is of size 16. Initially, there are three decision points at flip-flops 1's, 0's, and 7's endpoints. (f) The maximal clique at the first decision point in X' is $\{0, 1, 2, 3, 5, 7\}$. (g) The corresponding Y'. Flip-flop 1's maximal clique is $\{0, 1, 5, 7\}$. The appropriate MBFF cell is a 4-bit flip-flop, so $K_1 = \{0, 1, 5, 7\}$. (h) Runtime decision points are at flip-flops 1's, 4's, and 3's endpoints.

```
Algorithm INTEGRA
// Initialization
1.  lexicographically sort the MBFF library
2.  collapse MBFFs
3.  X' ← sort {s_x(i), e_x(i): i = 1..n}, j ← 1, Q ← ∅
// Main body
4.  while (X' is not empty) do
5.    find a decision point
6.    Q ← Q + essential flip-flops and related flip-flops
7.    Y' ← sort {s_y(i), e_y(i): i ∈ Q}
8.    foreach essential flip-flop k do
        // Flip-flop clustering
9.      K_max ← max_clique(Y', k)
10.     find the appropriate MBFF cell of bit number B for |K_max|
11.     K_max ← sort {e_x(i): i ∈ K_max - {k}}
12.     K_j ← flip-flop k and the first (B-1) flip-flops in K_max
        // Flip-flop placement
13.     find bounding box B_b for K_j
14.     project B_b's corner and center points to F_r(K_j)
15.     find the projected point with min distance between B_b and F_r(K_j)
16.     legalize this point and assign it to MBFF K_j
17.     if legalization fails then go to line 9
18.     Q ← Q - K_j, X' ← X' - K_j
19.     j++
```

Figure 6: Our MBFF clustering algorithm.

5. INTEGRA places the clustered flip-flop at a legal location with routing cost and density consideration

6. INTEGRA repeats steps 2–5 until all flip-flops are investigated.

4.1 Phase 1: Initialization

INTEGRA preprocesses the design intent as follows: In line 1 in Figure 6, MBFF library cells are lexicographically sorted, e.g., $<(1, 100, 100), (2, 172, 192), (4, 312, 285)>$ for Table 1. In line 2, each pre-clustered b-bit MBFF is collapsed into b flip-flops; the collapsed flip-flops are temporarily placed at the MBFF's location. (If a pre-clustered MBFF is not allowed to be modified, it is preserved, and its location is marked as occupied.) In line 3, the feasible region of each flip-flop is computed, and X' is created.

4.2 Phase 2: Flip-Flop Clustering

We first introduce "decision points" and "essential flip-flops" as follows:

Definition 1. If there exist two consecutive points x'_k and x'_{k+1} in X', where $x'_k = s_{x'}(i), x'_{k+1} = e_{x'}(j), 1 \le i, j \le n$, a decision point is the coordinate of x'_{k+1}, i.e., $e_{x'}(j)$.

Definition 2. The essential flip-flops with respect to a decision point are the flip-flops whose end points ordered from this decision point to the next decision point or to the end of X' for the last decision point.

THEOREM 1. *Consider X', a decision point, and the corresponding essential flip-flop. The maximal clique containing the essential flip-flops in x' interval graph can be found at this decision point.*

COROLLARY 1. *A decision point corresponds to at least one essential flip-flop. Hence, the number of decision points is less than or equal to the number of flip-flops.*

For the instance in Figure 5(a), Figure 5(e) shows there exist 3 decision points at flip-flops 1's, 0's, 7's end points;

flip-flop 1 is the first one's essential flip-flop, flip-flops 0, 4, and 2 are the second one's, and flip-flops 7, 3, 5, and 6 are the third one's. This instance has 8 flip-flops but only 3 decision points.

A decision point in X' means at which there exist some flip-flops whom we should decide how to cluster; otherwise, they cannot be clustered any more. These flip-flops are essential flip-flops with respect to this decision point. Figure 5(f) shows that at flip-flop 1's end point (the first decision point), flip-flop 1 has to be clustered, i.e., it is essential. Flip-flops 0, 2, 3, 5, 7 are its related flip-flops, meaning $\{0, 1, 2, 3, 5, 7\}$ forms a maximal clique in x' interval graph. Moreover, considering the interval from the second decision point (flip-flop 0's end point) to the third decision point (flip-flop 7's end point), $[4, 8]$, the maximal clique containing flip-flops 0, 4, 2 can be found at flip-flop 0's end point, $\{0, 2, 3, 4, 5, 7\}$. Furthermore, by the theorem, to find *all* maximal cliques[2] we need check only decision points instead of all flip-flops' end points. Each maximal clique can be found at some decision point, so we can maintain the global view.

To form an MBFF, the maximal clique found in X' should be further verified by the y' intervals of its members. For the flip-flops in the maximal clique in X', the starting and end points of their y' intervals are sorted in ascending order and form the working sequence Y'. Similarly, the maximal clique in Y' of each essential flip-flop can be found by checking the decision points in Y' within the essential flip-flop's y' interval. Figure 5(g) shows that the decision points in Y' within flip-flop 1's y' interval are flip-flops 3's and 5's end points. Flip-flop 3 corresponds to a maximal clique $\{0, 1, 3\}$, while flip-flop 5 corresponds to $\{0, 1, 5, 7\}$. We choose $\{0, 1, 5, 7\}$ and check the sorted MBFF library to find an appropriate MBFF cell to cluster them. An appropriate MBFF cell means its bit number is the largest but not exceeding the clique size. With the MBFF library specified in Table 1, flip-flops 0, 1, 5, 7, hence, form a 4-bit MBFF and are removed from X'. If the clique size is larger than the bit number of the selected MBFF cell, we cluster the flip-flops in ascending order of their x' end points. The created MBFF is then placed at a legal grid point with routing cost and placement density consideration (see Section 4.3). The same process is repeated until all flip-flops are investigated, i.e., X' is empty. Since flip-flops are iteratively clustered and removed from X', the size of X' gradually shrinks, and the runtime decision points might be deferred or even disappear, e.g., Figure 5(h) indicates that runtime decision points are at flip-flops 1's, 4's, and 3's end points.

INTEGRA iteratively finds a decision point in line 5, records the essential and related flip-flops by queue Q in line 6, and creates the corresponding working sequence Y' in line 7. From line 8 to line 12, for each essential flip-flop, INTEGRA repeatedly extracts the maximal clique in Y' and selects an appropriate MBFF cell to cluster it. After placing it (see Section 4.3), INTEGRA removes the clustered flip-flops from X' and Q in line 18.

4.3 Phase 3: Flip-Flop Placement

Each MBFF generated by flip-flop clustering (see Section 4.2) is placed to a legal grid point as close to the opti-

---

[2] In [8], Ramalingam and Rangan sort the end points of intervals in ascending order. They can compute *one* maximal clique for each interval in linear time.

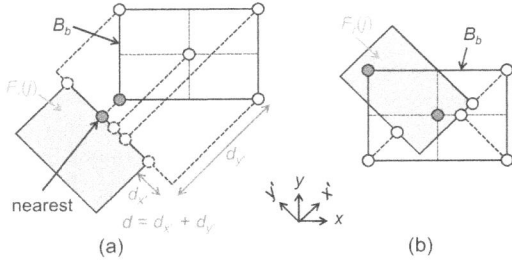

Figure 7: Flip-flop placement: The optimal location with minimum routing cost for two general cases.

mum location for the routing cost as possible. A legal grid point satisfies the following conditions:

1. It is a grid point (checked by Equation (6)).

2. It is not occupied by other gates or flip-flops.

3. It is density-safe.

The minimum routing cost of MBFF j occurs when j is located within the bounding box defined by low/high median x and y coordinates of its all fanin and fanout gates. To find the nearest point in the feasible region to the bounding box, [2] iteratively enlarges the bounding box until it reaches any point in the feasible region. Unlike [2], we do it efficiently. First of all, INTEGRA finds the x' and y' coordinates of the corner and center points of the bounding box in line 13. INTEGRA then projects these points into MBFF j's feasible region and finds the nearest point in lines 14 and 15. (Projecting the center point can prevent fanin/fanout gates from being selected.) The projected points and their distances can efficiently be computed by checking their x' and y' coordinates. Figure 7 shows two general cases that MBFF j's bounding box B_b may or may not overlap with its feasible region $F_r(j)$. In line 16, INTEGRA picks the nearest projected point and moves it within the feasible region until a legal point is found. INTEGRA then places the MBFF at the found point. If there exist no legal points in the feasible region, it goes back to line 9 to try another clique.

Table 1 indicates that the larger bit number, the smaller normalized area per bit. Once some flip-flops form an MBFF, the total area utilization is smaller, and the placement density constraint becomes looser. Because there are some induced empty space around, if MBFFs overlap with pre-placed cells (combinational elements) or macros, legalization can be applied to adjust these pre-placed cells; the displacement should be small, and the impact on timing is negligible.

5. EXPERIMENTAL RESULTS

We implemented INTEGRA in the C programming language and executed the program on a platform with an Intel Xeon 3.8 GHz CPU and with 16 GB memory under Ubuntu 10.04 OS. 6 pre-clustered benchmark circuits provided by [2] are listed in Table 2. The effective number of single-bit flip-flops (#FFs) ranges from 120 to 192,000. The bin size is 100×100 grids. The MBFF library is specified in Table 1. "Initial" lists the status of each input circuit.

Table 3 compares power, wirelength, and runtime among [11], [2] and INTEGRA. "Power ratio" (respectively "WL ratio") means the power (respectively wirelength) after MBFF

Table 2: Circuit Statistics.

Circuit	#FFs	Chip size (#Grids)	Initial	
			Power	Wirelength
C1	120	600×600	11,384	89,425
C2	480	1,200×1,200	46,404	348,920
C3	1,920	2,400×2,400	185,616	1,395,680
C4	5,880	4,200×4,200	566,972	4,290,655
C5	12,000	6,000×6,000	1,160,100	8,723,000
C6	192,000	24,000×24,000	18,561,600	139,568,000

Table 4: Comparison on FF Power and Clock Tree.

RISC32 CPU	[4]	Ours
# Single-bit FFs	3,689	75
# Dual-bit FFs	2,155	3,962
FF replacement rate	53.88%	99.06%
# Clock tree leaves	5,844	4,037
Clock tree synthesis report		
Normalized dynamic power for combinational ckt	1.000	1.009
Normalized dynamic power for clock buffers	1.000	0.789
Normalized dynamic power for FFs	1.000	0.933
# Clock subtrees	157	150
# Clock buffers	165	110
Depth of clock tree	5	5

clustering over that of the input design; "#Decision" represents the number of runtime decision points; "Avg. ratio" means the average speedup. "Lower bound" on power (respectively wirelength) is computed based on the 4-bit MBFF cell (respectively median fanin/fanout coordinates). We modify [11] to handle discrete MBFF libraries: When a clique in the intersection graph is found, we allocate an MBFF cell of the appropriate bit number and cluster flip-flops starting with the smallest degrees. Considering power saving, INTEGRA generates solutions with only 0.17% away from the lower bounds, while the modified version of [11] and [2] are 0.60% and 2.36% away, respectively. Keeping wirelength almost unchanged, INTEGRA can outperform modified [11] and [2] with 358.6X and 16.9X speedup. In particular, prior works suffer from long runtimes for $C6$, which is a large-scale design with numerous bins/grids. Moreover, the number of runtime decision points is significantly smaller than the number of flip-flops, on average 12% of flip-flop count.

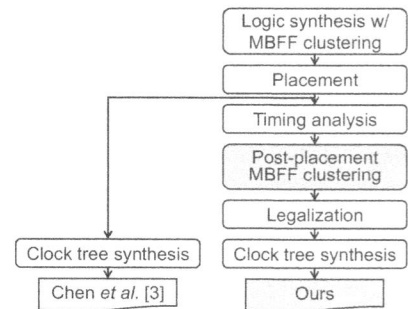

Figure 8: The experimental flow.

Table 3: Comparison on Power, Wirelength, and Runtime with [11] and [2].

Circuit	Lower bound		Modified [11]			[2]			Ours			
	Power ratio	WL ratio	Power ratio	WL ratio	Time (s)	Power ratio	WL ratio	Time (s)	Power ratio	WL ratio	#Decision	Time (s)
C1	82.2%	48.7%	82.8%	123.0%	0.03	85.2%	91.7%	< 0.01	82.8%	96.4%	28	< 0.01
C2	80.7%	49.9%	81.2%	124.8%	0.11	83.1%	94.7%	0.02	80.9%	102.0%	90	< 0.01
C3	80.7%	49.9%	81.3%	125.2%	0.53	82.9%	94.8%	0.07	80.8%	103.6%	229	< 0.01
C4	80.9%	49.7%	81.5%	124.7%	2.55	83.2%	94.5%	0.23	81.0%	104.1%	458	0.02
C5	80.7%	49.9%	81.3%	124.2%	8.01	82.9%	94.9%	0.52	80.7%	104.8%	690	0.05
C6	80.7%	49.9%	81.3%	124.4%	1994.61	82.8%	94.9%	76.94	80.7%	105.3%	3,007	1.11
Avg. ratio	-	-	-	-	358.61	-	-	16.87	-	-	-	1.00

To demonstrate the effectiveness of post-placement MBFF clustering, we apply the experimental flow shown in Figure 8 to a 32-bit RISC CPU design, which is pre-clustered at logic synthesis with 55 nm process by [3]. Logic synthesis is done by Synopsys Design Compiler [9], static timing analysis is reported by Synopsys Prime Time [9], and placement (including legalization) and clock tree synthesis are generated by Cadence SoC Encounter [1]. As listed in Table 4, the MBFF library contains only single-bit and dual-bit flip-flop cells [3]. 54% of the single-bit flip-flops in the initial design can be replaced by [3], while 99% can be replaced by IN-TEGRA. Our dynamic power consumed by flip-flops almost reaches the lower bound specified by the MBFF library. According to the clock tree synthesis report, compared with [3], we gain 21% saving on the clock buffer power. Only less than 1% overhead on the combinational dynamic power is paid due to MBFF clustering and legalization. Since clock power dominates the chip dynamic power, this overhead is negligible. The MBFF clustering results pass the static timing check, so our timing model is reasonable, and legalization just generates small displacement. Moreover, the number of leaves (5844 vs. 4037) is greatly reduced, the number of clock subtrees (157 vs. 150) and the number of clock buffers (165 vs. 110) are decreased to maintain the same clock skew, 300 ps. Hence, MBFF clustering can not only save the dynamic power and area of flip-flops but also reduce the power and area of the whole clock network. Compared with MBFF clustering at logic synthesis [3], it can be seen that the impact on combinational dynamic power is relatively small, while power of the whole clock network can considerably be improved by post-placement MBFF clustering, even under timing and placement density constraints.

6. CONCLUSIONS

In this paper, we present a fast post-placement multi-bit flip-flop clustering algorithm for clock power saving. Based on coordinate transformation and interval graphs, we adopt a pair of linear-size sequences as the representation. The concept of decision points helps us significantly reduce the times of clustering applied. Compared with prior work applying MBFF clustering at post-placement and early design stages, our results show the superior efficiency and effectiveness of our algorithm.

7. REFERENCES

[1] Cadence Design Systems, Inc. SoC Encounter.
[2] Y.-T. Chang, C.-C. Hsu, M. P.-H. Lin, Y.-W. Tsai, and S.-F. Chen. Post-placement power optimization with multi-bit flip-flops. In *Proc. ICCAD*, pages 218–223, 2010.
[3] L. Chen, A. Hung, H.-M. Chen, E. Y.-W. Tsai, S.-H. Chen, M.-H. Ku, and C.-C. Chen. Using multi-bit flip-flop for clock power saving by designcompiler. In *Proc. SNUG*, 2010.
[4] Y.-P. Chen, J.-W. Fang, and Y.-W. Chang. ECO timing optimization using spare cells and technology remapping. In *Proc. ICCAD*, pages 530–535, 2007.
[5] W. Hou, D. Liu, and P.-H. Ho. Automatic register banking for low-power clock trees. In *Proc. ISQED*, 2009.
[6] J. Kleinberg and E. Tardos. *Algorithm Design*. Addison Wesley, 2006.
[7] R. R. Pokala, R. A. Feretich, and R. W. McGuffin. Physical synthesis for performance optimization. In *Proc. ASIC*, pages 34–37, 1992.
[8] G. Ramalingam and C. P. Rangan. A unified approach to domination problems on interval graphs. *Information Processing Letters*, 27(5):271–274, 1998.
[9] Synopsys, Inc. Design Compiler / Prime Time.
[10] L.-T. Wang, Y.-W. Chang, and K.-T. Cheng, editors. *Electronic Design Automation: Synthesis, Verification, and Test*. Elsevier/Morgan Kaufmann, 2009.
[11] J.-T. Yan and Z.-W. Chen. Construction of constrained multi-bit flip-flops for clock power reduction. In *Proc. ICGCS*, pages 675–678, 2010.

Obstacle-aware Clock-tree Shaping during Placement

Dong-Jin Lee
University of Michigan
2260 Hayward Street
Ann Arbor, MI 48109-2121
ejdjsy@umich.edu

Igor L. Markov
University of Michigan
2260 Hayward Street
Ann Arbor, MI 48109-2121
imarkov@eecs.umich.edu

ABSTRACT

Traditional IC design flows optimize clock networks before signal-net routing and are limited by the quality of register placement. Existing publications also reflect this bias and focus mostly on clock routing. The few known techniques for register placement exhibit significant limitations and do not account for recent progress in large-scale placement and obstacle-aware clock-network synthesis.

In this work, we integrate clock network synthesis within global placement by optimizing register locations. We propose the following techniques: (1) obstacle-aware virtual clock-tree synthesis; (2) arboreal clock-net contraction force with virtual-node insertion, which can handle multiple clock domains and gated clocks; (3) an obstacle-avoidance force. Our work is validated on large-size benchmarks with numerous macro blocks. Experimental results show that our software implementation, called Lopper, prunes clock-tree branches to reduce their length by 30.0%~36.6% and average total dynamic power consumption by 6.8%~11.6% versus conventional approaches.

Categories and Subject Descriptors

B.7.2 [**Integrated Circuits**]: Design Aids—*placement and routing*

General Terms

Algorithms, Design

1. INTRODUCTION

Power consumption is one of the primary optimization objectives for modern IC designs [21]. It includes three basic components: *short-circuit* power, *leakage* power and *net-switching* power [13]. Net-switching power is usually the largest contributor, and clock networks are often responsible for over 30% of total power consumption due to their high capacitance and frequent switching [5,6,16,26]. The quality of clock networks is greatly affected by register placement, but mainstream literature on placement and most commercial EDA tools have largely overlooked this fact by focusing

on wirelength of signal nets [10], routability [29] and circuit timing [7]. As far as we know, high-quality register placement cannot be achieved by easy pre- or post-processing of existing techniques. To this end, most appropriate changes to cell locations that reduce the clock network may depend on the current structure of the clock network, which is not accounted for in existing placement tools.

Our analysis of prior work reveals serious limitations in published techniques Some methods coerce the placer into shortening the clock tree by capturing portions of the clock tree with the half-perimeter wirelength (HPWL) objective, which is usually applied only to signal nets [4,30]. This idea overlooks the fact that low-skew clock trees exhibit much greater wirelength than signal nets with the same bounding box. To make matters worse, the HPWL estimate does not offer much fidelity for clock-tree lengths, as we show in Figure 2. Furthermore, a handful of existing publications that optimize clock networks during placement (reviewed in Section 2) do not reflect recent progress in large-scale placement and clock-network synthesis, and do not compare their results with best-of-breed software. In most cases, they are evaluated on small benchmarks without routing/buffering obstacles rather than on modern ASIC or SoC designs with many macro blocks. *Our research addresses these gaps in the literature by developing a set of new techniques for clock-net optimization during placement and evaluating these techniques against leading academic software.* We extended the ISPD 2005 benchmark suite toward clock-network synthesis, with the largest benchmark including 2.1M standard cells and 327K registers. The benchmarks include numerous macros, which we interpret as routing obstacles.

To optimize the trade-off between clock network minimization and traditional placement objectives, we propose a new placement methodology based on *obstacle-aware virtual clock-tree synthesis* that extends force-directed placement by adding *a arboreal clock-net force* using virtual nodes. *A key challenge addressed in our work is preserving the quality of global placement when adding clock-net optimizations.* We also accommodate multiple clock domains and gated clocks. Our algorithms are integrated into the SimPL placer [9], which currently produces lowest-wirelength placements on the ISPD'05 benchmarks. The quality of register placement is evaluated by Contango 2.0 [12] – the winner of the ISPD 2010 contest. Experimental results show that our method can reduce clock-network capacitance by 30.0%~36.6% while reducing the overall dynamic power of the IC by 6.8%~11.6% compared to conventional approaches.

Permission to make digital or hard copies of all or part of this work for personal or classroom use is granted without fee provided that copies are not made or distributed for profit or commercial advantage and that copies bear this notice and the full citation on the first page. To copy otherwise, to republish, to post on servers or to redistribute to lists, requires prior specific permission and/or a fee.
ISPD'11, March 27–30, 2011, Santa Barbara, California, USA.
Copyright 2011 ACM 978-1-4503-0550-1/11/03 ...$10.00.

2. PRIOR WORK

Recent clock-network synthesis tools often construct initial trees with a simple delay model (e.g., Elmore) and then perform SPICE-accurate tuning [11, 12, 14, 22].

Clock-network optimization after placement can be performed by clustering nearby flip-flops [3, 20] to share inverters (inside flip-flops) and shorten the clock tree. This clustering does not adversely affect signal nets, but is rather limited by the locations of combinational gates. In high-performance CPUs flip-flops are often replaced by single latches, which reduces savings from clock-sink clustering.

Clock-network optimization during placement. To address the apparent conflict between clock-net optimization and traditional placement objectives, some researchers proposed techniques and algorithms for better register placement without intrusive interference in traditional placement objectives. Lu [15] proposed several techniques including Manhattan ring-based register guidance, center-of-gravity constraints for registers, pseudo-pins and register-cluster contraction. Cheon [4] proposed power-aware placement that performs both activity-based register clustering and activity-based net weighting to simultaneously reduce the clock and signal net-switching power. In order to reduce the clock network size, Wang [30] proposed dynamic clock-tree building (DCTB), multi level bounding box (MLBB) and multi level attractive force (MLAF), and integrated them into a force-directed placement (FDP) framework [28].

Limitations of existing techniques. Clock-net optimization during placement seeks better register locations but should not harm total wirelength of signal nets. A naive method is to increase the weight of the clock net and pull all registers together. Unfortunately, this method increases routing congestion and hot spots, and also leads to poor signal-net wirelength when dealing with more than several hundred registers [4, 30]. To definitively resolve the conflict between clock-net minimization and traditional placement objectives, careful problem formulation is essential.

Prior approaches to clock-net minimization in placement form two families. *Manhattan-ring guidance methods* commit registers to certain guidance locations and try to pull the registers close to the nearest such locations during placement [15]. However, such methods do poorly in the presence of numerous obstacles, e.g., macro-blocks, or when register locations found by the global placer are not uniformly distributed. In other words, guidance rings cannot accurately predict ideal locations for register clusters. Figure 1 illustrates how Manhattan-ring methods fail. In Figure 1(b), the sink group A is attracted by the closest Manhattan ring. The sinks in A are erroneously guided toward the obstacle. The sink group B and the related standard cells have heavy connections to the bottom macro block. However, the two bottom Manhattan rings encourage the sinks in B to move away from the center of B, which will likely increase signal-net wirelength significantly.

The second family of approaches performs clock-network synthesis using register locations from intermediate placement results. Specific techniques [4, 30] often simplify the structure of the clock network and bias the placement process to optimize such simplified networks. However, clock trees generated by those techniques are not realistic and very different from those generated by leading software. In the DCTB algorithm [30], the essential parameters of clock network synthesis, such as sink capacitance and wire capac-

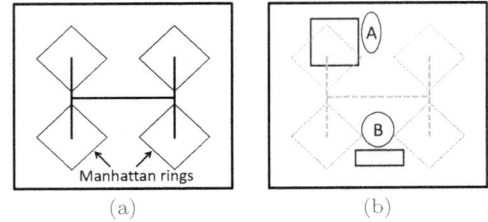

Figure 1: Two examples of Manhattan rings proposed in [15]. (a) Zero-skew Manhattan rings driven by an H-tree. (b) Manhattan rings on the design with obstacles. Obstacles are indicated by darker boxes, two sink groups (A, B) are represented as ellipses. The locations of the sinks are assumed to be decided based on wirelength-driven placement.

itance/resistance, are ignored, and the cost function is derived by only considering Manhattan length between sinks or nodes. The quick CTS algorithm in [4] is also much simpler than standard DME algorithms, which minimize wirelength with zero or bounded skew based on Elmore delay. Furthermore, all previous work ignores the presence of routing obstacles, common in modern IC designs, and this ignorance can undermine end results (Sections 4 and 6).

Previous publications that simplify clock-tree synthesis during placement [4, 30] cluster clock trees and represent these clusters with bounding boxes to model clock network reduction by placement objectives. Typically, registers are clustered at one or multiple levels based on the structure of the reference (simplified) clock tree, and bounding boxes are created for each cluster. The experimental results of [4, 30] show that bounding boxes are helpful for clock-net size reduction. However, we argue below that this method fails to represent clock-net reduction problem in placement.

Bounding boxes are represented by fake nets during placement and are optimized to reduce HPWL [9, 24]. The HPWL objective is relevant to placement because it estimates the lengths of signal routes reasonably well. However, clock routing is very different from signal-net routing and requires longer routes to ensure low skew. Therefore, HPWL does not offer accurate estimates of clock-tree lengths. Figure 2 shows that reducing HPWL of the clock net may increase the total length of the clock tree, demonstrating that the HPWL estimates lack not only accuracy, but also fidelity.

The authors of [30] adapted MLAF to compensate for the drawback of MLBB. However, we show in Section 4.2 that MLAF offers only a partial solution to this problem.

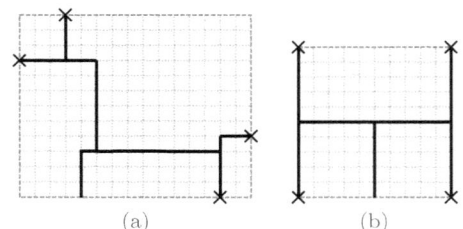

Figure 2: Bounding boxes of two partial ZST-DME clock trees. (a) HPWL of the bounding box is (15+12)=27. The total wirelength of the inside clock tree is 32. (b) HPWL is (10+10)=20 and the total wirelength of the clock tree is 35. The clock-net wirelength of (b) is greater than (a) although the bounding-box HPWL of (b) is notably smaller than (a) while the source-to-sink wirelength is 15 for all sinks.

3. OPTIMIZATION OBJECTIVE

Let $\mathcal{N}$ be the set of signal nets, and let $\mathcal{E}$ be the set of clock-net edges. To optimize clock networks in placement, we minimize the total switching power P_{sw}, defined as the sum of $\mathcal{N}$'s switching power $P_{\mathcal{N}}$ and $\mathcal{E}$'s switching power $P_{\mathcal{E}}$

$$P_{sw} = P_{\mathcal{N}} + P_{\mathcal{E}} \qquad (1)$$

If activity factors of signal nets and clock-net edges are available, then the total signal-net switching power is

$$P_{\mathcal{N}} = \sum_{n_i \in \mathcal{N}} \alpha_{n_i} HPWL_{n_i} C_n V^2 f \qquad (2)$$

and the total clock-net switching power is

$$P_{\mathcal{E}} = \sum_{e_i \in \mathcal{E}} \alpha_{e_i} L_{e_i} C_e V^2 f \qquad (3)$$

Here, α_{n_i} and α_{e_i} are the respective signal-net and clock-edge activity factors C_n and C_e are the respective unit capacitance for signal and clock wires, V is the supply voltage, f is the clock frequency, $HPWL_{n_i}$ is the HPWL of net n_i, and L_{e_i} is the Manhattan length of edge e_i. Activity factors of clock-net edges are required when multiple clock domains or gated clocks are utilized for given designs, otherwise $\alpha_{e_i} = 1$ as clock edges switch every clock cycle. The handling of gated clocks is discussed in Section 5 in more detail. If the activity factors of signal nets are not available, the computation of total switching power relies on *clock-power ratio* β, i.e., clock-net switching power divided by total switching power. In this case, the average activity factor of signal-net α_{avg} can be derived as

$$\alpha_{avg} = \frac{(1-\beta)\sum_{e_i \in \mathcal{E}} L_{e_i} C_e}{\beta \sum_{n_i \in \mathcal{N}} HPWL_{n_i} C_n} \qquad (4)$$

α_{avg} is utilized for the activity factors of all the signal nets.

4. PROPOSED TECHNIQUES

We propose a methodology and several new techniques to overcome limitations of prior work and reliably optimize large IC designs with numerous layout obstacles. Our approach consists of two major phases: (*i*) virtual clock-tree synthesis, (*ii*) arboreal clock-net contraction force, which is corrected by an obstacle-avoidance force.

4.1 Obstacle-aware virtual clock trees

Our virtual clock-tree synthesis handles macro blocks as wiring obstacles and produces obstacle-avoiding clock trees. The importance of utilizing obstacle-aware clock trees is illustrated in Figure 3 (the contraction forces are described in Section 4.2). Clock-net optimizations without obstacle handling pull clock sinks inside obstacles, which undermines global placement.

Experimental results in [12] show that the difference in total capacitance between initial zero-skew DME trees (based on Elmore delay) and the final SPICE-optimized trees is only 2.2% on average. Hence, initial trees produced by leading clock-network synthesis tools offer reasonably accurate capacitance estimates. To quickly construct a *virtual clock-tree* during placement, our methodology first performs traditional DME-based zero-skew clock-tree synthesis with Elmore delay model, subject to obstacle avoidance. Several techniques are known for this problem, including direct

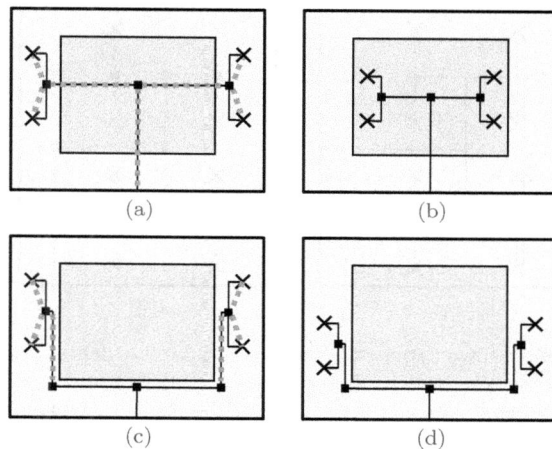

Figure 3: An example of clock-net optimization with an obstacle. (a) The virtual clock tree and corresponding contraction forces are created without considering the obstacle. (b) The result of a placement iteration with the forces in (a). (c) The obstacle is accounted during virtual clock-tree generation and when establishing additional forces. (d) The result of (c).

obstacle-avoiding clock-tree construction [8] and incremental repair of obstacle-unaware trees [11]. Each approach can be used in our methodology, but we found that incremental-repair techniques are simpler and yet produce high-quality trees.[1] Our clock trees target the $45\ nm$ technology used at the ISPD 2010 clock network synthesis contest [25].

4.2 Arboreal clock-net contraction force

If the virtual clock network connecting to current register locations faithfully represents a realistic clock network, then optimizing it directly should improve the final clock network produced by a specialized CTS tool after placement is complete. To this end, we extend force-directed placement with new, structurally-defined forces that seek to reduce individual edges of the virtual clock network. This technique communicates current clock-tree structure to the placement algorithm, and also allows the structure to change with placement.

Figure 4(a) illustrates a sample virtual clock tree. To reduce the length of ϵ_1 directly, all sinks downstream from e_1 can be moved in the direction of reducing the length of e_1. For each downstream sink of e_1, a force vector needs to be assigned. The force vectors created for e_1 should not affect other tree edges.

The sum of magnitudes of force vectors induced by e_1 ($F_{e_i}^{sum}$) needs to be carefully controlled to avoid excessive increase in signal-net wirelength. $F_{e_i}^{sum}$ may vary when the activity factors of clock edges differ (e.g., in gated clocks). Figure 4(a) illustrates force vectors. The force from e_1 is weaker than the force from e_2, $F_{e_1} < F_{e_2}$ since the sum of magnitudes should be same.

The main problem with this method is that the relative

---

[1]Extensive empirical studies and the experience of ISPD clock-network synthesis contests suggest that when clock sinks are placed outside the obstacles, the overlaps caused by obstacle-unaware trees can often be fixed with minimal impact on skew and total capacitance, compared to obstacle-aware trees.

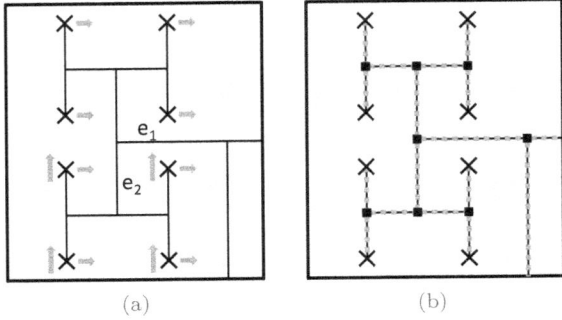

Figure 4: Two types of forces for clock-net optimization. Registers are indicated by crosses. (a) For each edge, the corresponding downstream registers are given force vectors. Right arrows are the force vectors for reducing e_1, and up arrows are the force vectors for reducing e_2. (b) Virtual nodes are inserted (squares), and forces are created between each pair of connected nodes (dotted lines).

locations of branching nodes from sinks are assumed to be same when the force vectors are created. However, optimal relative locations of the branching nodes change during the optimization. Therefore, placement iterations with fixed force vectors for sinks do not produce optimal locations.

To shorten clock wires, we propose *a arboreal clock-net contraction force with virtual-node insertion*. Our approach creates forces between clock-tree nodes and structurally transfer the forces down to registers. Virtual nodes represent branching nodes in the clock tree and split the clock tree into individual edges, seen as different nets by the placement algorithm. The virtual nodes have zero area and do not create overlap with real cells, so they do not affect the spreading process in force-directed placers. Zero-area nodes may or may not be allowed to overlap with obstacles (if such a node is placed over an obstacle, its overlap has zero area). In our case, virtual nodes should not be placed over obstacles to avoid routing over obstacles.

Compared to the fixed force vectors applied exclusively to sinks, our technique creates forces between flexible nodes and each force seeks to reduce the length of the corresponding clock edge. Unlike in the bounding-box based method, each force is integrated into the placement instance as a two-pin pseudo net, as shown in Figure 4(b).

To reduce dynamic power consumption of the IC, contraction forces are calculated based on the activity factors of the signal nets. When activity factors of signal nets are available, the average activity factor α_{avg} over all nets is

$$\alpha_{avg} = \frac{\sum_{n_i \in \mathcal{N}} \alpha_{n_i} HPWL_{n_i}}{\sum_{n_i \in \mathcal{N}} HPWL_{n_i}} \quad (5)$$

Otherwise, Equation 4 is utilized to compute α_{avg}. A two-pin net representing clock-net contraction forces for clock edge e_i is given a weight

$$w_{e_i} = \frac{C_e \alpha_{e_i}}{C_n \alpha_{avg}} \quad (6)$$

and the HPWL of a two-pin net from e_i is equal to the Manhattan length of e_i,

$$L_{e_i} = HPWL_{e_i} \quad (7)$$

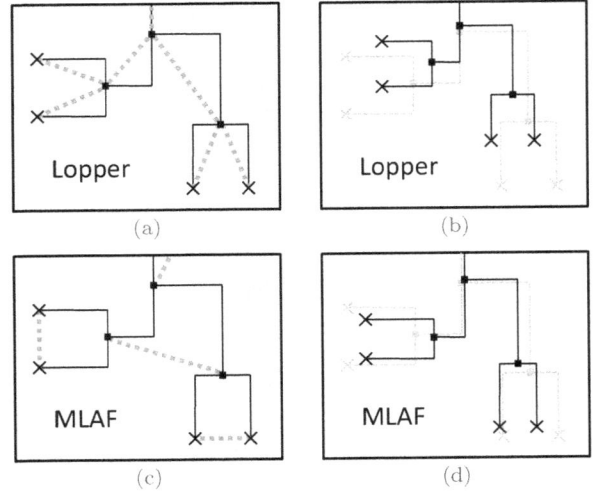

Figure 5: Comparison between our arboreal clock-net contraction force and MLAF of [30]. (a) Arboreal clock-net contraction forces are generated. (b) The modified register and virtual clock-node locations when forces in (a) are utilized. (c) The forces created by the MLAF algorithm. (d) The modified register and virtual clock-node locations when forces in (c) are utilized. We can observe that the edges between parents and children nodes are poorly handled for the force creation in (c), and our method is more efficient on non H-tree structures (which is common in modern designs).

Combining Equations 2, 3, 5 and 7 yields

$$\left(\sum_{n_i \in \mathcal{N}} \alpha_{avg} HPWL_{n_i} C_n + \sum_{e_i \in \mathcal{E}} \alpha_{e_i} HPWL_{e_i} C_e \right) V^2 f \quad (8)$$

By substituting α_{e_i} in terms of w_{e_i} (Equation 6), Equation 8 can be rewritten as

$$\left(\sum_{n_i \in \mathcal{N}} \alpha_{avg} HPWL_{n_i} C_n + \sum_{e_i \in \mathcal{E}} \alpha_{avg} w_{e_i} HPWL_{e_i} C_n \right) V^2 f \quad (9)$$

Let K be $\alpha_{avg} C_n V^2 f$, $\mathcal{M} = \mathcal{N} \cup \mathcal{E}$ and the weight value of signal net n_i be $w_{n_i} = 1$. Then,

$$P_{sw} = P_{\mathcal{N}} + P_{\mathcal{E}} = K \sum_{m_i \in \mathcal{M}} w_{m_i} HPWL_{m_i} \quad (10)$$

In other words, our techniques capture the switching-power minimization problem, which can be solved by any high-quality wirelength-driven placer capable of net weighting. Figure 5 compares our technique and MLAF from [30]. MLAF is ineffective in shortening clock nets that significantly differ from H-trees.

4.3 Obstacle-avoidance force

Given an obstacle-avoiding tree, we modify arboreal clock-net contraction forces to promote obstacle avoidance. Contraction forces based on an obstacle-avoiding clock tree do not necessarily improve every tree edge. as shown in Figure 6. In Figure 6(a), five edges are derived from a virtual obstacle-aware tree built as in Section 4.1. If we create forces for all the edges, subsequent optimization will produce the tree in Figure 6(b). The force f_4 associated with edge e_4 is rendered ineffective by the obstacle. Our force-modification algorithm for obstacle avoidance detects these obstacle-detouring edges and eliminates the contraction forces for them.[2] In this example, e_4 and e_5 are excluded from force construction, and the result is illustrated in Figure 6(c).

5. PROPOSED METHODOLOGY

We integrate our techniques into SimPL, a flat, force-directed quadratic placer [9]. Recall that analytic placers first minimize a function of interconnect length, neglecting overlaps between standard cells and macros. This initial step places many cells in densely populated regions. Clock-net contraction forces are ineffective at this step for two reasons: (i) the current virtual clock network may differ greatly from the final clock network. (ii) the contraction forces may restrict the spreading of the registers at the center of the design due to their high net weight. Therefore, our techniques are invoked between signal-net wirelength-driven global placement and detailed placement (including legalization).

Our clock-net optimization during placement is referred to as *Lopper*, and described in Figure 7.

5.1 The Lopper flow

At each iteration of Lopper, a new virtual clock tree is generated based on current register locations. We discard the previous virtual clock tree based on the following observation. The topology of a clock tree and the embedding of its wires minimize (i) skew as the primary objective, (ii) total wirelength as the secondary objective. When an iteration of Lopper is performed, the locations of the registers are modified in order to reduce the total wirelength of the given virtual clock tree. Since registers are displaced by different amounts (due to different connectivities), keeping the previous clock-tree structure would risk a large increase in skew. Therefore we regenerate the virtual clock tree for each iteration to obtain an optimal virtual clock tree with the current register locations. The tree topology typically undergoes only moderate changes, while branching nodes relocate to reduce skew.

Early placement iterations may greatly displace the registers, moving them over the obstacles in some cases. Therefore, Lopper ignores obstacles until average displacement of registers becomes small.

Global placement typically continues while HPWL continues improving, but clock-tree reduction in Lopper re-

---

[2]Consider a clock-tree edge that does not cross a given obstacle. The edge *detours* the obstacle if the straight line connecting the ends of the edge crosses the obstacle.

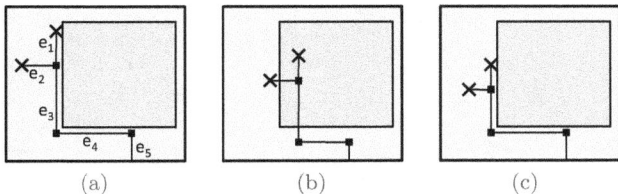

Figure 6: Obstacle-avoidance force. (a) Five edges of an obstacle-aware virtual clock tree. (b) The result when all the edges are utilized for contraction forces. (c) The result when e_4 and e_5 are excluded from force construction.

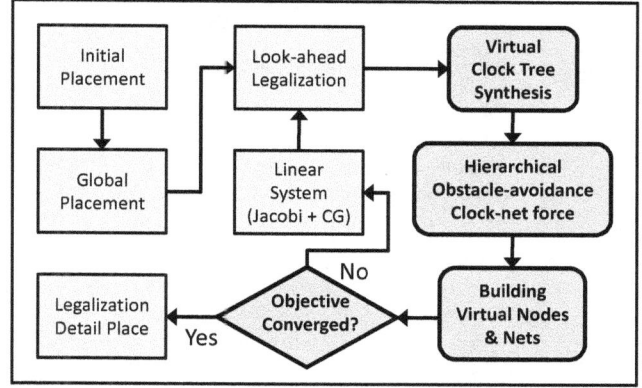

Figure 7: Key steps of Lopper integrated into the SimPL placer, as indicated with darker rounded boxes and a lozenge. Plain boxes represent the SimPL steps.

quires a different convergence criterion. After each iteration, total switching power is calculated and compared to previous values. Lopper is invoked repeatedly until total switching power (Equation 1) stops reducing.

Legalization and detailed placement are applied after Lopper is complete. It is important to preserve the virtual nodes and two-pin nets that represent the clock-net contraction forces during detailed placement because detailed placement algorithms usually optimize wirelength and would not have preserved clock-optimized register locations if guided only by signal nets.

5.2 Trade-offs and additional features

Quality control. Our techniques reduce the size of clock networks, but are likely to increase signal-net wirelength. The activity factor of each signal-net α_{n_i} or clock-power ratio β are required for Lopper to reduce total switching power. However, even clock-power ratio β is hard to estimate before the design is completed and can vary with various applications running on a CPU. Therefore, in our implementation the trade-off between clock-net and signal-net switching power can be easily controlled with a single parameter β. This simple quality control allows an IC designer to achieve intended total switching power of a chip without changing the algorithm or its internal parameters. Relevant trade-offs are illustrated in Table 2.

Gated clocks and multiple clock domains. Clock gating is a well-known and often the most effective approach to reduce clock network power dissipation [19]. To extend our techniques to gated clocks and multiple clock domains, each register s_i is given an activity factor α_{s_i} and the activity factors are propagated through the tree. The activity factor of an edge is the highest activity factor of its child edge or register (see Figure 8).

Once activity factors are propagated to tree edges in each clock tree, they are used to calculate net weights that represent clock-net contraction forces in Equation 6. Registers that switch less frequently due to clock gating will be more affected by signal nets than normal registers without clock gating. Our technique does not track the locations of gators assuming that the final clock tree and the gators are constructed after register placement. While we have not experimented with gator placement, we do not believe that it will affect results reported in our work.

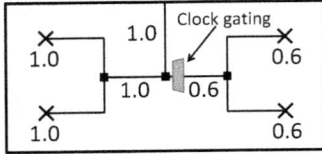

Figure 8: Activity-factor propagation for gated clocks. Registers are indicated with crosses. Tree edges and registers are labeled with activity factors.

Flexible integration. Through the Lopper flow, forces for clock-net optimization are represented in placement instances by virtual nodes and nets. No support for clock-net optimization is required in the placement algorithm. Therefore, Lopper can integrate any fast obstacle-aware clock-tree synthesis technique into any iterative high-performance wirelength-driven placer capable of net weighting.

6. EMPIRICAL VALIDATION

The benchmarks used in prior publications on clock-tree optimization during placement exhibit the following problems: (1) Empirical validation of each existing publication relies on one benchmark suite which is not utilized by any other work. Most of the benchmarks are inaccessible to public, therefore comparisons to new techniques are impossible. (2) The benchmark designs are based on unrealistically small placement instances. (3) Macro blocks became essential components, and many IC designs include more than hundreds of macros with fixed locations after floorplanning [1]. However, prior publications used the benchmarks without macro blocks or ignored macro blocks present in the benchmarks [30]. (4) Reference placement tools used for comparison are often outdated [15] or self-implemented [30]. Such comparisons risk not being representative of state-of-the-art EDA tools.

In this section, we propose a new benchmark set that addresses the above pitfalls. Our experimental results offer full comparisons with leading academic wirelength-driven placers and a known technique for register placement (MLAF). The quality of register locations is validated by a leading academic clock-network synthesis tool.

6.1 Experimental setup

The ISPD 2005 placement contest benchmark suite is being used extensively in placement research, and the academic community consistently advanced physical design techniques using the ISPD'05 benchmarks. These benchmarks are directly derived from industrial ASIC designs, with circuit sizes ranging from 210K to 2.1M placeable objects. We adapted eight designs from the ISPD'05 benchmarks and created register lists in which 15% of standard cells are selected to be registers. We selected the number 15% based on the industrial designs introduced in [4], where the average 14.65% of cells are registers. The largest benchmark has 327K registers. Fixed macro blocks are viewed as routing blockages during clock-network synthesis. The benchmarks are mapped to the Nangate 45 nm open cell library [18] to facilitate clock-network synthesis with parameters from ISPD 2010 CNS contest. The standard-cell height (or row height) is set to 1.4 μm according to the 45 nm library. Clock-power ratio β is set to 0.3 for clock network optimization during placement based on the industrial circuits from [4], where clock power is responsible for 31.9% of total power on average. The unit-wire capacitances for signal-net and clock-

Name	Cells	Regs	Macros	CoreX (mm)	CoreY (mm)	Area (mm^2)
clkad1	210K	32K	56	1.247	1.246	1.554
clkad2	255K	38K	177	1.640	1.638	2.686
clkad3	451K	68K	721	2.706	2.722	7.363
clkad4	494K	74K	1329	2.706	2.722	7.363
clkbb1	278K	42K	30	1.247	1.246	1.554
clkbb2	535K	84K	923	2.181	2.192	4.781
clkbb3	1095K	165K	666	3.231	3.242	10.47
clkbb4	2169K	327K	639	3.756	3.772	14.16

Table 1: The new CLKISPD'05 benchmarks.

net (C_n, C_e) are set to $0.1fF/\mu m$, $0.2fF/\mu m$ respectively based on the 45 nm technology model from the ISPD'10 contest [25] and the Nangate open-cell library [18]. Supply voltage and clock frequency are set to $1.0V$ and $2GHz$. The coordinate of clock source is set to the bottom left corner of core area except when it is blocked by macros. When the desired location is blocked, we move the clock source to the closest unblocked coordinate. Since many academic placers handle the ISPD'05 benchmarks, a direct comparison of clock-network quality and signal-net wirelength is possible. The new benchmarks (referred to as *CLKISPD'05*) are described in Table 1.

The quality of clock networks based on the final register locations of each placer is evaluated by Contango 2.0 [12]. Contango 2.0 is the winner of the ISPD 2009 and 2010 Clock Network Synthesis (CNS) contests and produces clock trees with less than 7.5 ps skew in the presence of variation on the ISPD'10 CNS benchmarks. During our experiments, we exclude SPICE-accurate tuning in Contango 2.0 for two reasons: (1) the designs from the ISPD'05 benchmarks are too large to run SPICE simulations, (2) the average added capacitance during the SPICE-driven optimization on the ISPD'10 CNS benchmarks is 2.2% of total clock-net capacitance (including sink, wire and buffer capacitance), suggesting that the initial trees optimized for Elmore delay provide good estimates of power consumption.

6.2 Empirical results

Table 5 compares results of our methodology to the leading academic placers on the CLKISPD'05 benchmarks. The results of SimPL [9] are used as reference for comparison. α_{avg} is computed for each benchmark based on the given $\beta = 0.3$, and total wire-switching power is calculated based

β	α_{avg}	Orig. P (mW)	ClkWL (mm)	HPWL (m)	Pwr (mW)	(Rel)
Orig	-	-	209.1	8.968	-	-
0.1	0.420	837.0	184.2	9.073	**835.8**	0.999
0.15	0.264	557.2	173.5	9.128	**551.3**	0.990
0.2	0.187	419.1	165.7	9.188	**409.9**	0.978
0.25	0.140	334.8	158.0	9.225	**321.5**	0.960
0.3	0.109	279.9	152.3	9.233	**262.2**	0.939
0.35	0.087	239.7	151.0	9.280	**221.9**	0.925
0.4	0.070	209.2	144.8	9.305	**188.2**	0.900
0.45	0.057	185.9	144.5	9.316	**164.0**	0.882
0.5	0.047	168.0	139.5	9.342	**143.6**	0.854
0.55	0.038	151.8	135.7	9.343	**125.3**	0.826
0.6	0.031	139.3	128.0	9.425	**109.6**	0.787

Table 2: The results on *clkad1* with various clock power ratios β. The specifications of the reference placement produced by SimPL are in the row *Orig*. α_{avg} is calculated based on β and reference placement produced by SimPL. Total wire-switching power values of the reference placement with the corresponding β are represented in the column *Orig. P*. The relative power ratios are indicated with *Rel*.

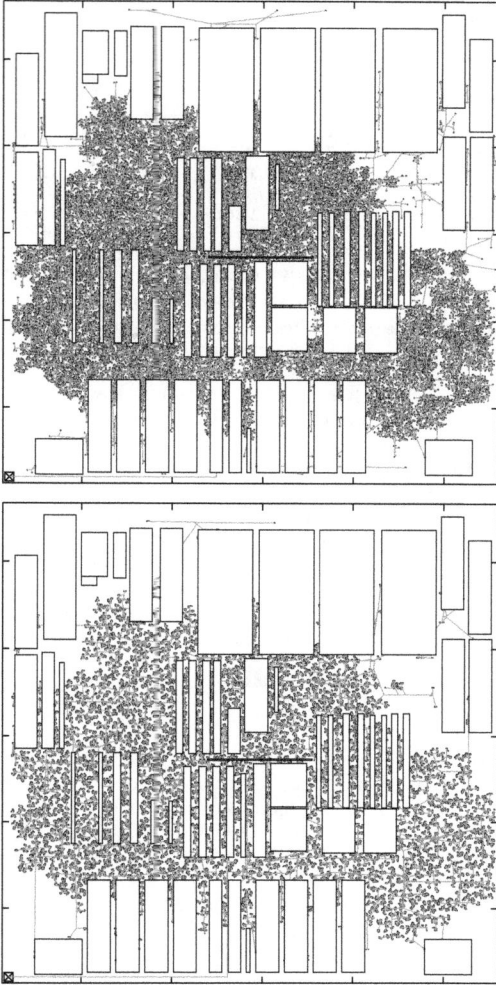

Figure 9: Clock trees for clkad1, based on a SimPL register placement (top) and produced by proposed techniques (bottom). The respective clock-tree wirelengths based on SimPL and our method are 209.13 mm **and 152.27** mm**. The total switching power of SimPL and our method are 279.9** mW **and 263.0** mW **respectively.**

Bench	Orig. Flow		w/o OAVCT		w/o OAF	
	ClkWL (mm)	Pwr (mW)	ClkWL (mm)	Pwr (mW)	ClkWL (mm)	Pwr (mW)
clkad1	152.3	263.0	165.7	267.8	158.5	265.3
clkad2	161.0	278.4	170.9	285.5	163.7	278.7
clkad3	326.9	583.0	362.1	595.1	340.8	587.4
clkad4	354.4	640.4	403.1	657.2	379.8	649.4
clkbb1	166.3	295.7	172.6	297.4	169.1	296.4
clkbb2	371.2	661.4	411.2	673.8	389.9	666.7
clkbb3	602.2	1085	663.1	1104	627.2	1093
clkbb4	1266	2279	1412	2331	1328	2102
Avg	$1.0\times$	$1.0\times$	+9.5%	+1.8%	+4.1%	+0.7%

Table 3: Impact of excluding obstacle-aware virtual clock trees (OAVCT), obstacle avoidance forces (OAF). OAVCT and OAF are excluded in the columns under "w/o OAVCT" and only the OAF step is removed in "w/o OAF"

switching power. For example, the result for $\beta = 0.6$ consumes 109.6 mW for total wire-switching power, but if the same circuit is used for the applications with $\beta = 0.1$, the total wire-switching power computed by Equations 1 - 3 is 842.9 mW, which is greater than the switching power of the reference placement 836.9 mW. This implies that clock-net optimization must utilize activity factors of signal nets or clock-power ratios to reduce total switching power.

Table 3 shows the impact of obstacle-aware virtual clock trees (OAVCT) and obstacle avoidance forces (OAF). When OAVCT is excluded, DME trees without obstacle handling are utilized for the remaining flow. The results indicate that 9.5% of clock-net wirelength can be reduced on average by utilizing obstacle-aware trees. The advantage of OAVCT is reduced on benchmarks with a few obstacles such as *clkbb1* where a few obstacles exist at the top left corner of the chip. Obstacle-avoidance forces reduce clock-net length by 4.1% and total switching power by 0.7%.

Table 4 compares results of our technique to the technique called MLAF on MLBB [30]. We re-implemented their MLAF algorithm and integrated it into the SimPL placer [9] instead of the FDP framework [28] they utilized. Since their DCTB algorithm cannot process obstacles, our obstacle-aware virtual clock-tree generation algorithm in Section 4.1 is utilized for the MLAF algorithm. In terms of clock-net wirelength and net-switching power, the average gain from the MLAF technique is limited by 43.5%, 30.6% of the improvement of our technique respectively, which means that our arboreal clock-net contraction force is 3.3$\times$ more effective for switching-power reduction than MLAF.

Bench	SimPL+MLAF		
	ClkWL (mm)	HPWL (m)	Pwr (mW)
clkad1	182.4 (46.9%)	9.194 (85.3%)	274.2 (33.7%)
clkad2	200.9 (35.8%)	10.76 (76.2%)	293.0 (24.0%)
clkad3	402.5 (46.6%)	24.71 (76.9%)	609.8 (35.7%)
clkad4	449.5 (42.4%)	22.24 (86.9%)	676.6 (30.7%)
clkbb1	203.8 (47.9%)	11.48 (84.9%)	309.7 (36.1%)
clkbb2	473.8 (36.7%)	17.16 (80.0%)	699.3 (23.4%)
clkbb3	743.5 (46.5%)	40.81 (91.0%)	1139 (22.9%)
clkbb4	1587 (45.5%)	94.77 (80.2%)	2399 (38.1%)
Avg	(43.5%)	(82.7%)	(30.6%)

Table 4: Results of the MLAF technique integrated into SimPL with comparison to our technique. The numbers in parentheses represent the amount of reduction(ClkWL, Pwr)/increase(HPWL) when the amount of reduction/increase of our technique is 100%.

on α_{avg}. On average, the combination of SimPL and Lopper reduces total clock-tree length by 30.0%, total wire-switching power by 6.8% while the total HPWL of the signal nets only increases by 3.1% compared to SimPL. Compared to FastPlace3 [27] and mPL6 [2], our methodology reduces the total clock-net wirelength by 32.1%, 36.6%, total wire-switching power by 10.5%, 11.6% respectively, while the total signal-net HPWL is smaller than that produced by Fast-Place3 by 1.4% and very similar to that produced by mPL6. Figure 9 compares two clock trees based on different register placements from SimPL and our method.

To further study the relative significance of clock-power ratio β, we show in Table 2 the impact of varying β on the benchmark *clkad1*. The average activity factor of signal nets α_{avg} is computed based on the reference layout and utilized for computing the total wire-switching power. The performance of Lopper is improved when clock networks consume a greater portion of total power. Table 2 also shows that reducing clock networks does not necessarily reduce the total

Bench	α_{avg}	FastPlace3			mPL6			SimPL 101			SimPL+Lopper			
		ClkWL (mm)	HPWL (m)	Pwr (mW)	ClkWL (mm)	HPWL (m)	Pwr (mW)	ClkWL (mm)	HPWL (m)	Pwr (mW)	ClkWL (mm)	HPWL (m)	Pwr (mW)	⊙ (min)
clkad1	0.109	214.7	9.119	285.5	248.2	9.092	298.3	209.1	8.968	279.9	**152.3**	9.233	**263.0**	4.30
clkad2	0.099	236.2	10.92	310.1	267.0	10.74	318.9	223.1	10.54	297.6	**161.0**	10.83	**278.4**	7.11
clkad3	0.091	469.3	24.95	640.8	467.6	24.99	640.8	468.5	24.08	624.7	**326.9**	24.90	**583.0**	13.4
clkad4	0.112	540.9	23.12	732.9	615.6	22.62	751.6	519.4	21.70	692.6	**354.4**	22.32	**640.4**	14.1
clkbb1	0.099	250.5	11.24	323.6	245.1	11.29	322.5	238.2	11.18	317.6	**166.3**	11.53	**295.7**	6.32
clkbb2	0.149	539.2	18.07	752.6	514.1	17.77	733.6	533.2	16.75	710.9	**371.2**	17.26	**661.4**	31.9
clkbb3	0.103	892.6	42.65	1236	1032	40.15	1240	866.3	39.22	1155	**602.2**	40.97	**1085**	35.3
clkbb4	0.093	1907	97.32	2575	2119	96.77	2650	1855	92.96	2473	**1266**	95.21	**2279**	110
Avg		1.03×	1.05×	1.04×	1.11×	1.03×	1.06×	1.00×	1.00×	1.00×	**0.70×**	1.03×	**0.93×**	

Table 5: Results on the CLKISPD'05 benchmark suite. ClkWL represents total wirelength of a clock network synthesized by the initial phase of Contango 2.0 [12]. HPWL is total HPWL of signal nets. Pwr is total net-switching power. SimPL+Lopper is 2.57× faster than mPL6 and 2.05×, 2.50× slower than FastPlace3, SimPL respectively.

7. CONCLUSIONS

Despite the increasing significance of power optimization in VLSI, state-of-the-art placement algorithms only optimize signal-net switching power and ignore clock-network switching responsible for over 30% of total power. We propose new techniques and a methodology to optimize total dynamic power during placement for large IC designs with macro blocks. To this end, we advocate obstacle-aware virtual clock-tree synthesis, a arboreal clock-net contraction force with virtual nodes that can handle gated clocks, and an obstacle-avoidance force for clock edges. Our methodology is integrated into the SimPL placer [9], and the total switching power is measured by utilizing Contango 2.0 [12] — both programs are leading academic software. A new set of 45 nm benchmarks is proposed to better represent modern IC designs. Experimental results show that our method lowers the overall dynamic power by significantly reducing clock-net switching power. Other benefits of our optimizations (not explicitly evaluated in this paper) include smaller insertion delay in clock trees, diminished sensitivity to process variations, and reduced supply voltage noise.

8. REFERENCES

[1] C. J. Alpert, D. P. Mehta, S. S. Sapatnekar, "Handbook of Algorithms for Physical Design Automation," *CRC Press*, 2009.

[2] T. F. Chan et al, "mPL6: Enhanced Multilevel Mixed- Size Placement," *ISPD'06*, pp. 212-214.

[3] Y.-T. Chang et al, "Post-Placement Power Optimization with Multi-Bit Flip-Flops," *ICCAD'10*, pp. 218-223.

[4] Y. Cheon, P.-H. Ho, A. B. Kahng, S. Reda and Q. Wang, "Power-Aware Placement," *DAC'05*, pp. 795-800.

[5] M. Donno, E. Macci, and L. Mazzoni, "Power-Aware Clock Tree Planning," *ISPD'04*, pp. 138-147.

[6] P. E. Gronowski, W. J. Bowhill, R. P. Preston, M. K. Gowan, and R. L. Allmon, "High-Performance Microproccesor Design," *IEEE JSSC*, 33(5) (1998), pp. 676-686.

[7] W. Hou, X. Hong, W. Wu and Y. Cai, "A path-based timingdriven quadratic placement algorithm," *ASPDAC'03*, pp. 745-748.

[8] A. B. Kahng,C.-W. Tsao, "Practical Bounded-Skew Clock Routing," *J. VLSI Signal Proc.* 16(1997), pp.199-215.

[9] M.-C. Kim, D.-J. Lee and I. L. Markov, "SimPL: An Effective Placement Algorithm," *ICCAD'10*, pp. 649-656.

[10] J. M. Kleinhans, G. Sigl, F. M. Johannes and K. J. Antreich, "GORDIAN: VLSI Placement by Quadratic Programming and Slicing Optimization," *IEEE Trans. on CAD*, 10(3): 356-365, 1991.

[11] D.-J. Lee, I. L. Markov, "Contango: Integrated Optimization of SoC Clock Networks," *DATE'10*, pp. 1468-1473.

[12] D.-J. Lee, M.-C. Kim and I. L. Markov, "Low-Power Clock Trees for CPUs," *ICCAD'10*, pp. 444-451.

[13] Y. Liu, X. Hong, Y. Cai, W. Wu, "CEP: A Clock-Driven ECO Placement Algorithm for Standard-Cell Layout," *ASIC'01*, pp. 118-121.

[14] J. Lu et al, "A Dual-MST Approach for Clock Network Synthesis," *ASPDAC'10*, pp. 467-473.

[15] Y. Lu et al, "Navigating Registers in Placement for Clock Network Minimization," *DAC'05*, pp. 176-181.

[16] N. Magen, A. Kolodny, U. Weiser, and N. Shamir, "Interconnect-power Dissipation in a Microprocessor," *SLIP'04*, pp. 7-13.

[17] G. J. Nam, C. J. Alpert, P. Villarrubia, B. Winter and M. Yildiz, "The ISPD2005 Placement Contest and Benchmark Suite," *ISPD'05*, pp. 216-220.

[18] Nangate Inc. Open Cell Library v2009 07, 2009. Downloadable from http://www.nangate.com/openlibrary

[19] J. Oh and M. Pedram, "Gated Clock Routing for Low-Power Microprocessor Design", *IEEE Trans. on CAD*, Vol. 20, No. 6, pp. 715-722, 2001.

[20] R.P. Pokala, R.A. Feretich and R.W. McGuffin, "Physical Synthesis for Performance Optimization", *ASIC'92*, pp. 34-37.

[21] J. M. Rabaey, A. Chandrakasan, B. Nikolic, "Digital Integrated Circuits: A Design Perspective," *Prentice Hall*, Second Edition, 2003.

[22] X.-W. Shih et al, "Blockage-Avoiding Buffered Clock-Tree Synthesis for Clock Latency-Range and Skew Minimization," *ASPDAC'10*, pp. 395-400.

[23] G. Sigl, K. Doll and F. M. Johannes, "Analytical Placement: A Linear or a Quadratic Objective Function?" *DAC'91*, pp. 427-431.

[24] P. Spindler, U. Schlichtmann, F. M. Johannes, "Kraftwerk2 - A Fast Force-Directed Quadratic Placement Approach Using an Accurate Net Model," *IEEE Trans. on CAD*, 27(8) 2008, pp. 1398-1411.

[25] C. N. Sze, "ISPD 2010 High-Performance Clock Network Synthesis Contest: Benchmark Suite and Results," *ISPD'10*, pp. 143-143.

[26] V. Tiwari, D. Singh, S. Rajgopal, G. Mehta, R. Patel, and F. Baez, "Reducing Power in High-Performance Microprocessors," *DAC'98*, pp. 732-737.

[27] N. Viswanathan, M.Pan, C.Chu, "FastPlace 3.0: A Fast Multilevel Quadratic Placement Algorithm with Placement Congestion Control," *ASPDAC'07*, pp. 135-140.

[28] K. P. Vorwerk, A. Kennings and A. Vannelli, "Engineering details of a stable force-directed placer," *ICCAD'04*, pp. 795-800.

[29] M. Wang and M. Sarrafzadeh, "Congestion minimization during placement," *ISPD'99*, pp. 145-150.

[30] Y. Wang, Q. Zhou, X. Hong and Y. Cai, "Clock-Tree Aware Placement Based on Dynamic Clock-Tree Building," *ISCAS'07*, pp. 2040-2043.

Timing Slack Aware Incremental Register Placement with Non-uniform Grid Generation for Clock Mesh Synthesis

Jianchao Lu
ECE Dept.
Drexel University
Philadelphia, PA
jl597@drexel.edu

Xiaomi Mao
ECE Dept.
Drexel University
Philadelphia, PA
xm27@drexel.edu

Baris Taskin
ECE Dept.
Drexel University
Philadelphia, PA
taskin@coe.drexel.edu

ABSTRACT

A novel clock mesh network synthesis approach is proposed in this paper which generates an improved mesh size with registers placed incrementally considering the timing slack on the data paths and the non-uniform grid wire placement. The primary objective of the method is to reduce the power dissipation without a global skew degradation, which is achieved through a sparse and non-uniform mesh implementation with registers incrementally placed in close vicinity to the mesh grids. The incremental register placement is based on the timing information in order to preserve the timing slack of the circuit. Experimental results show that the total wirelength (mesh grid wires and stub wires) as well as the power dissipation is reduced significantly on the clock mesh network. Specifically, the wirelength of the mesh network and the power dissipation of the clock network are reduced by 52% and 48% on average, respectively. Moreover, the global clock skew and the non-negative timing slack are preserved.

Categories and Subject Descriptors

B.7.2 [**Hardware**]: Integrated circuits—*Design aids*

General Terms

Algorithms, Design, Performance

Keywords

VLSI CAD, Physical Design, Clock Network, Clock Mesh, Register Placement

1. INTRODUCTION

In a typical high performance microprocessor design, the clock distribution network is synthesized with redundancy in order to reduce on-chip variations. Clock mesh [11, 16, 17, 20, 23], cross links [12, 14, 15] and clock tree with spines are the commonly used clock structures with redundancy. Among these structures, the clock mesh is getting increasingly popular nowadays and often adopted in the high-end microprocessor design as the clock mesh provides high reliability [11, 16, 17, 20, 23].

Permission to make digital or hard copies of all or part of this work for personal or classroom use is granted without fee provided that copies are not made or distributed for profit or commercial advantage and that copies bear this notice and the full citation on the first page. To copy otherwise, to republish, to post on servers or to redistribute to lists, requires prior specific permission and/or a fee.
ISPD'11, March 27–30, 2011, Santa Barbara, California, USA.
Copyright 2011 ACM 978-1-4503-0550-1/11/03 ...$10.00.

The redundancy in a typical clock mesh network permits a low global clock skew variation. The low global clock skew is achieved at the expense of high power consumption compared to tree or other structures because of the excessive wires (mesh grid and stub wires) and buffers used as well as the short circuit power introduced. Due to the popularity of clock mesh in very large scale microprocessors, design automation efforts are made in the area of clock mesh synthesis and optimization [2, 10, 13, 18, 22]. In [22], buffer driver insertion and sizing are studied as well as the mesh reduction for power savings. In [13], an optimal mesh size selection method under skew constraints is proposed. The method in [13] encompasses buffer placement and sizing as well as reduction of mesh wires. In [18], steiner tree like connections between registers and meshes are created — instead of connecting registers to the mesh individually — such that the total stub wires are reduced. A uniform clock mesh is assumed in these works in [13, 18, 22]. In [2] and [10], non-uniform clock meshes are explored in order to reduce mesh wirelength and the power consumption. In [2], the timing delays of the combinational logic paths are considered when building the clock mesh such that the grid density can be adjusted based on the timing criticality. In [10], the stub wires connected between sink registers and the clock meshes are reduced by allowing an incremental movement of the mesh grid.

In the traditional integrated circuit design flow, the placement and clock network synthesis stages are performed sequentially. The primary objectives of placement do not include the optimization of register placement for the succeeding clock network synthesis stage. This may result in a low quality (e.g. power consuming) clock network after synthesis. It is desirable to combine the placement and clock network synthesis stages to provide a better physical design. The novel approach of synthesizing clock mesh network proposed in this paper combines the placement and clock network synthesis stages of the traditional IC physical design flow through incremental placement. In this method, both the registers and the mesh wires are incrementally placed towards each other considering the timing slack such that the total stub wirelength is significantly reduced. Moreover, a more favorable (sparse and non-uniform) grid is selected automatically with limited skew degradation. The proposed method integrates the clock network synthesis stage with the incremental placement of registers. The advantages of the clock mesh network generated by the proposed method are the following:

1. The power consumption of the clock mesh network is reduced compared to previous clock mesh design methods due to the sparse mesh network and the reduced stub wirelength,

2. The non-negative timing slack of the circuit is preserved after the incremental register placement,

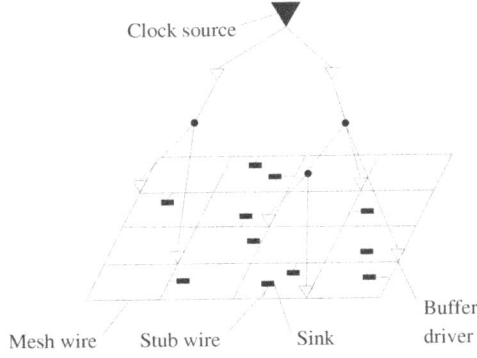

Figure 1: An illustration of the clock mesh network.

3. The skew variation of the clock network is comparable to a uniform, full mesh grid clock network, despite the sparser mesh grid.

The trade-offs of the proposed method are the changing timing slack and the logic routing wirelength. The timing slack is guaranteed to be non-negative by the proposed method with minimal slack decrease. The logic routing wirelength increase is only 5.9% on average on the benchmark circuits, which is very limited. The proposed method is thus highly practical.

The rest of the paper is organized as follows. In Section 2, the preliminaries about clock mesh and the static timing constraints are presented. In Section 3, the proposed methods are introduced. In Section 4, the experimental results are summarized. The paper is finalized in Section 5.

2. PRELIMINARIES

In physical IC design, uniform clock meshes are preferred since the mesh grid can be placed between uniform power rails to prevent crosstalk. This paper follows this principle of utilizing uniformly placed tracks as potential candidates for clock mesh grid synthesis. To this end, the full redundancy of the clock mesh is used in optimization. The proposed optimization selects a subset of the potential tracks for a non-uniform clock grid synthesis.

The two primary metrics when designing clock meshes are the total wirelength (impacting power) and the clock skew (caused by variations). In Section 2.1 and Section 2.2, the wirelength of the mesh and stub wires and the clock skew are discussed, respectively.

2.1 Clock mesh wirelength

An illustration of the clock mesh network is presented in Figure 1. The total wirelength L_{total} on a mesh network can be calculated as:

$$L_{total} = L_{mesh} + L_{stub}, \qquad (1)$$

where L_{mesh} and L_{stub} are the mesh grid wirelength and the stub wirelength, respectively.

The mesh and the stub wires contribute to the total power consumption through dynamic power dissipation. The dynamic switching capacitance of the clock network contributed by the mesh and stub wires is directly proportional to the total wirelength. In order to reduce the power consumption on the clock mesh, the proposed methods in [2, 13, 22] reduce the mesh wirelength while the methods in [10, 18] reduce the stub wirelength. The method proposed in this paper reduces both the mesh and stub wires through the inte-

grated placement and clock network synthesis approach. In particular, the stub wirelength is further reduced than [10, 18] as most of the stub wires are eliminated through incremental placement. The mesh wires are reduced through the mesh size selection and the mesh wire placement methods.

2.2 Clock skew in clock mesh networks

In [13], the global clock skew t_{skew} on a mesh network is estimated as:

$$t_{skew} = t_{skew}^{buf} + D_{mesh}(d_{max}) + D_{stub}(L_{stub}^{max}), \qquad (2)$$

where t_{skew}^{buf}, $D_{mesh}(d_{max})$ and $D_{stub}(L_{stub}^{max})$ are the skew introduced by the buffer drivers of the mesh, the maximum delay on the mesh from a buffer driver to a stub wire tapping point and the maximum delay from a tapping point of a stub wire to a sink register, respectively. In Equation (2), the skew introduced by the buffer driver t_{skew}^{buf} can be compensated using the prescribed skew tree generation method [3] when synthesizing the top level clock tree. Increasing the number of buffer drivers also improves t_{skew}^{buf} through improving the driving strength of the mesh, however, with penalty in increased power consumption. Inserting more buffer drivers also reduces the second term $D_{mesh}(d_{max})$. The third term can be reduced by reducing the stub wirelength. In a traditional, uniform clock mesh, the third term is reduced by using a denser clock mesh, where stub wires are typically shorter.

According to Equation (2), power and clock skew are two contradicting objectives. In order to achieve a low clock skew, a dense mesh network is required to reduce $D_{mesh}(d_{max})$ and $D_{stub}(L_{stub}^{max})$. However, more power is consumed as more mesh grids and buffers are necessary. If low power is preferred, a sparse network will be synthesized. However, the clock skew will be degraded due to the long stub wires and larger $D_{stub}(L_{stub}^{max})$. The trade-off between the clock skew and the total power consumption is broken by incrementally moving registers on to the mesh such that the stub wirelength is still limited with a sparser mesh grid.

2.3 Static timing constraints

A local data path $R_i \rightarrow R_f$ as shown in Figure 2 consists of two registers $R_{i(nitial)}$ and $R_{f(inal)}$ and a combinational logic block. The minimum and maximum propagation delays on the combinational block are denoted by D_{PMin}^{if} and D_{PMax}^{if}, respectively. The clock-to-output delay of a register R_i is denoted by D_{CQ}^i, whereas S_f is the setup time of the register R_f. The parameters t_i and t_f represent the clock delays to registers R_i and R_f, respectively and the clock period is denoted by T.

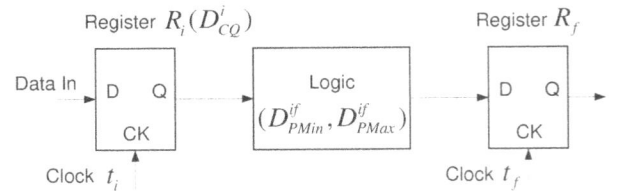

Figure 2: An illustration of the register to register timing path.

The timing analysis of a synchronous circuit is performed by satisfying the *setup* and *hold* timing constraints for each local data path:

$$Setup: \quad t_i + D_{CQ}^i + D_{PMax}^{if} \leq t_f + T - S_f - L_{if}, \quad (3)$$

$$Hold: \quad t_i + D_{CQ}^i + D_{PMin}^{if} \geq t_f. \quad (4)$$

$$(5)$$

For zero clock skew systems, which are the norm for synchronous circuits, clock delays t_i and t_f are identical, simplifying the timing constraints. Thus, if the sum of the maximum data propagation time D_{PMax}^{if} and the clock-to-output delay D_{CQ}^i of the register R_i minus the setup time S_f of the register R_f is greater than the operating clock period, a timing violation occurs [4]. After the clock period T is chosen, the placement and routing of the circuit should guarantee the setup and hold constraints for each datapath are satisfied with a non-negative timing slack [9]. The setup timing slack is more critical since the hold violations can always be fixed by inserting delays on a datapath [21]. The timing slack L_{if} on each timing path $R_i \rightarrow R_f$ of the circuit can be calculated as:

$$L_{if} = T - S_f - D_{CQ}^i - D_{PMax}^{if}. \quad (6)$$

3. METHODOLOGY

The proposed method flow is illustrated in Figure 3. The proposed method takes an existing placement result as the input. A static timing analysis is performed to identify the timing slack of each data path. Based on this information, the feasible moving regions of each register are created. The non-uniform clock mesh is generated and placed in order to simultaneously reduce the mesh wires and stub wires according to the feasible moving regions of the registers. The registers are then incrementally moved based on the timing and the clock mesh placement. To finalize, the buffer drivers of the clock mesh are inserted and the top level clock tree is generated. The three stages of the proposed methodology are introduced through Section 3.1 to Section 3.3, respectively.

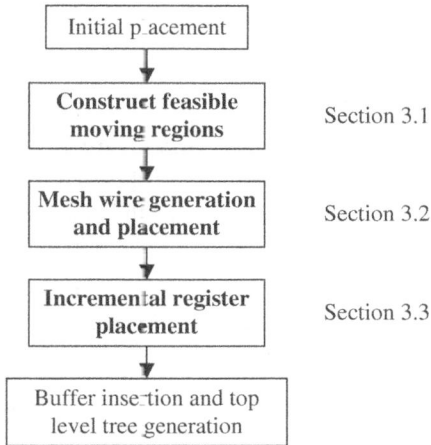

Figure 3: The methodology flow.

3.1 Building the feasible moving regions

The proposed method suggests the incremental placement of the registers towards the clock meshes. The timing slack is considered during the incremental placement (movement) of the registers in order to guarantee the functionality correctness of the design. The feasible regions for incremental register placement are defined

based on these slack, timing path, and physical path definitions. Note that the timing slack of a register-to-register path $R_i \rightarrow R_f$ is associated with all the physical paths on the register-to-register (timing) path. The incremental placement of the registers affects the locations of the registers but not the combinational logic gates constituting the physical paths. Consequently, incremental register placement changes the slack of the entire timing path, however, only the physical paths at the fanout of the initial register R_i and the fanin of the final register R_f are affected. The remaining physical paths between the combinational gates remain unaffected. To this end, the datapath delay D_{PMax}^{if} is decomposed to three parts:

$$D_{PMax}^{if} = D_{fo_k}^i + D_{m_k}^{if} + D_{fi_k}^f, \quad (7)$$

as illustrated in Figure 4. $D_{fo_k}^i$ is the wire delay from the output of the register R_i to the input of the kth fanout gate of the register R_i. $D_{m_k}^{if}$ is the gates and wire delay from the input of the kth fanout gate of the register R_i to the input of the fanin gate of register R_f. $D_{fi_k}^f$ is the gate and wire delay from the input of the fanin gate of register R_f to the input of the register R_f.

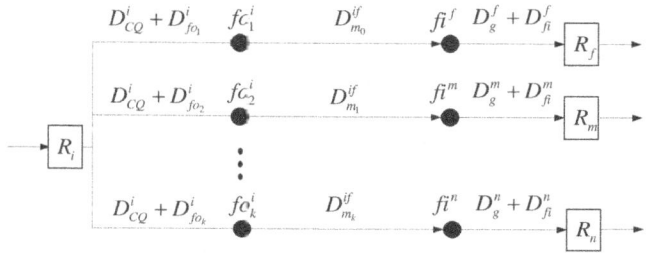

Figure 4: The illustration of the delay decomposition.

For each fanout physical datapath of register R_i on local datapath $R_i \rightarrow R_f$, the timing constraints presented as Equation (3) can be rewritten in order to reflect the location change of the registers as the delay change on the timing path. The fanout wirelength $w_{fo_k}^i$ of the register R_i will change through the incremental placement of the register R_i. The change in the fanout wirelength will affect the delay on the kth fanout wire and the clock to output delay of the register D_{CQ}^i as the load capacitance changes. The setup timing constraints are thus rewritten as:

$$D_{CQ}^i + D_{fo_k}^i \leq T - S_f - L_{if} - D_{m_k}^{if} - D_{fi_k}^f. \quad (8)$$

With similar reasoning, the change in the fanin wirelength w_{fi}^f affects the delay $D_{fi_k}^f$ of the fanin gate and the fanin wire. The timing constraints must satisfy:

$$D_{fi_k}^f \leq T - S_f - L_{if} - D_{m_k}^{if} - D_{fo_k}^i - D_{CQ}^i. \quad (9)$$

At the post placement stage, the clock period T and each part of the original datapath delay $D_{m_k}^{if}$, $D_{fo_k}^i$ and $D_{fi_k}^f$ are known. In order to guarantee the functionality correctness under variation, the timing slack L_{if} of each register timing path $R_i \rightarrow R_f$ should be non-negative. For the inequalities (8) and (9), the left hand side delay functions are monotonically increasing functions of fanout wirelength $w_{fo_k}^i$ and fanin wirelength w_{fi}^f, respectively. Given a positive slack, a maximum fanout wirelength $W_{fo_k}^i$ and fanin wirelength W_{fi}^i can be calculated for each register R_i. As long as the manhattan distance of the register R_i is less than the maximum

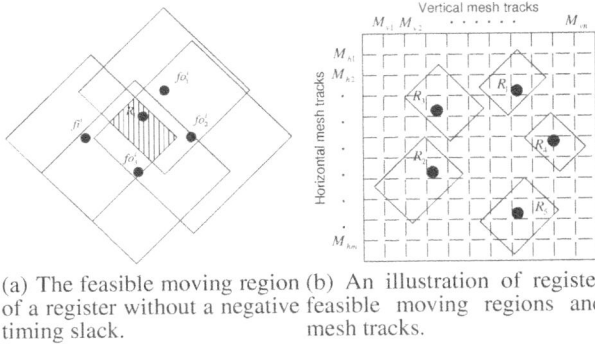

(a) The feasible moving region (b) An illustration of register of a register without a negative feasible moving regions and timing slack. mesh tracks.

Figure 5: The feasible moving regions and mesh tracks.

fanout wirelength $W_{fo_k}^i$ to the corresponding fanout gates and the maximum fanin wirelength W_{fi}^i to the fanin gate, the timing slack of the registers are guaranteed to be feasible.

The feasible moving regions for each fanout and fanin gates of register R_i are created based on $W_{fo_k}^i$ and W_{fi}^i, respectively. For instance, at the location of the kth fanout gate of register R_i, a tilted rectangle with radius $W_{fo_k}^i$ is created as shown in Figure 5(a) such that the manhattan distance from the register R_i to the gate is equal to $W_{fo_k}^i$ on the boundary of the region. As long as the register is placed within the created tilted rectangle region, the timing slack on the kth fanout path is satisfied. For each fanout and fanin gate of register R_i, a feasible rectangle region is created. The shaded overlapping region of all the tilted rectangle regions of the fanin and fanout gates of register R_i is defined as the feasible moving region of register R_i as shown in Figure 5(a). Note that the moving region construction is over optimistic as when one register is moved, the slack of the other register which has a datapath to the moved register changes. In other words, the feasible region of movement generated at this stage for each register R_i is valid when the rest of the registers are unmoved. Thus, in this stage, only the so called feasible moving region is created; the register placement will be performed at a later stage guaranteeing the positive timing slack.

3.2 Mesh wire generation and placement

The clock mesh network consists of the mesh network with grids and the stub wires connecting all the registers and the top level clock tree. In the proposed method, the mesh grid and the stub wires are generated after building the feasible moving regions of each register R_i. The objective of the mesh wire generation is to generate a sparse clock mesh which guarantees a low clock skew. The proposed method allows a non-uniform clock mesh in order to reduce the stub wirelength as well as the number of grids.

On the chip area, the mesh tracks are created, where each mesh track represents a possible placement of the clock mesh grids as shown in Figure 5(b). The mesh tracks M_{hi} and M_{vi} represent the possible horizontal and vertical mesh locations, respectively. The mesh tracks are defined at the floorplanning stage through the uniformly distributed power rails as described in Section 2. In this work, the mesh generation and placement problem is formulated as a weighted set cover problem. The weighted set cover problem is defined as [7]:

Given a universe U and a family of subsets S_i where each subset S_i has a positive weight, find the series of sets whose union is U and the total weights of these sets are minimized.

In the mesh wire generation problem, each horizontal and vertical mesh track M_{hi} and M_{vi} is considered as a set. The registers are the elements and the universe of the problem is all the registers. A register R_i is included in a set M_{hi} (M_{vi}) if the feasible moving region FR_i of the register overlaps with the mesh track M_{hi} (M_{vi}). It is assumed that if a feasible moving region FR_i of a register has overlapping with a mesh track, the register can be moved close to the mesh track. The objective is to find the minimum number of mesh wires such that the all the registers can be moved close to these mesh wires. The problem is equivalent to finding the minimum weight set cover for the register universe given the subsets M_{hi} and M_{vi}.

Each set M_{hi} (M_{vi}) has a weight $|M_{hi}|$ ($|M_{vi}|$) defined as:

$$|M_{hi}|\ (|M_{vi}|) = \sum_{\forall R_k \in M_{hi}\ (M_{vi})} WT_{R_k}^{M_{hi}\ (M_{vi})}, \quad (10)$$

where $WT_{R_k}^{M_{hi}\ (M_{vi})}$ is the weight of each register. The weight of each register when connecting to M_{hi} (M_{vi}) is defined as:

$$WT_{R_k}^{M_{hi}\ (M_{vi})} = Cdist(R_k, M_{hi}\ (M_{vi})), \quad (11)$$

where C is a scaling constant. Equation (11) suggests that the weight of the register R_i in set $M_{hi}(M_{vi})$ is proportional to the distance of the register R_i to the mesh track. As such, the solution favors less incremental movement of the registers. Solving the set cover problem identifies the mesh tracks which lead to the minimum cost (total weight) in incremental register placement. The scaling constant of the registers C is defined for the ease of computation and in the experiments it is identical for each register.

The set cover problem is a well-known NP-complete problem [7]. In this work, a simple yet effective greedy approximation algorithm is applied [7]. The algorithm greedily adds the set (horizontal or vertical mesh wire) with the minimum amortized cost into the solution at each iteration until the sets in the solution include all the registers. The amortized cost is defined as the cost of the set $|M_{hi}|$ ($|M_{vi}|$) divided by the number of new elements added when the set M_{hi} (M_{vi}) is chosen. The sets (M_{hi}s and M_{vi}s) in the solution are the mesh wires that are generated and placed for the non-uniform clock mesh network.

3.3 Timing slack aware incremental register placement avoiding overlapping

Although the feasible moving regions for the registers are generated as explained in Section 3.2, aggressively moving the registers onto a mesh segment inside the moving region does not always guarantee the timing slack requirement. This is because moving one register may negatively affect the timing slack, and thus, the feasible moving region of the other registers which have a timing path to it. Moreover, moving the registers incrementally close to the mesh may introduce overlapping. In the proposed method, the incremental register placement is formulated as linear programming formulation and solved optimally without timing violation and register overlapping.

Each register is moved towards the mesh wire assigned in the set cover solution. If more than one mesh wire has overlapping with one register, the register will be assigned to the closest mesh wire in order to reduce the moving distance of the incremental placement. During this incremental register placement process, the setup timing constraint (8) has to be satisfied, where $D_{fo_k}^i$, D_{CQ}^i and $D_{fi_k}^f$ are functions of the wire capacitance and the capacitive load. The wire capacitance and the capacitive load of the gate depends on the wire length. In this work, the delay change on the wire and the gate are conservatively modeled using linear functions:

$$D_{fo_k}^i = K_w C_0 w_{fo_k}^i, \tag{12}$$

$$D_{CQ}^i = D_{R0}^i + K_r^i C_0 w_{fo_k}^i, \tag{13}$$

$$D_{fi_k}^f = K_w C_0 w_{fi_k}^f + D_{G0}^f + K_G^f C_0 w_{fi_k}^f, \tag{14}$$

where C_0 is the unit wire capacitance. The parameters K_w, K_r^i and K_G^f are the slopes of the wire delay versus wire capacitance curve, register delay versus capacitive load curve and the fanin gate delay of register R_f versus the capacitive load, respectively. The parameters D_{R0}^i and D_{G0}^f are the clock-to-output delay and the gate delay when the capacitive load is zero (0). The wirelength $w_{fo_k}^i$ and $w_{fi_k}^f$ can be estimated using the distances of the fanout and fanin gate to the register:

$$w_{fo_k}^i = |x_{R_i} - x_{fo_k}^i| + |y_{R_i} - y_{fo_k}^i|, \tag{15}$$

$$w_{fi}^f = |x_{R_f} - x_{fi}^f| + |y_{R_f} - y_{fi}^f|, \tag{16}$$

where x_{R_i} and y_{R_i}, $x_{fo_k}^i$ and $y_{fo_k}^i$ and x_{fi}^f and y_{fi}^f are the x and y locations of the register R_i, the x and y locations of the kth fanout gate of register R_i, the x and y locations of the fanin gate of the register R_f, respectively. The linear approximation of the delay change on the wire and the gates is conservative because the delay typically is modeled quadratically. e.g. Elmore delay [8]. The linear approximation guarantees that the estimated (linear) delay is always higher than the higher order models. Alternative delay modeling can be performed for accuracy. However, the proposed design method is based on linear programming, thus, the linear delay model is selected to generate linear constraints. The conservative modeling estimates the wire and gate delays to be higher, which provides conservative but guaranteed timing constraints. The difference in overestimation is available as a positive timing slack after incremental placement, which is favorable for a practical operation.

The assignment of registers to mesh wires is obtained from the set cover solution in Section 3.2. If a register R_i is in the set $M_{h,j}$ or M_{vj} of the set cover solution, register R_i connects to the mesh wire M_{hj} or M_{vj}, respectively. If the register R_i resides in more than one set, the register R_i will connect to the mesh wire which has the minimum stub wirelength at its original location. The stub wirelength w_{stub}^i of each register R_i is the minimum distance from the location of the register R_i to its corresponding mesh wire:

$$w_{stub}^i = \begin{cases} |x_{R_i} - x_{M_{vj}}|, & \text{if } R_i \text{ connects to } M_{vj}, \\ |y_{R_i} - y_{M_{hj}}|, & \text{if } R_i \text{ connects to } M_{hj}, \end{cases} \tag{17}$$

The objective of the formulation is to minimize the total stub wirelength:

$$Objective : \min \sum_{\forall R_i} w_{stub}^i. \tag{18}$$

Simultaneous with these requirements in timing, the physical requirement in preventing the overlapping of registers is considered. As shown in Figure 6 let the length and width of a register be L_r and W_r, respectively. One of the following four overlapping avoidance constraints has to be satisfied in order to guarantee there is no overlapping between each pair of registers R_i and R_j:

$$x_{R_i} - x_{R_j} \geq W_r, \tag{19}$$

$$x_{R_j} - x_{R_i} \geq W_r, \tag{20}$$

$$y_{R_i} - y_{R_j} \geq L_r, \tag{21}$$

$$y_{R_j} - y_{R_i} \geq L_r. \tag{22}$$

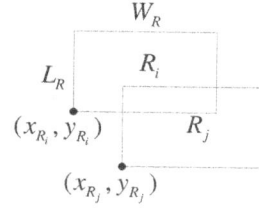

Figure 6: The overlap illustration.

These constraints prevent the horizontal and vertical overlapping of register placement based on the register length and width. The constraints (19) and (20) are mutually exclusive, similar to the constraints (21) and (22). In order to form an LP formulation for the problem, only one of the fours constraint is placed in the LP formulation between each pair of registers R_i and R_f. To this end, the following flow is proposed to generate and reduce the overlapping avoidance constraints to one:

1. Construct the stub wire minimization problem considering timing slack without the overlapping avoidance constraints. In this formulation, Equation (18) is the objective. Inequalities (8,12–17) are the constraints. Solve the formulation to obtain the incremental register placement results.

2. In the incremental register placement result, if two registers R_i and R_j are non-overlapping, add the constraint which has the maximum left hand side value among Equations (19–22) to the overlapping avoidance formulation.

3. In the incremental register placement result, if two registers R_i and R_j are overlapping, add the constraint which has the maximum left hand side value for the *original* locations of the registers R_i and R_j in Equations (19 – 22) to the overlapping avoidance formulation.

4. If two registers are placed on two different grids which do not have any intersection (e.g. two parallel horizontal or vertical grids), the overlapping avoidance constraints between the two registers can be eliminated.

In brief, the LP is solved without considering the physical overlapping avoidance constraints first (step 1). If the registers are placed to non-intersecting grids, the overlapping avoidance constraint is unnecessary as the registers will never overlap (step 4). Otherwise, the most conservative constraint is added to the LP and solved for the optimal placement without physical overlapping of register with each other (steps 2 and 3).

The linear programming formulation is presented in Table 1. The objective of the formulation is to minimize the total stub wirelength connecting the registers to the mesh wires by incrementally moving the registers. The timing constraints and the overlapping constraints are generated. Note that $xdist(a,b)$ and $ydist(a,b)$ represents the distance between nodes a and b on the horizontal direction and vertical direction, respectively. The constraints about $xdist(a,b)$ and $ydist(a,b)$ are used to linearize the distance constraints. For each register pair, at most one constraint among the last four constraints presented as "or" appears in the LP formulation. By solving the formulation, the optimal locations $(\hat{x}_{R_i}, \hat{y}_{R_i})$ of each register R_i and the corresponding total stub wirelength are obtained.

Table 1: The linear programming formulation for incremental register placement.

Minimize the total stub wirelength.

$$\min \quad \sum_{\forall R_i} w^i_{stub}$$

$$\text{s.t.} \quad D^i_{fo_k} + D^i_{CQ} + D^f_{fi_k} \leq T - S_f - L_{if} - D^{if}_{m_k},\ \forall(R_i \to R_f)$$
$$D^i_{fo_k} = K_w C_0 w^i_{fo_k},\ \forall R_i$$
$$D^i_{CQ} = D^i_{R0} + K^i_r C_0 w^i_{fo_k},\ \forall R_i$$
$$D^f_{fi_k} = K_w C_0 w^f_{fi_k} + D^f_{G0} + K^f_G C_0 w^f_{fi_k},\ \forall R_f$$
$$w^i_{stub} = xdist(R_i, M_{vj})\ (or\ ydist(R_i, M_{hj})|),\ \forall R_i$$
$$xdist(R_i, M_{vj}) \geq x_{R_i} - x_{M_{vj}},\ \forall R_i$$
$$xdist(R_i, M_{vj}) \geq x_{M_{vj}} - x_{R_i},\ \forall R_i$$
$$ydist(R_i, M_{hj}) \geq y_{R_i} - y_{M_{hj}},\ \forall R_i$$
$$ydist(R_i, M_{hj}) \geq y_{M_{hj}} - y_{R_i},\ \forall R_i$$
$$w^i_{fo_k} = xdist(R_i, fo_k) + ydist(R_i, fo_k),\ \forall R_i$$
$$xdist(R_i, fo_k) \geq x_{R_i} - x_{fo_k},\ \forall R_i$$
$$xdist(R_i, fo_k) \geq x_{fo_k} - x_{R_i},\ \forall R_i$$
$$ydist(R_i, fo_k) \geq y_{R_i} - y_{fo_k},\ \forall R_i$$
$$ydist(R_i, fo_k) \geq y_{fo_k} - y_{R_i},\ \forall R_i$$
$$w^f_{fi} = xdist(R_f, fi) + ydist(R_f, fi),\ \forall R_f$$
$$xdist(R_f, fi) \geq x_{R_f} - x_{fi},\ \forall R_f$$
$$xdist(R_f, fi) \geq x_{fi} - x_{R_f},\ \forall R_f$$
$$ydist(R_f, fi) \geq y_{R_f} - y_{fi},\ \forall R_f$$
$$ydist(R_f, fi) \geq y_{fi} - y_{R_f},\ \forall R_f$$
$$x_{R_i} - x_{R_j} \geq W_r,$$
$$or\ x_{R_j} - x_{R_i} \geq W_r,$$
$$or\ y_{R_i} - y_{R_j} \geq L_r,$$
$$or\ y_{R_j} - y_{R_i} \geq L_r.$$

3.4 Discussions

The buffer driver insertion process of the proposed method adopts the similar set cover solution in [22]. The set-cover problem proposes a clock driver placement and sizing solution in order to drive the registers on the mesh for a given global clock skew requirement. In the proposed method, since the registers are moved close to the clock meshes, the resulting optimal mesh grids are sparser. In order to reduce the clock skew introduced by the second item in Equation (2), the buffers are allowed to be inserted not only on the intersections of the mesh grids, but also in the middle of the mesh wires. In the proposed method, the top level tree is generated using the method in [5].

The overlapping between the registers is considered in the LP formulation. In the proposed method, the overlapping between the registers and the logic gates due to the incremental placement is resolved by placement legalization using `IC Compiler`. However, the overlapping between registers and gates can be avoided by allowing the registers to be placed on white space only.

4. EXPERIMENTAL RESULTS

The proposed algorithm flow is implemented in C++. The top level clock tree is generated using a buffered DME algorithm [5, 6] to drive the mesh grid. The clock mesh networks with a top level tree are translated into the ISPD10 clock network contest format [19] and simulated using *Ngspice* with a 45nm PTM model card. *IC Compiler* of *Synopsys* is used to perform the initial placement, timing slack analysis and routing. The LP formulations are solved by the online solver Feaspump and SCIP from [1]. Since the benchmark circuits provided by the ISPD'10 clock network contest do not have any logic gate information, the benchmark circuit used in the experiments are the five largest circuit from the ISCAS'89 benchmark. As reference, note that the register count for the largest

ISCAS'89 circuit is in the same level with the ISPD'10 contest benchmark (1728 vs. 2249). The runtime of the proposed method on all the benchmark circuits are less than 7 minutes.

The proposed mesh generation and incremental register placement methods are compared against the non-uniform mesh placement method proposed in [10]. The *iterative k-means* method in [10] is implemented in C++. Since the mesh size selection and mesh wire placement are integrated into the proposed set cover solution, the mesh network generated by the proposed method using the mesh size in [10] is not available. Thus, two sets of comparison are performed:

Set 1 The mesh network generated by the proposed method is compared to the mesh network generated by the method in [10] using the optimized mesh size in [10].

Set 2 The mesh network generated by the proposed method is compared to the mesh network generated by the method in [10] using the mesh size optimized by the proposed method.

These comparisons are performed in order to demonstrated the wire reduction effects of the proposed method.

The result of the first set of comparison from the above list is summarized in Table 2. For the same circuits, the proposed method typically generates a sparser mesh network. The stub wirelength is reduced by 89.6% and the mesh wirelength is reduced by 24.4%, on average. The total wire reduction on the mesh network is 51.9% on average.

The results for the second set of comparison from the above list is summarized in Table 3. The stub wirelength reduction is more significant at 91.9% on average than the method in [10]. This is such as the method proposed in this paper suggest incremental register placement towards the mesh grids whereas the method in [10] keeps the placement intact. Due to the experimental setup of choosing the same mesh grid sizes as in [10], there is no mesh reduction. The total wire reduction is 50.8% on average.

The power and global clock skew are compared between the proposed method and the method in [10], and the results are summarized in Table 4. The power reduction is presented as the total switching capacitance including the wire capacitance, buffer capacitance on the clock mesh and the top level clock tree. Comparing to the mesh generated using the method in [10] with the mesh size in [10] (Set 1), a 48.3% reduction on the total capacitance is observed. This reduction is achieved through the wire reduction and the top level clock tree altogether (less buffer drivers and thus smaller clock tree). The clock skew on the proposed mesh network is similar (0ps change) to the previous work in [10]. Comparing to the mesh generated using the method in [10] but with the mesh size optimized by the proposed method (Set 2), the power reduction is 28.1% on average. The average clock skew is reduced by 0.8ps on average using the proposed method due to the less stub wire.

The trade-offs of the proposed method in logic wire routing and timing slack are performed and analyzed using `IC Compiler`. The trade-offs are summarized in Table 5. It is observed that the timing slack is reduced by 22ps (7.3%) on average using the proposed method, which is very limited compared to the original timing slack before applying the proposed clock mesh synthesis. A non-negative timing slack is guaranteed in the proposed formulation. The logic wire routing is only increased by 5.9% on average due to the incremental placement of the registers, which is very limited compared to the power saving on clock network.

The placement result for the circuit ISCAS'89 s35932 before and after the proposed clock mesh network synthesis are illustrated in Figure 7(a) and Figure 7(b), respectively. The optimal mesh grid

Table 2: Experimental Set 1: Wirelength comparison using the optimized mesh size in [10].

Circuit	[10] w/ same mesh size in [10] (Set 1)				Proposed method				Improvement		
	Grid	Stub (μm)	Mesh (μm)	Total (μm)	Grid	Stub (μm)	Mesh (μm)	Total (μm)	Stub	Mesh	Total
s13207	8*8	3281	4848	8129	6*7	389	3938	4327	88.1%	18.8%	46.8%
s15850	8*8	2226	4062	6288	5*4	178	2285	2463	91.9%	43.7%	60.8%
s35932	12*12	10112	10871	20983	11*7	985	8157	9142	90.3%	25.0%	56.4%
s38417	12*12	8839	10794	19633	10*9	1252	8546	9798	85.8%	20.8%	50.1%
s38584	11*11	8533	12668	21201	12*7	674	10941	11615	92.1%	13.6%	45.2%
Average									89.6%	24.4%	51.9%

Table 3: Experimental Set 2: Wirelength comparison using the proposed optimized mesh size.

Circuit	[10] w/ proposed mesh size (Set 2)				Proposed method				Improvement		
	Grid	Stub (μm)	Mesh (μm)	Total (μm)	Grid	Stub (μm)	Mesh (μm)	Total (μm)	Stub	Mesh	Total
s13207	6*7	4012	3938	7950	6*7	389	3938	4327	90.3%	-	45.6%
s15850	5*4	3683	2285	5968	5*4	178	2285	2463	95.2%	-	58.7%
s35932	11*7	12867	8157	21024	11*7	985	8157	9142	92.3%	-	56.5%
s38417	10*9	11440	8546	19986	10*9	1252	8546	9798	89.1%	-	50.9%
s38584	12*7	9173	10941	20114	12*7	674	10941	11615	92.6%	-	42.3%
Average									91.9%	-	50.8%

Table 5: Trade-off analysis.

Ckt.	Slack Information				Logic Wire Incr.
	Pre-syn (ps)	Post-syn (ps)	Slack decr. (ps)	Decr. %	
s13207	297	272	25	8.4%	7.7%
s15850	213	180	33	15.5%	8.3%
s35932	277	265	12	4.3%	3.6%
s38417	647	612	35	5.4%	9.0%
s38584	113	110	3	2.7%	0.7%
Average			22	7.3%	5.9%

calculated by the proposed method is 7×11. It is visually observed that the registers (highlighted as dark blue boxes, the light green boxes represents the logic gates) are placed on a 7×11 grid wires.

5. CONCLUSIONS

A clock mesh network synthesis flow is proposed in this paper. In this flow, the timing slack aware incremental register placement with the non-uniform clock mesh generation methods are proposed. The clock mesh synthesis flow is demonstrated to have significantly (48%) less total switching capacitance on the clock distribution network than the previous work. The method is able to reduce the power dissipation without skew degradation. The proposed method is the first work known in literature that combines the incremental register placement with clock mesh network synthesis to reduce the on-chip power dissipation. The experimental results show that the methodology flow is effective and can be easily integrated into existing industrial physical design flow.

6. REFERENCES

[1] *NEOS Solvers*. http://neos.mcs.anl.gov/neos/solvers/.

[2] A. Abdelhadi, R. Ginosar, A. Kolodny, and E. G. Friedman. Timing-driven variation-aware nonuniform clock mesh synthesis. In *Proceedings of the Great Lakes Symposium on VLSI (GLSVLSI)*, pages 15–20, May 2010.

[3] R. Chaturvedi and J. Hu. An efficient merging scheme for prescribed skew clock routing. *IEEE Transactions on Very Large Scale Integration (VLSI) Systems*, 13(6):750–754, June 2005.

[4] W.-K. Chen, editor. *The VLSI Handbook*. CRC Press, 1st edition, 1999.

[5] Y. Chen and D. F. Wong. An algorithm for zero-skew clock tree routing with buffer insertion. In *Proceedings of the European Conference on Design and Test (ED&TC)*, pages 230–236, March 1996.

[6] J. Cong, A. B. Kahng, C.-K. Koh, and C.-W. Tsao. Bounded-skew clock and steiner routing. *ACM Transactions on Design Automation of Electronic Systems (TODAES)*, 3(3):341–388, 1998.

[7] T. H. Cormen, C. E. Leiserson, R. L. Rivest, and C. Stein. *Introduction to Algorithms*. MIT Press, 2nd edition, 2001.

[8] W. Elmore. The transient response of damped linear networks with particular regard to wideband amplifiers. *Journal of Applied Physics (AIP)*, 19(1):55–63, January 1948.

[9] E. G. Friedman. *Clock Distribution Networks in VLSI Circuits and Systems*. IEEE Press, 1995.

[10] M. R. Guthaus, G. Wilke, and R. Reis. Non-uniform clock mesh optimization with linear programming buffer insertion. In *Proceedings of the ACM/IEEE Design Automation Conference (DAC)*, pages 74–79, June 2010.

[11] N. Kurd, J. Barkarullah, R. Dizon, T. Fletcher, and P. Madland. A multigigahertz clocking scheme for the pentium(r) 4 microprocessor. *IEEE Journal of Solid-State Circuits (JSSC)*, 36(11):1647–1653, Nov. 2001.

[12] A. Rajaram, J. Hu, and R. Mahapatra. Reducing clock skew variability via cross links. In *Proceedings of the ACM/IEEE Design Automation Conference (DAC)*, pages 18–23, June 2004.

Table 4: Total capacitance and clock skew of the synthesized mesh network.

Circuit	Set 1		Set 2		Proposed method		Impro. over Set 1		Impro. over Set 2	
	Skew (ps)	Cap (fF)	Skew (ps)	Cap (fF)	Skew (ps)	Cap (fF)	Skew (ps)	Cap	Skew (ps)	Cap
s13207	0.7	5656	1.2	4138	0.3	3127	-0.4	44.7%	-0.9	24.4%
s15850	0.8	4935	1.2	2748	0.8	1837	0.0	62.8%	-0.4	33.1%
s35932	0.7	13169	2.1	9975	1.8	7112	1.1	46.0%	-0.3	28.7%
s38417	1.4	12791	1.2	9803	0.9	7280	-0.5	43.1%	-0.3	25.7%
s38584	1.1	13531	2.9	10417	0.8	7428	-0.3	45.1%	-2.1	28.7%
Average							0.0	48.3%	-0.8	28.1%

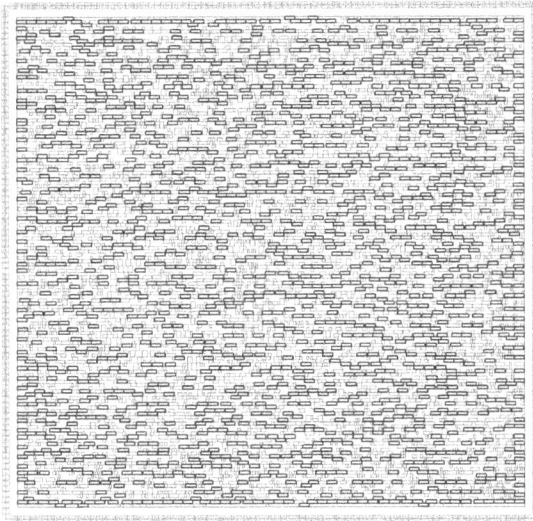

(a) Before incremental register placement.

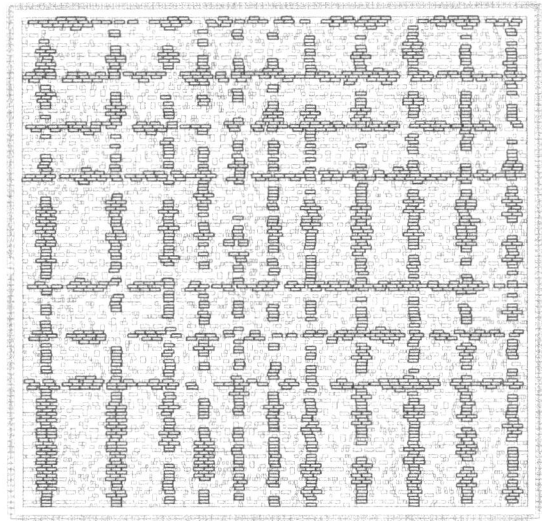

(b) After incremental registers placement.

Figure 7: The incremental register placement illustration for ISCAS'89 s35932.

[13] A. Rajaram and D. Pan. Meshworks: An efficient framework for planning, synthesis and optimization of clock mesh networks. In *Asia and South Pacific Design Automation Conference (ASPDAC)*, pages 250–257, Jan. 2008.

[14] A. Rajaram and D. Z. Pan. Variation tolerant buffered clock network synthesis with cross links. In *Proceedings of the International Symposium on Physical Design (ISPD)*, pages 157–164, 2006.

[15] A. Rajaram, D. Z. Pan, and J. Hu. Improved algorithms for link-based non-tree clock networks for skew variability reduction. In *Proceedings of the International Symposium on Physical Design (ISPD)*, pages 55–62, 2005.

[16] P. Restle, C. Carter, J. Eckhardt, B. Krauter, B. McCredie, K. Jenkins, A. Weger, and A. Mule. The clock distribution of the power4 microprocessor. In *Proceedings of the IEEE International Solid-State Circuits Conference (ISSCC)*, volume 1, pages 144–145, Feb. 2002.

[17] P. Restle, T. McNamara, D. Webber, P. Camporese, K. Eng, K. Jenkins, D. Allen, M. Rohn, M. Quaranta, D. Boerstler, C. Alpert, C. Carter, R. Bailey, J. Petrovick, B. Krauter, and B. McCredie. A clock distribution network for microprocessors. *IEEE Journal of Solid-State Circuits (JSSC)*, 36(5):792–799, May 2001.

[18] R. S. Shelar. An algorithm for routing with capacitance/distance constraints for clock distribution in microprocessors. In *Proceedings of the International Symposium on Physical Design (ISPD)*, pages 141–148, Mar. 2009.

[19] C. N. Sze. Ispd 2010 high performance clock network synthesis contest: benchmark suite and results. In *Proceedings of the International Symposium on Physical Design (ISPD)*, pages 143–143, 2010.

[20] S. Tam, J. Leung, R. Limaye, S. Choy, S. Vora, and M. Adachi. Clock generation and distribution of a dual-core xeon processor with 16mb l3 cache. In *Proceedings of the IEEE International Solid-State Circuits Conference (ISSCC)*, pages 1512–1521, Feb. 2006.

[21] B. Taskin and I. S. Kourtev. Delay insertion method in clock skew scheduling. In *Proceedings of the International Symposium on Physical Design (ISPD)*, pages 47–54, April 2005.

[22] G. Venkataraman, Z. Feng, J. Hu, and P. Li. Combinatorial algorithms for fast clock mesh optimization. *IEEE Transactions on Very Large Scale Integration Systems (TVLSI)*, 18(1):131–141, Jan. 2010.

[23] T. Xanthopoulos, D. Bailey, A. Gangwar, M. Gowan, A. Jain, and B. Prewitt. The design and analysis of the clock distribution network for a 1.2 ghz alpha microprocessor. In *Proceedings of the IEEE International Solid-State Circuits Conference (ISSCC)*, pages 402–403, Feb. 2001.

Impact of Manufacturing on Routing Methodology at 32/22 nm

Alexander Volkov

Mentor Graphics

Fremont, CA, USA

Alexander_Volkov@mentor.com

Abstract

As the IC industry accelerates adoption of the 45 nm and 32 nm process nodes, designers are facing significant new challenges in meeting quality and manufacturability targets. The new challenges of nanometer routing—including very large (1B transistor) designs, complex DRC/DFM, and multiple optimization objectives—are stressing the fundamental ability of digital routing tools to solve the layout topology.

To ensure that physical designs can be reliably manufactured, foundries are greatly expanding the number and complexity of design rules and DFM requirements at advanced nodes. The number of DRC and DFM rules has roughly doubled between the 90- and 32-nm nodes, depending on the foundry. The rule complexity, which is measured by the number of operations required to verify the rules, has grown even faster.

DFM checks, which used to be voluntary, are now becoming mandatory just like traditional DRCs—with the type and complexity of checks being foundry dependent. DFM violations can cause issues ranging from chip failure to reduced reliability and decreased performance. If they aren't made carefully, however, changes to improve manufacturability can reduce performance, increase power consumption, or otherwise compromise the design..

20 nm brings one more aspect from manufacturing side to the routing. It's double patterning (DP) what is completely new term both for process and Place&Route system. Contrary to DRC/DFM which is local effect double pattern violation is global one. Improper geometry on one side of a chip could result in the problem on other side of the chip. Such formulation contradicts to the local Search&Repair approach.

To fully realize the advantages of moving to a new process node, while also maintaining turn-around-time, designers need a new routing platform that can more accurately model the numerous and complex DRC and DFM rules from the earliest stages of the routing process, natively invoke signoff DRC/DFM models and engines, use timing-driven routing to optimize critical paths, and manage huge design sizes with predictable runtimes.

Categories & Subject Descriptors: B.7.2 Placement and routing

General Terms: Algorithms, Design.

Bio

The speaker is Principal Engineer at Mentor Graphics and Technical Leader of Olympus-SoC routing team. Having has 15 years experience in EDA industry. Last decade his primary focus is the development of commercial ASIC detail routing systems. He is the author of Olympus-SoC polygon library, DRC/LVS/Antenna verification, DFM optimization techniques and integration with Calibre tools. He actively participated in development of detail routing algorithms and 45/32nm process qualifications of Mentor's routing solution. He received M.S. of Computer Design and Organization from Moscow Institute of Physics and Technology.

Copyright is held by the author/owner(s).
ISPD'11, March 27–30, 2011, Santa Barbara, California, USA.
ACM 978-1-4503-0550-1/11/03.

The ISPD-2011 Routability-Driven Placement Contest and Benchmark Suite

Natarajan Viswanathan, Charles J. Alpert, Cliff Sze, Zhuo Li, Gi-Joon Nam, Jarrod A. Roy

IBM Corporation, 11501 Burnet Road, Austin, TX 78758
{ nviswan, alpert, csze, lizhuo, gnam }@us.ibm.com
jarrod.a.roy@gmail.com

ABSTRACT

The last few years have seen significant advances in the quality of placement algorithms. This is in part due to the availablity of large, challenging testcases by way of the ISPD-2005 [17] and ISPD-2006 [16] placement contests. These contests primarily evaluated the placers based on the half-perimeter wire length metric. Although wire length is an important metric, it still does not address a fundamental requirement for placement algorithms, namely, the ability to produce routable placements.

This paper describes the ISPD-2011 routability-driven placement contest, and a new benchmark suite that is being released in conjunction with the contest. All designs in the new benchmark suite are derived from industrial ASIC designs, and can be used to perform both placement and global routing. By way of the contest and the associated benchmark suite, we hope to provide a standard, publicly available framework to help advance research in the area of routability-driven placement.

Categories and Subject Descriptors

B.7.2 [**Hardware, Integrated Circuits, Design Aids**]: Placement and routing

General Terms

Algorithms, Design, Experimentation, Performance

Keywords

Placement, Benchmarks, Physical Design

1. INTRODUCTION

Physical synthesis is one of the most important steps in the design of large-scale integrated circuits, and has a significant impact on design closure [1, 14]. One of the key challenges for modern physical synthesis flows is that of routing congestion [2]. There are multiple factors that contribute

to the issue of routing congestion in advanced process technologies like $65nm$ and below. A few of them being, increased use of embedded IPs or memories on the die that block metal layers, more layer stacks to achieve higher performance, reduced die size to control manufacturing cost, and complicated logic structures such as cross-bars. As a result, physical synthesis needs to consider congestion during the entire design flow.

Previous ISPD contests [16, 17] have been instrumental in bringing significant advances in the quality of global placement algorithms [5, 6, 8, 11, 12, 20–22]. These contests evaluated the placers using the half-perimeter wire length (ISPD-05) and their spreading capability by way of a placement target_density in addition to wire length and runtime (ISPD-06). Although wire length and spreading ability are important metrics, they still do not address a fundamental requirement for placement algorithms, namely, the ability to produce routable placements. An excessively packed design with good wire length or one that is well spread out, is of no consequence if the design is ultimately unroutable. In addition to the advances in placement, there has also been significant progress in routing algorithms, both in terms of speed and solution quality [7, 9, 10, 13, 15, 18, 23–25, 27].

Despite the advances in placement and routing, only a handful of techniques combine the two (e.g., [19, 26]) to handle routability-driven placement. Hence, the focus of this year's contest is to determine the top performing placers in terms of routability, as evaluated by a global router.

The key objectives of the ISPD-2011 contest are:

- Release more advanced benchmarks derived from industrial ASIC designs, that can be used to perform both placement and global routing. These benchmarks are representative of today's designs with numerous placement blockages, more metal layers, varying metal width and spacing across layers, etc. As a result, they complicate both the placement and routing steps.

- Spur academic research in the area of fast, yet reasonably accurate routing congestion analysis.

- Motivate research to integrate placement and global routing to handle routability-driven placement.

- Coming up with standard metrics to evaluate the routability of a given placement.

We hope that a standard, publicly available framework will lead to some interesting developments in the area of routability-driven placement. By way of this contest, academic placement algorithms would be solving the same problem as industrial place-and-route algorithms, where routability happens to be one of the key objectives of placement.

Permission to make digital or hard copies of all or part of this work for personal or classroom use is granted without fee provided that copies are not made or distributed for profit or commercial advantage and that copies bear this notice and the full citation on the first page. To copy otherwise, to republish, to post on servers or to redistribute to lists, requires prior specific permission and/or a fee.
ISPD'11, March 27–30, 2011, Santa Barbara, California, USA.
Copyright 2011 ACM 978-1-4503-0550-1/11/03 ...$10.00.

The rest of this paper is organized as follows: Section 2 gives a brief description of the contest. Section 3 describes the ISPD-2011 benchmark suite. Section 4 presents the floorplan layouts for all the designs in the benchmark suite. Finally, Section 5 provides concluding remarks.

2. THE ISPD-2011 CONTEST

One of the requirements for the contest was the availability of a set of "golden" routers to evaluate the solution quality of the placements. Prior to the contest, the coordinators worked with the academic routing teams of GRIP [23, 24], NTHU-Route [7, 13], FGR [10, 18], FastRoute [25, 27] and BoxRoute [9, 15] to qualify their routers to be used for the contest.

In addition, prior to the contest, four of the eight benchmarks and the source code or library files of the qualified golden routers were released to all the competing teams. This was done for the following reasons: (a) to give the teams an idea of the complexity of the designs, (b) ensure that the placers were able to handle the file formats and generate a valid placement solution, and (c) give the teams an opportunity to incorporate the routers within their placers and/or evaluate the routability of their placements.

Approximately three weeks prior to the 2011 ISPD symposium, the teams were required to submit their placement tool binaries and any run-scripts. The contest coordinators then ran the tools on all eight benchmarks and used the golden routers to evaluate the solution quality of the placements.

The placements were evaluated on the following metrics:

- Legality.
- Routability.
- Overall run-time to generate a placement solution.

A placement is considered legal if it satisfies the following criteria:

- The fixed nodes are in their original locations.
- The movable nodes are placed within the placement region.
- The movable nodes are aligned to the circuit rows.
- Within each row, the movable nodes are placed in valid placement sites.
- There is no overlap among the nodes (movable and/or fixed).

The qualified golden routers were used to evaluate the routability of the placement solutions. The metric used was the total overflow (TOF) number as reported by the golden routers. For any tile edge whose demand exceeds its available capacity, the overflow (OF) of the tile edge on a particular metal layer is defined as the excess routing tracks that span the tile edge on that layer multiplied by the sum of the wire width and wire spacing for the layer. The TOF is the summation of the OF values for all the tile edges across all metal layers. We used the TOF metric for the contest as it is a reasonable first-order metric to determine routability.

To encourage reasonable turn-around-time, we used the run-time of the placers to scale the overflow values. Thus, for the same overflow value, the placer with the faster run-time would get a higher score.

3. THE ISPD-2011 BENCHMARK SUITE

Table 1 summarizes the characteristics of the designs released as part of the ISPD-2011 benchmark suite. For each design, the reported statistics are:

- *Total Nodes*: The total number of nodes in the design.
- *Movable Nodes*: The number of movable nodes in the design.
- *Terminal Nodes*: The number of fixed "terminal" nodes in the design. No overlap is allowed between the movable and terminal nodes.
- *Terminal_NI Nodes*: The number of fixed "terminal_NI" nodes in the design. Overlap is allowed between the movable and terminal_NI nodes.
- *Total Nets*: The total number of nets in the design.
- *Total Pins*: The total number of pins in the netlist.
- *Design Util.*: The design utilization in percentage. Design utilization is defined as the ratio of the area sum of the movable and terminal nodes to the area of the placement region.
- *Design Den.*: The design density in percentage. Design density is defined as the ratio of the area sum of the movable nodes to the available free-space in the design. Where, free-space is given by: (area of the placement region – area sum of the terminal nodes).

All the designs in the benchmark suite were translated from the IBM internal data format to the GSRC Bookshelf placement format [3, 4] using the IBM CPLACE tool. To preserve the integrity of the industrial designs during translation and present the data as-is, two new features were introduced for the ISPD-2011 contest. The following subsections give a brief description of these features.

3.1 Non-rectangular Fixed Nodes

A subset of the fixed nodes in the design are not rectangular. Place and route tools need to handle them in an appropriate manner as these nodes would have an impact on various metrics like placement density and routing edge capacity. An example non-rectangular node is depicted in Figure 1(a). From Figure 1(b), this non-rectangular node is represented as: (a) an enclosing rectangle (blue box with the solid line), and (b) a set of rectangular component shapes (red boxes with dashed lines). Finally, Figure 1(c) shows the corresponding representation in the benchmark file format. From Figure 1(c): (a) the *circuit.nodes* file gives the dimensions of the enclosing rectangle, (b) the *circuit.pl* file gives the lower-left coordinate of the enclosing rectangle, (c) the *circuit.shapes* file gives the component shape definitions for the non-rectangular node, and (d) the *circuit.nets* file (not shown in the figure) gives the pin-offsets from the center of the enclosing rectangle.

3.2 Macro-pins on "terminal_NI" Nodes

The use of a hierarchical design flow is becoming more prevalent in the industry due to the increasing complexity of present-day designs. To provide interconnection between the different levels of hierarchy, designs often use fixed pins that are not restricted to the periphery of the placement region.

Design	Total Nodes	Movable Nodes	Terminal Nodes	Terminal_NI Nodes	Total Nets	Total Pins	Design Util. (%)	Design Den. (%)
superblue18	483452	442405	25063	15984	468918	1864306	67	47
superblue4	600220	521466	40550	38204	567607	1884008	70	44
superblue5	772457	677416	74365	20676	786999	2500306	77	37
superblue1	847441	765102	52627	29712	822744	2861188	69	35
superblue2	1014029	921273	59312	33444	990899	3228345	76	28
superblue15	1123963	829614	252053	42296	1080409	3816680	73	60
superblue10	1129144	914921	153595	60628	1085737	3665711	75	35
superblue12	1293433	1278084	8953	6396	1293436	4774069	56	44

Table 1: Design statistics of the ISPD-2011 benchmark suite.

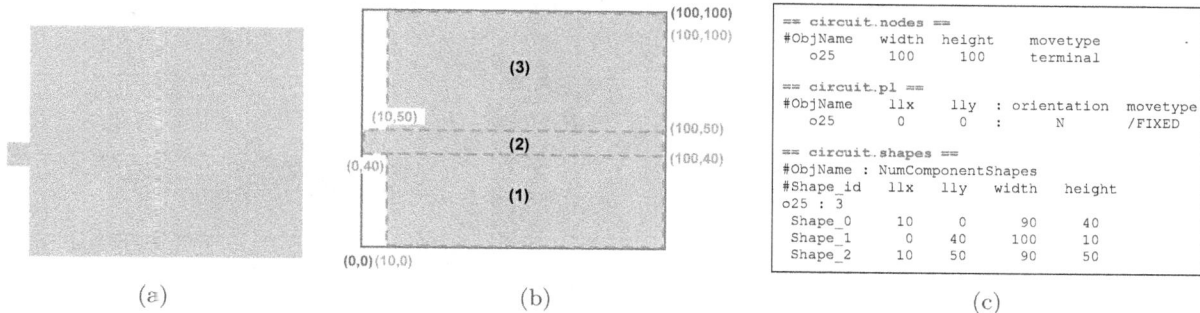

```
== circuit.nodes ==
#ObjName    width    height    movetype
    o25      100      100      terminal

== circuit.pl ==
#ObjName    llx    lly  : orientation  movetype
    o25      0      0   :      N        /FIXED

== circuit.shapes ==
#ObjName : NumComponentShapes
#Shape_id   llx    lly    width    height
o25 : 3
   Shape_0   10     0      90       40
   Shape_1    0     40     100      10
   Shape_2   10     50     90       50
```

(a)   (b)   (c)

Figure 1: (a) A non-rectangular fixed node, (b) Corresponding representation as an enclosing rectangle (blue box with the solid line) and a set of rectangular component shapes (red boxes with the dashed lines), (c) Benchmark file format description of the non-rectangular node.

We denote such pins as "macro-pins". Macro-pins reside on a metal layer above the ones used within a standard-cell for its internal pins and routing. To represent such pins, we introduce a new node class called "terminal_NI". Each terminal_NI node is associated with one macro-pin. For placement, these nodes appear to reside above the placement region. In other words standard-cells can be placed below the terminal_NI nodes without resulting in an overlap. For routing, the pin associated with the terminal_NI node is present on a higher metal layer (e.g., M4) as opposed to M1 or M2.

3.3 Benchmark Files

To introduce the new features described in the previous sub-sections, and to provide routing information for each design, two new files were added to the GSRC Bookshelf placement format. As a result, each design in the benchmark suite has the following files:

1. *circuit.aux*: Auxiliary file with a list of the files that need to be read in by the placer.

2. *circuit.nodes*: For each node in the design, specifies node attributes: name, dimensions, movetype (movable, terminal or terminal_NI).

3. *circuit.nets*: Specifies the set of nets and associated pins in the design.

4. *circuit.pl*: Specifies the location and orientation for all the nodes.

5. *circuit.scl*: Specifies the circuit row information to place the nodes.

6. *circuit.shapes*: Specifies the set of component rectangular shapes for all the non-rectangular nodes.

7. *circuit.route*: Provides the information to perform global routing, given a placement of the nodes in the design.

8. *circuit.wts*: Currently not used as all the nodes and nets in the design have the same weight.

4. FLOORPLAN LAYOUTS

Figure 2 and Figure 3 depict the floorplans for all the designs in the benchmark suite. The figures show only the fixed "terminal" nodes. The light-red shaded boxes with blue boundary represent the rectangular fixed nodes and the gray shaded boxes represent the non-rectangular fixed nodes in the design. These figures demonstrate the varying characteristics and associated challenges of modern ASIC floorplans. For example, all the fixed nodes in *superblue18* and *superblue12* are pushed to the periphery of the placement region. This gives the placement tool a large amount of free-space in and around the center of the placement region. On the other hand, the fixed nodes in *superblue15* and *superblue10* fragment the placement region into multiple sub-regions. In addition, the tall thin fixed nodes in most of the designs create "alleys". Alleys in the placement region can lead to significant congestion if a net gets split across the corresponding fixed nodes.

5. CONCLUSIONS

This paper describes the ISPD-2011 routability-driven placement contest, and a new benchmark suite that is being released in conjunction with the contest. These benchmarks

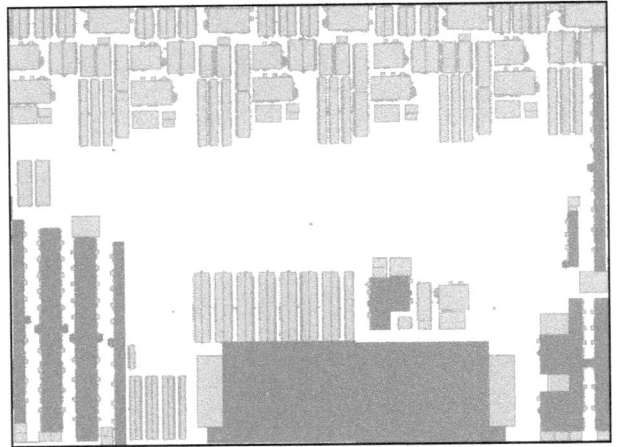

superblue18   superblue4

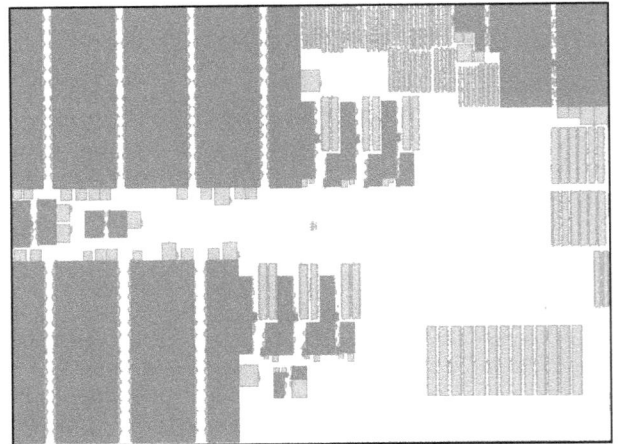

superblue5   superblue1

Figure 2: Flcorplan layout figures of designs superblue18, superblue4, superblue5 and superblue1. The light-red shadɘd boxes with blue boundary represent the rectangular fixed nodes and the gray shaded boxes represent the non-rectangular fixed nodes in the design.

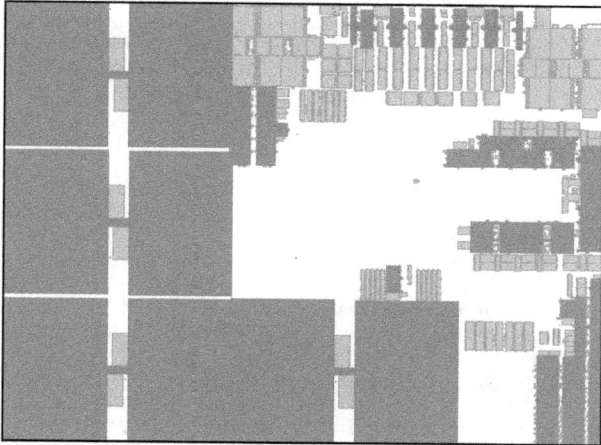

superblue2

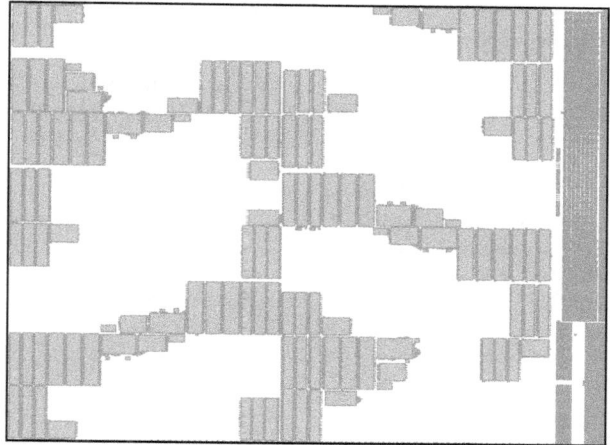

superblue15

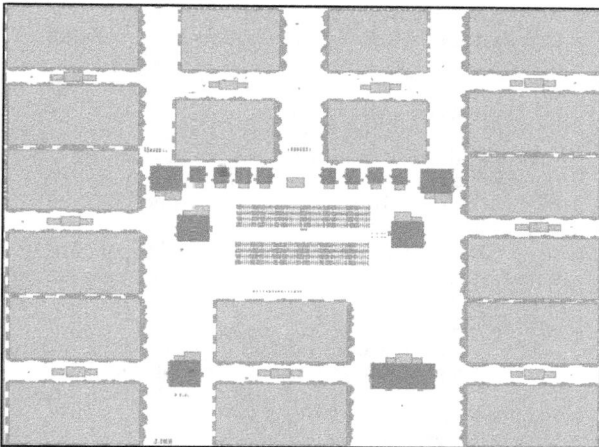

superblue10

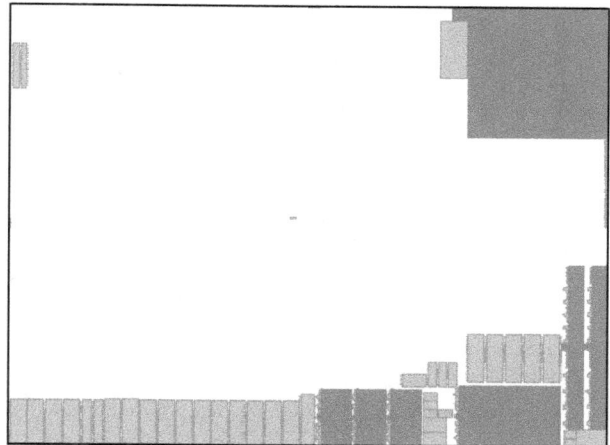

superblue12

Figure 3: Floorplan layout figures of designs superblue2, superblue15, superblue10 and superblue12. The light-red shaded boxes with blue boundary represent the rectangular fixed nodes and the gray shaded boxes represent the non-rectangular fixed nodes in the design.

reflect the complexities of modern ASIC designs with numerous placement blockages, more metal layers, varying metal width and spacing across layers, etc. As a result, they complicate both the placement and routing steps. In addition, the contest provides standard metrics and routers to evaluate the routability of placement algorithms. We hope that such a standard, publicly available framework will lead to some interesting developments in the area of routability-driven placement.

6. ACKNOWLEDGMENTS

One of the main requirements for the contest was a set of routers to evaluate the placement solutions. The authors would like to thank Hamid Shojaei, Azadeh Davoodi, Chou Hsueh Ju, Yen-Jung Chang, Yu-Ting Lee, Tsung-Hsien Lee, Jhih-Rong Gao, Pei-Ci Wu, Ting-Chi Wang, and other members of the NTHURoute team, Jin Hu, Igor Markov, Yanheng Zhang, Chris Chu, Jhih-Rong Gao and David Pan for taking the time to qualify their routers to be used for the ISPD-2011 contest. Thanks also go to Bertram Bradley and Randy Darden at IBM for their help in releasing the benchmarks.

7. REFERENCES

[1] C. J. Alpert, S. Karandikar, Z. Li, G.-J. Nam, S. T. Quay, H. Ren, C. N. Sze, P. G. Villarrubia, and M. Yildiz. Techniques for fast physical synthesis. *Proceedings of IEEE*, 95(3):573–599, Mar. 2007.

[2] C. J. Alpert, Z. Li, M. D. Moffitt, G.-J. Nam, J. A. Roy, and G. Telleze. What makes a design difficult to route. In *Proc. Intl. Symp. on Physical Design*, pages 7–12, 2010.

[3] A. E. Caldwell, A. B. Kahng, and I. L. Markov. Placement formats, rev. 1.2. In *http://vlsicad.ucsd.edu/GSRC/bookshelf/Slots/Placement/plFormats.html*.

[4] A. E. Caldwell, A. B. Kahng, and I. L. Markov. Toward CAD-IP reuse: The MARCO GSRC Bookshelf of fundamental CAD algorithms. In *IEEE Design and Test*, pages 72–81, 2002.

[5] T. Chan, J. Cong, and K. Sze. Multilevel generalized force-directed method for circuit placement. In *Proc. Intl. Symp. on Physical Design*, pages 185–192, 2005.

[6] T. F. Chan, J. Cong, J. R. Shinnerl, K. Sze, and M. Xie. mPL6: Enhanced multilevel mixed-size placement. In *Proc. Intl. Symp. on Physical Design*, pages 212–214, 2006.

[7] Y.-J. Chang, Y.-T. Lee, and T.-C. Wang. NTHU-Route 2.0: A fast and stable global router. In *Proc. IEEE/ACM Intl. Conf. on Computer-Aided Design*, pages 338–343, 2008.

[8] T.-C. Chen, Z.-W. Jiang, T.-C. Hsu, H.-C. Chen, and Y.-W. Chang. A high-quality mixed-size analytical placer considering preplaced blocks and density constraints. In *Proc. IEEE/ACM Intl. Conf. on Computer-Aided Design*, pages 187 – 192, 2006.

[9] M. Cho and D. Z. Pan. Boxrouter: A new global router based on box expansion and progressive ilp. In *Proc. ACM/IEEE Design Automation Conf.*, pages 373–378, 2006.

[10] J. Hu, J. A. Roy, and I. L. Markov. Completing high-quality routes. In *Proc. Intl. Symp. on Physical Design*, pages 35–41, 2010.

[11] A. B. Kahng, S. Reda, and Q. Wang. Architecture and details of a high quality, large-scale analytical placer. In *Proc. IEEE/ACM Intl. Conf. on Computer-Aided Design*, pages 890–897, 2005.

[12] M.-C. Kim, D.-J. Lee, and I. L. Markov. SimPL: An effective placement algorithm. In *Proc. IEEE/ACM Intl. Conf. on Computer-Aided Design*, pages 649–656, 2010.

[13] T.-H. Lee and T.-C. Wang. Robust layer assignment for via optimization in multi-layer global routing. In *Proc. Intl. Symp. on Physical Design*, pages 159–166, 2009.

[14] Z. Li and C. J. Alpert. What is physical synthesis. *ACM/SIGDA E-Newsletter*, 40(12), Dec.

[15] K. Y. D. Z. P. M. Cho, K. Lu. Boxrouter 2.0: Architecture and implementation of a hybrid and robust global router. In *Proc. IEEE/ACM Intl. Conf. on Computer-Aided Design*, pages 503–508, 2007.

[16] G.-J. Nam, C. J. Alpert, and P. Villarrubia. ISPD 2006 placement contest: Benchmark suite and results. In *Proc. Intl. Symp. on Physical Design*, pages 167–167, 2006.

[17] G.-J. Nam, C. J. Alpert, P. Villarrubia, B. Winter, and M. Yildiz. The ISPD2005 placement contest and benchmark suite. In *Proc. Intl. Symp. on Physical Design*, pages 216–220, 2005.

[18] J. A. Roy and I. L. Markov. High-performance routing at the nanometer scale. In *Proc. IEEE/ACM Intl. Conf. on Computer-Aided Design*, pages 496–502, 2007.

[19] J. A. Roy, N. Viswanathan, G.-J. Nam, C. J. Alpert, and I. L. Markov. CRISP: Congestion reduction by iterated spreading during placement. In *Proc. IEEE/ACM Intl. Conf. on Computer-Aided Design*, pages 357–362, 2009.

[20] P. Spindler and F. M. Johannes. Fast and robust quadratic placement combined with an exact linear net model. In *Proc. IEEE/ACM Intl. Conf. on Computer-Aided Design*, pages 179 – 186, 2006.

[21] N. Viswanathan, G.-J. Nam, C. J. Alpert, P. Villarubia, H. Ren, and C. Chu. RQL: Global placement via relaxed quadratic spreading and linearization. In *Proc. ACM/IEEE Design Automation Conf.*, pages 453–458, 2007.

[22] N. Viswanathan, M. Pan, and C. Chu. Fastplace 3.0: A fast multilevel quadratic placement algorithm with placement congestion control. In *Proc. Asia and South Pacific Design Automation Conf.*, pages 135–140, 2007.

[23] T.-H. Wu, A. Davoodi, and J. T. Linderoth. GRIP: scalable 3D global routing using integer programming. In *Proc. ACM/IEEE Design Automation Conf.*, pages 320–325, 2009.

[24] T.-H. Wu, A. Davoodi, and J. T. Linderoth. A parallel integer programming approach to global routing. In *Proc. ACM/IEEE Design Automation Conf.*, pages 194–199, 2010.

[25] Y. Xu, Y. Zhang, and C. Chu. Fastroute 4.0: Global router with efficient via minimization. In *Proc. Asia and South Pacific Design Automation Conf.*, pages 576–581, 2009.

[26] Y. Zhang and C. Chu. CROP: Fast and effective congestion refinement of placement. In *Proc. IEEE/ACM Intl. Conf. on Computer-Aided Design*, pages 344–350, 2009.

[27] Y. Zhang, Y. Xu, and C. Chu. FastRoute 3.0: A fast and high quality global router based on virtual capacity. In *Proc. IEEE/ACM Intl. Conf. on Computer-Aided Design*, pages 344–349, 2008.

Vertical Slit Transistor Based Integrated Circuits (VeSTICs): Feasibility Study

Wojciech Maly
ECE Department, Carnegie Mellon
Pittsburgh, PA 15213, USA
maly@ece.cmu.edu

Abstract

In this presentation feasibility of Vertical Slit Transistor Based Integrated Circuits (VeSTICs) is evaluated. VeSTICs paradigm has been conceived as a response to the rapidly growing complexity of the traditional CMOS-based approach to challenges posed by the nano-scale era. This paradigm has been constructed using notion of a strict layout regularity imposed on VeSTIC layouts. The central element of the proposed vision is new junction-less Vertical Slit Field Effect Transistor (VeSFET) with twin independent gates. It is expected that VeSTICs will enable much denser, much easier to design, test and manufacure ICs, as well as, will be 3D-extendable and OPC-free. (More exhaustive description of this paradigm one can find at: *http://vestics.org*).

This talk reviews all of the above VeSTICs characteristics. The focus of the talk, however, is on first silicon results (just obtained with simple SOI-like process), extremely low power and dependency between actual achievable transistor density and layout design rules. These characteristics are compared to a variety of traditional CMOS-based paradigms using the same infrastructure.

Categories and Subject Descriptors
B.7.1 [**Integrated Circuits**]: Design Style – *Advanced Technologies*

General Terms
Design, Economics.

Authors Keywords
VeSFET, VeSTICs, Twin Gate, Double Gate, Dual Gate, Junction-less, Strict Layout Regularity, Regular Layout, 3D ICs, OPC-free, SOI, Extremely Low Power, DFM .

Bio

Wojciech Maly received his education in Poland. Since September 1983, he has been with Carnegie Mellon University, where he is a Whitaker Professor of Electrical and Computer Engineering.

Dr. Maly's research interests have been focused on the interfaces between IC design, testing and manufacturing with the stress on the stochastic nature of phenomena relating these three domains.

Copyright is held by the author/owner(s).
ISPD'11, March 27–30, 2011, Santa Barbara, California, USA.
ACM 978-1-4503-0550-1/11/03.

Litho and Design: Moore Close than Ever

Vivek Singh

Intel Fellow

Intel Corporation

Hillsboro, OR, USA

vivek.singh@intel.com

Abstract

As the gap between the lithography wavelength and critical feature size has continued to increase, the semiconductor industry has had to adjust. Previously, scaling along Moore's Law had relied on improvement in lithography equipment, occasionally by reducing the wavelength and frequently by improving the effective numerical aperture. For a few years now, this scaling has relied more on optimization of the lithography process and on deeper co-optimization between process and design. Recently, a key element of lithography optimization has been computational lithography, which in turn has consisted of two advanced features: inverse lithography, and source-mask optimization. The second prong of scaling, co-optimization between process and design, has involved close cooperation between the two camps, using tools and methods that have been given the umbrella label of DFM. While some of the mystique surrounding DFM continues to linger, we at Intel have always believed that DFM is defined in terms of the results it produces. In that sense, DFM is a broad set of practices that helps produce compelling products at high yield levels on a competitive schedule. While the fundaments of these practices have existed for many years, the volume of effort and degree of sophistication have increased with each technology generation. By co-optimizing design and process early in the development cycle for a given node, we arrive at a set of design rules that meet the process and design requirements. The thoroughness of this early work results in these rules being stable through the development cycle. The application of these practices to 32 nm and 22 nm nodes, with ongoing refinements to meet new challenges, is ensuring the continued march of Moore's Law.

This paper will cover the spectrum of lithography options that lie ahead, computational lithography solutions that are helping drive Moore's Law, and the general implications of both on design.

Categories & Subject Descriptors: B.7.2 [Hardware]: Integrated Circuits-Design Aids

General Terms: Algorithms

Bio

Vivek Singh is an Intel Fellow and Director of Computational Lithography in Intel's Technology and Manufacturing Group. He is responsible for Intel's CAD and modeling tool development in full chip OPC, lithography verification, rigorous lithography modeling, next-generation lithography selection, inverse lithography technologies and double patterning. Singh joined Intel in 1993 as a modeling applications engineer. He was appointed team leader for the Resist and Applications Group in 1996, and became the Manager of the Lithography Modeling Group in 2000. He was appointed Intel Fellow in 2008. Singh graduated from the Indian Institute of Technology in Delhi with a bachelor's degree in chemical engineering in 1989. He earned a master's degree in chemical engineering in 1990, a Ph.D. minor in electrical engineering in 1993, and a Ph.D. in chemical engineering in 1993, all from Stanford University.

Copyright is held by the author/owner(s).

ISPD'11, March 27–30, 2011, Santa Barbara, California, USA.

ACM 978-1-4503-0550-1/11/03.

E-Beam Lithography Stencil Planning and Optimization with Overlapped Characters

Kun Yuan and David Z. Pan
ECE Dept. Univ. of Texas at Austin, Austin, TX 78712
{kyuan, dpan}@cerc.utexas.edu

ABSTRACT

Electronic Beam Lithography (EBL) is an emerging maskless nanolithography technology which directly writes the desired circuit pattern into wafer using e-beam, thus it overcomes the diffraction limit of light in current optical lithography system. However, low throughput is its key technical hurdle. In conventional EBL system, each rectangle in the layout will be projected by one electronic shot, through a Variable Shaped Beam (VSB). This would be extremely slow. As an improved EBL technology, Character Projection(CP) shoots complex shapes, so called characters, by putting them into a pre-designed stencil to increase throughput. However, only a limited number of characters can be put on the stencil due to its area constraint. For those patterns not in the stencil, they still need to be written by VSB. A key problem is how to select an optimal set of characters and pack them on the CP stencil to minimize total processing time. In this paper, we investigate a new problem of EBL stencil design with overlapped characters. Different from previous works, besides selecting appropriate characters, their placements on the stencil are also optimized in our framework. Our experimental results show that compared to conventional stencil design methodology without overlapped characters, we are able to reduce total projection time by 51%.

Categories and Subject Descriptors

B.7.2 [**Hardware, Integrated Circuit**]: Design Aids

General Terms

Algorithms, Design, Performance

Keywords

Electronic Beam Lithography, Stencil Design

1. INTRODUCTION

As aggressive scaling continues, the conventional 193nm optical photolithography technology is facing the great challenge of printing sub-32nm. For near future, double/multiple patterning lithography has been developed as temporary solution for 32nm, 22nm, even 16nm, technology [1–3]. In the longer future, the semiconductor industries and researchers have been actively pushing on alternative emerging nanolithography to print finer feature size below 16nm, such as Electronic Beam Lithography (EBL), Extreme Ultra Violet (EUV) and nanoimprint.

Permission to make digital or hard copies of all or part of this work for personal or classroom use is granted without fee provided that copies are not made or distributed for profit or commercial advantage and that copies bear this notice and the full citation on the first page. To copy otherwise, to republish, to post on servers or to redistribute to lists, requires prior specific permission and/or a fee.
ISPD'11, March 27–30, 2011, Santa Barbara, California, USA.
Copyright 2011 ACM 978-1-4503-0550-1/11/03 ...$10.00.

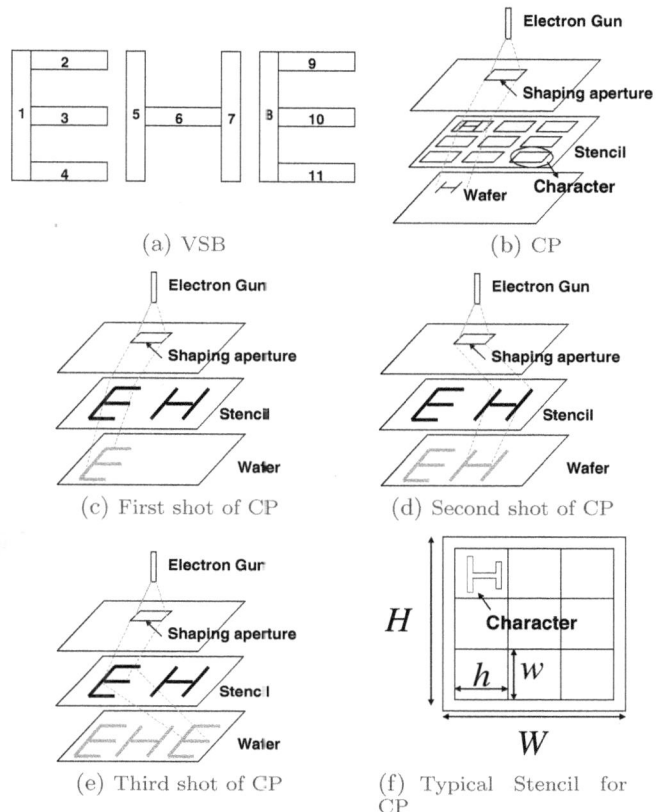

Figure 1: Electron Beam Lithography

(a) VSB (b) CP (c) First shot of CP (d) Second shot of CP (e) Third shot of CP (f) Typical Stencil for CP

EBL [4–6] is a maskless technology which shoots desired patterns directly into a silicon wafer, with charged particle beam. The primary advantage is that it is one of the ways to beat the diffraction limit of light of current well-adopted optical lithography [7]. However, the key limitation of electron beam lithography is low throughput.

The conventional type of EBL system is Variable Shaped Beam (VSB). In VSB, the layout is usually decomposed into a set of rectangles, and each one would be shot into resist by dose of electron sequentially. As Fig. 1 (a) shows, the pattern of "EHE" is divided into eleven rectangles and needs total eleven shots. The whole processing time of this technique increases with number of beam shots. This makes its throughput very low for modern complicated design, which is commonly composed of significant number of small rectangles.

The Character Projection (CP) technology [4–6] has been invented for improving the throughput of VSB methods. The key idea is to print some complex shapes in one electronic beam shot, rather than writing multiple small rectangles. This reduces manufacturing time significantly. In detail, as the projection system of CP in Fig. 1 (b) illustrates, a library of layout configurations, called **Characters**, or **Templates**, are prepared on a **stencil** first. During manufacturing, if any character exists in the targeted design, it will be chosen in the system and projected into the wafer. To print the example of Fig. 1 (a), suppose two characters "E" and "H" are pre designed for the stencil. By adjusting and aligning the shaping aperture and stencil, we can print the patterns of "E", "H", "E" in sequential manner, as Fig. 1 (c)-(e) shows. Totally, it only takes three shots.

Due to less beam shots for the same layout, CP system is much faster than VSB. However, the number of characters is limited due to the area constraint of the stencil. As in the example of Fig. 1 (f), there are only maximum $\lfloor W/w \rfloor \lfloor H/h \rfloor$ characters. For modern design, it is not practical to fully make use of CP, due to numerous distinct circuit patterns. Those patterns which do not match any character are still required to be written by VSB.

Several methodologies have been proposed to design and select group of circuit patterns as characters for minimizing total projection time of both CP and VSB. In [8], frequently-used standard cells are greedily chosen as characters, processed by CP technology. M.Sugihara at el. [9–12] employ integer linear programming to optimize the throughput, given a set of character candidates. Recently, EDA vendor D2S inc [4–6] proposes improving stencil design from a new point of view, but with no detailed algorithm presented. They show that, in practice, when individual character/template is designed, blanking area is usually reserved around its boundaries. By sharing blanks between adjacent templates, more characters can be placed on the stencil than the regular design of Fig. 1 (f), better improving the throughput.

The work of [4–6] implies that, to fully minimize the total projection time of EBL, besides selecting appropriate characters as [8–12], their relative locations on the stencil should be taken into account at the same time due to possible overlapping. In the paper, we will investigate on this new problem of electronic beam lithography stencil design with overlapped characters. One/two dimensional problem is researched separately, depending on whether the available overlapping space of characters is non-uniform in either horizontal or both directions. The main contributions of our work are stated as follows.

1. We co optimize the selection process of characters and their physical placements on stencil for effective EBL throughput improvement.

2. We propose a four-phase iterative refinement process to conduct one-dimensional stencil design optimization. A Hamilton-path based approach has been developed to solve single-row reordering efficiently and effectively.

3. We develop a Sequence Pair (SP) based simulated annealing framework to optimize general two-dimensional stencil design. Two SP-related techniques have been proposed to ensure correct and fast character placement evaluation, and two specialized perturbation methods have been developed for robust solution improvement of simulated annealing process.

2. PRELIMINARY AND PROBLEM FORMULATION

2.1 Overlapped Character

Electronic Beam Lithography (EBL) is a maskless technique, which shoots desired patterns directly into a silicon wafer, and

can potentially combat device parameter variations [13–15]. Various investigation [9–12] have been conducted on the optimization of character selection for EPL technology, where no intersection is allowed between templates on the stencil, as shown by Fig. 1 (f). Recently, the work of [4–6] shows that the design of stencil can be further improved by overlapping adjacent characters, which allows more templates to be put and increase the throughput.

As pointed out by [4–6], when individual character is designed, blanking space is usually reserved around its enclosed rectangular circuit pattern, shown by Fig. 2 (a). The reason is that, when the electron beam is scatted from the shaping aperture of Fig. 1 (b), it could span larger area on the stencil than the layout to be printed. In order to avoid projecting any unwanted image, the white space should be preserved. These blanking areas offer great opportunity for character sharing.

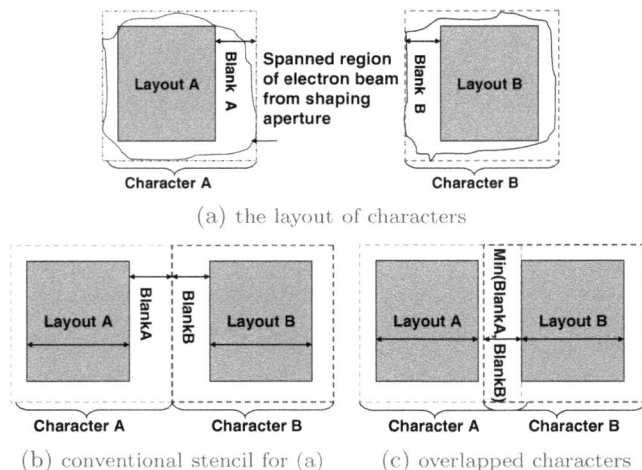

(a) the layout of characters

(b) conventional stencil for (a)   (c) overlapped characters

Figure 2: Overlapped characters for improving the stencil densities.

Suppose the required white space around layout A and B are $BlankA$ and $BlankB$ respectively, in Fig. 2 (a). If the characters are conventionally aligned by edge as Fig. 2 (b), it results in a waste of area. The space between layout A and B is actually $BlankA + BlankB$, which is more than required for both patterns. By contrast, we would greatly reduce the total area of character A and B by sharing an amount of $min(BlankA, BlankB)$ space. In this case, $max(BlankA, BlankB)$ white width is still reserved between layout A and B, which is sufficient for ensuring correct printing image.

2.2 Stencil Design Challenge

The main challenge of stencil design with overlapped characters comes from the fact that, for each character, the amount of required blanking space is not uniform, strongly depending on its enclosed layout patterns. In consequence, for different placements of characters, the area reduction from template overlapping may vary a lot. Therefore, unlike the traditional design of Fig. 1 (f), the number of maximum allowable characters in the stencil is not fixed. To achieve high quality solution, the detailed physical placement information of all the characters must be taken into account. This makes the problem of stencil design with overlapped characters not only different from but also more difficult than conventional non-overlapping one, addressed in [9–12].

As the example of Fig. 3 (a) illustrates, suppose there are three character candidates A-C, and we would like to pack them into a simple stencil of Fig. 3 (b) for minimum projection time. As easily seen, their blanking spaces are quite different. In conventional design where overlapping is not considered, at most two of them can fit. On the other side, when the blanking space is shared by adjacent characters, the result is correlated with the detailed physical implementation of stencil, and could be different from traditional design. If these three candidates are tried out by the order of A-B-C like Fig. 3 (c), only A and B can be put in. Patterns C is out of bound and has to be processed by VSB technique. This does not lead to higher throughput than conventional non-overlapped methodology. In contrast, if rearranged as C-B-A as Fig. 3 (d), all of these three patterns can be used as CP characters. Obviously, it is a better stencil optimization.

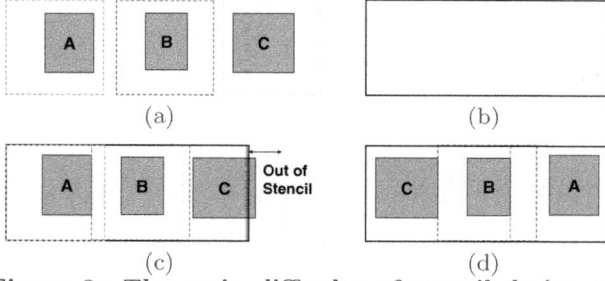

Figure 3: The main difficulty of stencil design with overlapped characters

2.3 Problem Formulation

In this subsection, we will formulate the problem of EBL stencil design with overlapped characters.

Similar to previous work [9–12], we assume a set of character candidates have already been given. To model overlapping information, as Fig. 4 (a) illustrates, assume the blanking spaces of each candidate c_i, from left, right, top and bottom boundaries, are l_i, r_i, t_i and b_i respectively. The orientation of these candidates are not allowed to be flipped, since it actually becomes a different template, as explained in [10]. When two candidates c_i and c_j are put adjacent to each other horizontally, their maximum allowed overlap is set as o_{ij}^H, which is $min(r_i, l_j)$ as shown by Fig. 4 (b). Similarly, Fig. 4 (c) defines the maximum vertical overlapping margin o_{ij}^V. o_{ij}^H and o_{ij}^V vary for different i and j.

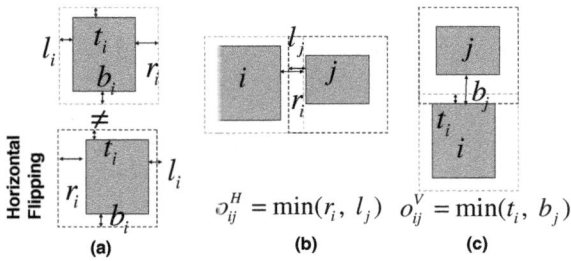

Figure 4: The dimensional variable of character candidates

Moreover, since the manufacturing time of EBL is dominantly determined by electronic beam shooting, in our work, we make use of total number of shots as the measurement of projection time. Suppose each candidate c_i is referred r_i^c times in the chip. For each of its appearance, the candidate c_i will be projected by either CP or VSB method, with a number of shots n_i^{CP} and n_i^{VSB}. The total processing time (number of shots) of the entire circuit is computed by following equation.

$$\sum_{c_i \in C^{CP}} r_i^c n_i^{CP} + \sum_{c_i \in (C^C \setminus C^{CP})} r_i^c n_i^{VSB} \qquad (1)$$

C^C is the set of all the character candidates. C^{CP} is the union of selected candidates processed by CP method, which is a subset of C^C.

In our work, for simplification purpose, we only design and optimize the stencil for single design. The general case of multiple chips can be easily extended, where the characters would be reused by different designs. Based on above description, our optimization problem can be stated as below:

Problem Formulation: Given a design and its set of character candidate C^C, select a subset C^{CP} out of C^C as characters, and place them on the stencil S. The objective is to minimize the total projection time (number of shots) of this design expressed by Equation (1), while the placement of C^{CP} is bounded by the outline of S. The width and height of stencil is W and H, respectively, and all the candidate has unique width w and height h. The maximum overlapping margin between adjacent characters is given by o_{ij}^H and o_{ij}^V.

In this paper, we will first investigate on the special case of one dimensional stencil design in Section 3, where the amount of blanking spaces differs only in either horizontal or vertical direction. Then, in Section 4, the algorithm, for generalized two dimensional problem, will be developed.

3. ONE DIMENSIONAL STENCIL DESIGN

Normally, each template implements one standard cell. That is to say, the enclosed circuit patterns of all the characters have the same height, and their layouts near top and bottom boundary edges are mostly regular power rails. As a result, illustrated by Fig. 5 (a), the required blanking spaces on the top t and bottom b are nearly identical for these candidates.

Therefore, in such case, characters are usually be placed on the stencil in a row-based manner, shown by Fig. 5 (b). All rows have a unique height h. The overlapped blanking margin h_o between adjacent rows are also the same, which is $min(t_i, b_i)$. In consequence, as Fig. 5 (c) shows, the overlapping-aware stencil design becomes a one-dimensional problem. The number of character rows can be pre determined as $\lfloor (H - h_o)/(h - h_o) \rfloor$. The candidates would be packed into these rows with maximum width W.

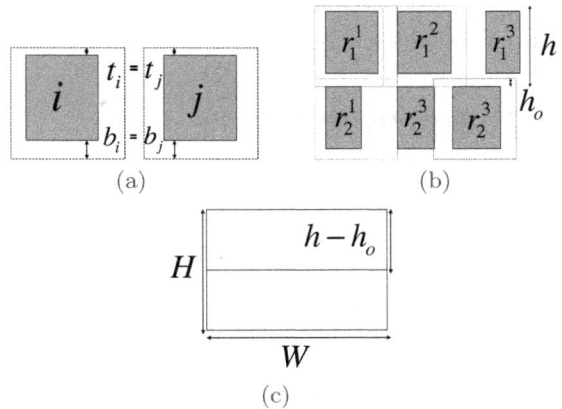

Figure 5: One-dimensional Stencil Design.

The overview of our four-phase iterative refinement algorithm for this special one-dimension problem is given in Fig. 6, and the details will be discussed in following subsections.

3.1 Greedy One Dimensional Bin Packing

To construct a reasonable good starting point, we adopt a descending best-fit bin packing algorithm to push the character candidates into stencil, until there is no enough capacity.

Note that the overall projection time (number of shots) of Objective (1) can also be represented as

$$\sum_{c_i \in C^C} r_i^c n_i^{VSB} - \sum_{c_i \in C^{CP}} r_i^c (n_i^{VSB} - n_i^{CP}) \qquad (2)$$

where the first part is independent with stencil design. To reduce the processing time, $\sum_{c_i \in C^{CP}} r_i^c (n_i^{VSB} - n_i^{CP})$ should be made as large as possible, during greedy bin-packing.

Therefore, as preprocessing, we first assign each candidate c_i a profit value p_i, $r_i^c (n_i^{VSB} - n_i^{CP})$. The bigger p_i is, the larger amount of projection efforts can be saved by printing c_i using CP than VSB method. For getting good greedy optimization result, the c_i with larger profit should be given higher priority to be placed on the stencil. Guided by this heuristic, in the second step, the character candidates, which have not been on the stencil yet, will be sorted decreasingly based on their profits and packed in a sequential manner.

Next, these sorted candidates will be pushed into stencil by a best-fit packing strategy. When c_i is to be packed, the row, which has the most amount of capacities left *after* accommodating c_i, will be picked. This is to consider the possible shared space between adjacent objects, when we are computing the remaining room in each row. As Fig. 7 (a) illustrates, suppose only two rows are available and candidate C is to be packed next. It appears that row R1 has more capacity left. However, as Fig. 7 (b) illustrates, when we try out C in both rows, it is R2 which has larger remaining room. As a result, candidate C is packed into R2, shown by Fig. 7 (c).

3.2 Single Row Reordering

After greedy bin packing, there is no room left to accommodate more candidates. However, as motivated by Fig. 3, we can adjust the relative locations of already-placed characters in each row to shrink its occupied width and increase remaining capacity. This allows pushing in more candidates, which further reducing the overall projection time. Therefore, in this phase, our goal is to minimize the total width of its characters in each row for maximizing remaining capacity.

Suppose row r contains a set of $c_0^r .... c_n^r$ characters from left

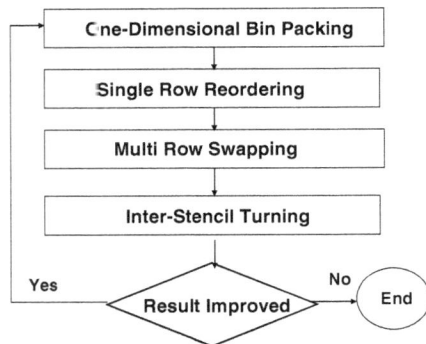

(a)　　　　　　　　　(b)

(c)

Figure 7: This figure illustrates the procedure of best-fit bin packing with overlapping awareness.

to right, its total occupied width can be computed as $\sum_{i=0}^{n} w - \sum_{i=0}^{n-1} o_{i,i+1}^H$. It is not difficult to see that $\sum_{i=0}^{n} w$ is a constant as long as the number of characters is not changed. Therefore, to minimize the total occupied width, the overall overlapped blanking margin $\sum_{i=0}^{n-1} o_{i,i+1}^H$ should be maximized.

To compute optimal character permutation for maximum amount of shared blanking width, we formulate a *minimum cost Hamiltonian path* problem. First of all, a graph G is constructed as follows: Each c_i^r is represented by a vertex v_i^r. For each pair of v_i^r and v_j^r, we add two directed edges e_{ij} and e_{ji}. The associated costs are $o_{big}^H - o_{ij}^H$ and $o_{big}^H - o_{ji}^H$, respectively. o_{ij}^H / o_{ji}^H is the shared space when c_i is put left/right adjacent to c_j, and o_{big}^H is a constant value, bigger than any of o_{ij}^H. To maximize $\sum_{i=0}^{n-1} o_{i,i+1}^H$, it suffice to find a path visiting each node of G exactly once such that the total edge weighs ($\sum_{e \in Path} (o_{big}^H - o_{ij}^H)$) along this path is minimized. As Fig. 8 (a) illustrates, a graph for three character placement (A,B,C) is given. Suppose the minimum cost Hamiltonian path is found as Fig. 8 (b), Fig. 8 (c) shows its corresponding character placement.

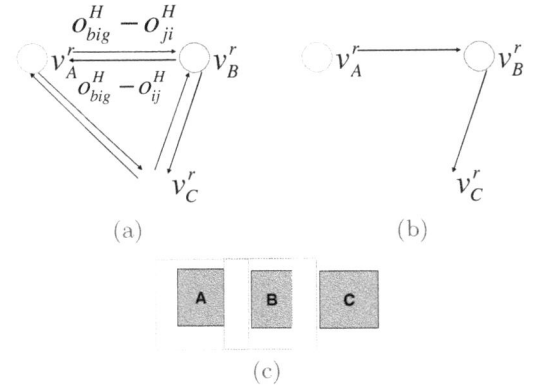

(a)　　　　　　　　(b)

(c)

Figure 8: This figure shows how to optimize the occupied-width of each row as min-cost Hamiltonian path problem

Practically, since the problem of minimum cost Hamiltonian path is NP-hard, it may be expensive to solve the whole row in one time. In that case, our heuristic is to partition the row into

Figure 6: The overview of one dimensional stencil design with overlapped characters.

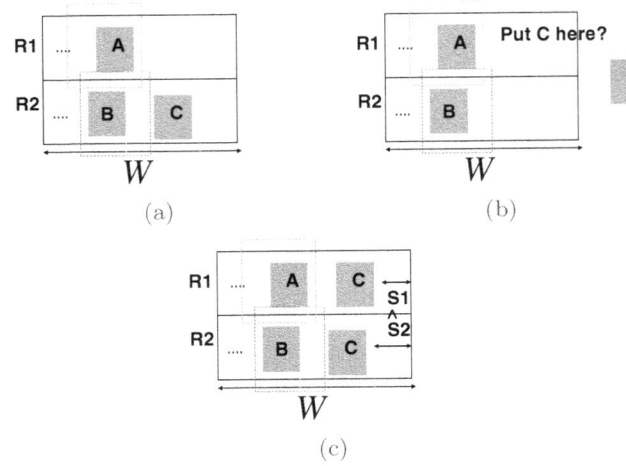

multiple overlapped smaller segments, and solve each segment by Hamiltonian path-based method.

3.3 Multiple Row Swapping

After single row reordering, the character permutation within each row has been extensively optimized. However, it is still possible to increase their remaining capacities, by swapping characters from different rows. As Fig 9 illustrates, by swapping r_1^2 and r_2^2, both characters find "better" neighbors with more overlapped blanking space. For row R1 and R2, their remaining rooms are both increased.

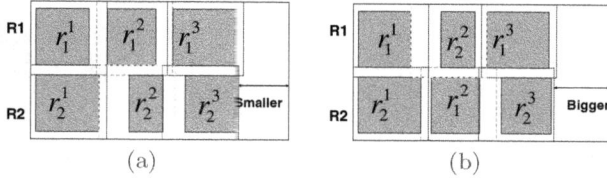

Figure 9: This figure explains the motivation of multi-row swapping.

The algorithm is briefly explained as follows. We test every pair of characters from different rows. Only when the remaining capacities of both rows are increased after swapping, it is considered as a *reasonable* swap. This ensures, the modified placement is definitely better than original one. The reason is that, after swapping, if one row gains more room but another has less, it is possible that the following optimization is hurt by the row with shrunk capacity.

After all the *reasonable* swap pairs are found, they are sorted increasingly by capacity gains, and performed one by one. When certain swap is done, the associated characters and their neighbors are locked. Any swap in the later trials is not allowed to move these locked characters as well as their neighbors. This honors previous optimization result.

3.4 Inter Stencil Tuning

The previous single and multiple row optimization are conducted based on the initial solution of bin-packing algorithm in Section 3.1. This may limit the optimization space. To get out of local optima, as the last step of ech iteration, we would like to exchange the placed characters with those which have not been selected.

Our approach is to randomly pick and exchange two character candidates, where one is from the stencil and another is not. The swapping will be accepted, only if the overall projection time, number of shots, Objective (1) is reduced and the remaining capacity of any row is not shrunk.

4. TWO DIMENSIONAL STENCIL DESIGN

In this section, we investigate on the general case of EBL stencil design with overlapped characters. The blanking spaces of templates are non-uniform along both horizontal and vertical directions. Due to NP-completeness of this problem, we adopt a simulated-annealing based heuristic approach to perform a robust iterative improvement.

4.1 Sequence Pair Representation

To represent the character placement solution, we make use of sequence pair (SP) proposed in [16].

Given a set of character candidates C^C, its SP consists in two permutations $\overline{X}\&\overline{Y}$ of these templates $(c_0, c_1...c_n)$, which specifies their geometry relationships as below.

$$(\overline{X}:< .., c_i..c_j.. >, \overline{Y}:< ...c_i...c_j... >) : c_i \text{ is left to } c_j \quad (3)$$

$$(\overline{X}:< .., c_j..c_i.. >, \overline{Y}:< ...c_i...c_j... >) : c_i \text{ is below } c_j \quad (4)$$

Based these constraints, we can map any SP into a solution of character placement as following procedure:

Procedure 1:
Step1: Compute a packing solution of C^C, following similar methods of [16, 17]. The details will be described in Section 4.1.1 and 4.1.2.
Step2: Assuming the left-bottom coordinates of packing results and stencil are the same, the candidates, which are located completely within the outline of stencil, are considered as selected characters. □

The step1 is the critical one in above transformation. Due to specific properties of our problem, its implementation actually differs from the conventional approaches of [16, 17], explained as follows.

4.1.1 Correct Packing Algorithm

The key step of packing solution evaluation from SP is to determine the physical coordinates of each block. This problem has been well investigated, when overlapping is not considered between adjacent blocks. The original algorithm is proposed in [16], and improved by [17] with new solution pruning technique. The work of [16] is extensible for our overlapping-enabled character placement problem. However, the key speed-up idea in [17] does not apply, although it is much faster.

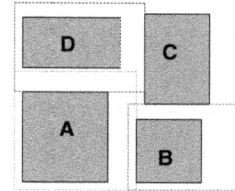

$$SF : \overline{X} = (D\ A\ C\ B)$$
$$\overline{Y} = (A\ B\ D\ C)$$

(a)

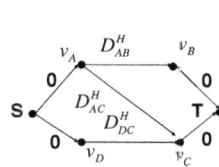

$$D_{AB}^H = \frac{1}{2}w_A + \frac{1}{2}w_B - o_{A,B}^H$$
$$D_{AC}^H = \frac{1}{2}w_A + \frac{1}{2}w_C - o_{A,C}^H$$
$$D_{DC}^H = \frac{1}{2}w_D + \frac{1}{2}w_C - o_{D,C}^H$$

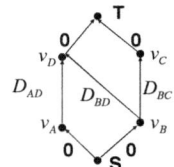

$$D_{BC}^V = \frac{1}{2}h_B + \frac{1}{2}h_C - o_{B,C}^V$$
$$D_{BD}^V = \frac{1}{2}h_B + \frac{1}{2}h_D - o_{B,D}^V$$
$$D_{AD}^V = \frac{1}{2}h_A + \frac{1}{2}h_D - o_{A,D}^V$$

(b) H graph   (c) V graph

Figure 10: This figure explains packing evaluation of [16] based on sequence pair

The method of [16] is based on longest path algorithm, and starts from constraint graph construction. Given a SP, a H/V graph is built first to capture the horizontal/vertical relationship between different blocks. Assume there are totally C^C candidates, the H/V graph has $|C^C|+2$ vertexes, one v_i for each candidate c_i plus a source s and sink t. If c_j is (left adjacent to)/(below) c_k, a directed edge e_{jk} is added from v_j to v_k. The weight of e_{jk} is the minimum possible horizontal/vertical distance between the centers of c_j and c_k. Beside these, there is a zero-weight edge from source to every v_i, and a zero-weight edge from every v_i to sink. For the example of Figure 10 (a), Figure 10 (b) and (c) show the resulting H and V constraint graphs, respectively.

After that, the x/y coordinates of these candidates can be obtained by finding its weighted longest path algorithm from source. As easy to see, this methodology is also applicable for our problem, where overlapped space is allowed between adjacent vertexes. The only difference is that, when the weights of edge are assigned, the amount of shared blanking space must be considered, as highlighted by the red cycles in Figure 10 (b) and (c).

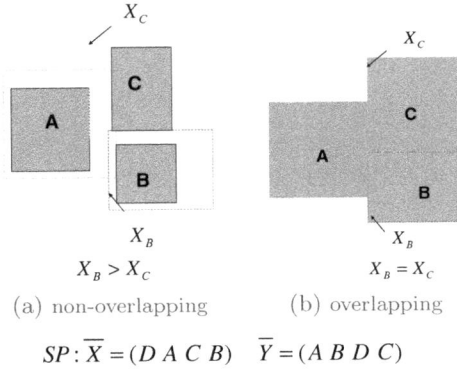

$$SP: \overline{X} = (D\ A\ C\ B) \quad \overline{Y} = (A\ B\ D\ C)$$

Figure 11: This figure illustrates the key idea of of [17]

On the other side, the work of [17] does not explicitly build the constraints graphs but depends on the longest common subsequence computations. They evaluate the placement of character candidates much faster than [16], depending on the following property.

Property 1 *Given two blocks B and C, if we put them (right adjacent to)/below a common component A, then the x/y coordinates of these two blocks should be same.*

The correctness of this property can be easily seen for the conventional packing, as shown by Figure 11 (a), while overlapping is not considered. However, it does not hold true, when the sharing of characters becomes possible. As Figure 11 (b) illustrates, due to different overlapping margins, the coordinates of B and C are not the same.

4.1.2 Fast Packing Evaluation

After evaluating packing solution, in the step2 of Procedure 1, the candidates outside the outline of stencil will not be taken as characters. This implies, the detailed locations of these candidates are not important, and do not have to be computed in the step1. Great speedup can be achieved by making use of this property.

In detail, in the implementation of SP-based minimum area packing, we stop placement evaluation as soon as the contour of already-packed character candidates is completely outside the outline of stencil by at least a margin of o_{max}, given that o_{max} is the maximum value of o_{ij}^H and o_{ij}^V.

This strategy will not effect the solution of character placement. For any of unpacked candidates by the stopping time, it can not be totally fit into the stencil no matter how to push it around the boundaries of already-packed character clusters.

4.2 Simulated Annealing

During simulated annealing, we continuously make small modification on sequence pair, and evaluate the resulting stencil design. The new SP/solution will be for sure adopted if reducing the total time of Objective (1). While it is actually a worse character placement, this non-improving result is accepted with probability decreasing over time.

In this subsection, we present two effective SP perturbation methods for better local search towards shorter projection time: throughput-driven swapping and slack-based insertion.

4.2.1 Throughput-driven Swapping

The first type of perturbation we perform is throughput-driven swapping. The basic idea is to try reducing overall projection time by swapping the positions of two candidates in the $\overline{X}\&\overline{Y}$ SP. This is equivalent to exchange their relative locations in the packing solution.

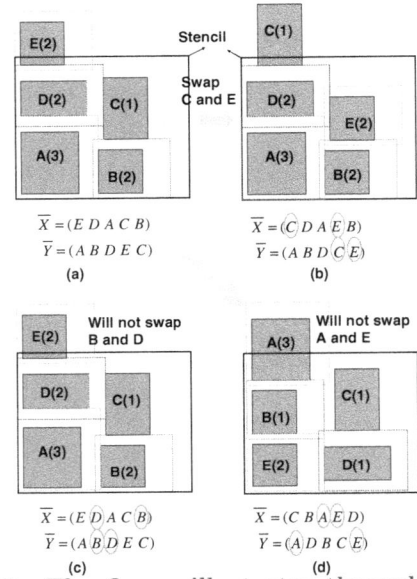

Figure 12: The figure illustrates throughput-driven swapping.

Fig. 12 illustrates a motivational example, which has five blocks A-E to be packed. The required number of shots, to project any of these candidates once, are assumed as 1 and 10 for CP (n_i^{CP}) and VSB (n_i^{VSB}) methods respectively. The digit in the parentheses denotes how many times r_i of each comp<onent, will be used and printed in the design.

Fig. 12 (a) gives a SP representation and its corresponding stencil design, based on the Procedure 1 in Section 4.1. Following the definition of Objective 1, the total processing time (number of shots) are $3+2+1+2+10\times2 = 28$, since A-D are selected as characters while E is not. If swapping the locations of C and E in SP as Fig. 12 (b), we would end up with a better stencil design with less amount processing time. It only takes a number of 19 shots, which is computed as $3+2+10+2+2 = 19$, in this case.

In the detailed implementation, we enforce two heuristic swapping constraints, to enable efficient and effective shot number reduction.

First of all, given a SP, out of the pair of elements to be changed, we require that one candidate c_s, should have been selected as characters by its corresponding stencil packing, while the other one, c_o is not. For the example of Fig. 12 (a), we only allow the exchange of the positions between E and any of A-D. The swapping among any two of A-D is not enabled. The reason is that if the two candidates to be swapped are both in or out of stencil already, most likely the new SP generates a stencil solution with same set of selected characters and just different geometrical ordering. As an example, if we swap candidate B and C< which are both already in the stencil, like from Fig. 12 (a) to Fig. 12 (c), the resulting packing result also selects A-D as characters, still requiring 28 shots totally.

Secondly, after randomly picking in-stencil candidate c_s and out-of-stencil one c_o for swapping, we will compute the difference of their profits p_o-p_s, to decide whether this swapping would be tried on. The profit p_o/p_s is defined as same as $r_i(n_i^{VSB} - n_i^{CP})$ in Section 3.1, which reflects the reduction of the shoot number by printing this candidate by CP rather than VSB. If we swap the locations of c_s and c_o, it is highly like that c_s will be pushed out of stencil but the c_o would be selected as character in turn. Assuming all the other candidates stay either in or outside the stencil, as the state before the swapping, the total shot reduction by this exchange can be approximated as p_o-p_s. Therefore, if the difference p_o-p_s is smaller than zero, it is in high possibility that the swapping under consideration will not lead to better packing result. For the example of Fig. 12 (a), suppose c_s and c_o are A and E, respectively, and it turns out p_o-p_s is -9. In this case, the corresponding stencil design indeed becomes worse, taking 35 shots as Fig. 12 (d) shows.

4.2.2 Slack-based insertion

Given a SP and its corresponding character solution, our purpose of slack-based insertion is to add-in a new candidate, which currently is not serving as character, into the stencil. To ensure robust throughput improvement, we would like to find a good strategy to insert such extra candidate, so that all the previously already-placed characters are still kept on the stencil in most trials. This equals to increase the number of usable templates. In this subsection, we make use of the concept of slack, applied in [18], to search such a good insertion location.

Given a character c_s on the stencil, its x/y slack is defined as the allowed movement range of x/y coordinates of c_s, under the constraint that none of all the other already-placed characters would be pushed outside the stencil after such move. Fig. 13 (a)-(b) illustrate a simple example, with four characters A-D. Their leftmost and rightmost packing solutions are shown by Fig. 13 (a) and (b), respectively. Based on this two extreme cases, the x slack of C, for example, can be computed as $X_c^{right} - X_c^{left}$.

Once slacks are known, we randomly pick a *base* character c_b, which has large slacks in both x and y directions, and insert a new candidate c_{new} before it. The reason is that the location of such *base* can be moved in relatively big amount to make space for additional character. In terms of SP operation, this can be done by simply changing the position of c_{new} right before c_b in $\overline{X}$ and $\overline{Y}$ permutations. As illustrated by Fig. 13 (c), suppose the c_b and c_{new} are candidate C and E, respectively. The resulting new SP is obtained by insert E right in front of the position of C, as shown by Fig. 13 (d) .

5. EXPERIMENTAL RESULTS

We implement our algorithm in C++ and test on Intel Core 2.0GHz Linux machine with 32G RAM. LKH [19] is chosen as the solver for min-cost Hamilton path. Moreover, Parquet [18] is adopted as our simulated annealing framework.

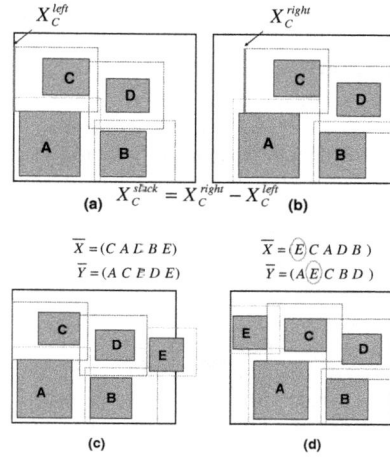

Figure 13: The simple example of slack-based insertion.

Table 1: Statistics on testcases.

ckts	csize	total area	total blanks	optimal area
1D-1	3.8x3.8	1.444	0.416	1.028
1D-2	4.0x4.0	1.6	0.479	1.121
1D-3	4.2x4.2	1.764	0.514	1.25
1D-4	4.4x4.4	1.936	0.569	1.367
2D-1	3.8x3.8	1.444	0.414	1.03
2D-2	4.0x4.0	1.6	0.529	1.071
2D-3	4.2x4.2	1.764	0.662	1.102
2D-4	4.4x4.4	1.936	0.774	1.162

To test the efficiency of proposed methods, we randomly generate eight benchmarks. The size of stencil is set as 100um x 100um, and a total number of 1000 character candidates with unique size are generated. The sharable blanking area within each candidate is randomly decided, uniformly distributed between 0%-50% character width. For the special case of one dimensional problem, the blanking space along vertical direction is set as a constant value. Moreover, for each candidate c_i, we randomly assign a triple of value $(r_i, n_i^{VSB}, n_i^{CP})$ as its referred time in chip, and respective number of shots by VSB and CP. n_i^{VSB} is made 5-10x larger than n_i^{CP}.

The detailed statistical data for individual testcase is shown in Table 1. The first column denotes the name of benchmarks, which "1D-x" and "2D-x" are applied for one and two dimensional problem, respectively. "csize" is the size of each character candidate, formatted by "um x um". The units of all the other columns are "$1e^4 um^2$". "total area" shows the total area of all the character candidates, and "total blanks" is the summation of their sharable blanking space. "optimal area" is computed as "total area" minus "total blanks", typically larger than the area of given stencil. This matches the fact that even under best possible case of stencil design, where all the blanking area are indeed shared by adjacent characters, the entire set of the candidates can not be fully pushed into the stencil.

For comparative reason, we implement two different stencil design approaches. The first one **NO-OVERLAP** is based on the work of [12], where no overlapped characters is allowed. A little difference is that, in its implementation, only one stencil with unique character size is considered. Moreover, for our problem, their algorithm is somewhat degenerated into a method of selecting the most profitable candidates, which profit is judged by $r_i(n_i^{CP} - n_i^{VSB})$. In the second comparative approach **GREEDY**, possible sharing is taken into account, but a greedy methodology is applied to chose character candidates. In 1D problem, only the first phase of heuristic Descending Best-Fit (DBF) packing in Section 3.1 is performed.

For two-dimensional problem, 2D DBF packing is conducted.

5.1 One-dimensional Stencil Design

Table 2 lists the comparison of stencil design in one-dimensional case. "#shot" shows the total processing time (number of shots) of the circuit by using corresponding stencil design methodologies, which is computed by the equation of Objective 1. "#char" is the number of characters that fits into stencils, and "#CPU" tells the runtime of these stencil optimization methods, in terms of seconds.

As we can see, compared to **NO-OVERLAP**, we are able to averagely put 42% more characters on the stencil, and reduce the total projection time (number of shots) by 51%. With respect to **GREEDY** algorithm, our approach still achieves averagely 14% more projection time reduction, by allowing 7% more characters placed. The CPU time of our approach is relatively large but its absolute value is only around 20s. These results show the effectiveness and efficiency of our proposed four-phase iterative refinement algorithm.

For this special one-dimension problem, **GREEDY** looks also quite useful. The reason is that the vertical blanking spaces of these candidates are uniform in this case, and have been fully shared during the stencil design.

Table 2: Result Comparison for 1D problem

ckts	NO-OVERLAP			GREEDY			our approach		
	#shot	#char	CPU(s)	#shot	#char	CPU	#shot	#char	CPU
1D-1	28654	676	1.2	13528	901	2.2	10083	951	22.3
1D-2	41727	625	1.1	17929	836	2.1	14921	880	21.8
1D-3	38460	529	0.9	25155	727	1.9	22503	768	20.6
1D-4	41260	484	0.8	29462	665	1.8	26756	702	20.1
total	150101	2314	4	86074	3129	8	74263	3301	84.8
ratio	2.0	1	0.05	1.16	1.35	0.10	1	1.42	1

5.2 Two-dimensional Stencil Design

Table 3 lists the comparison of stencil design in general two-dimensional case. The meaning of labels are the same as Table 2. Compared to **NO-OVERLAP** and **GREEDY** methods, in average, our proposed SP-based algorithm places 28% and 24% more characters on stencil, which reduces the projection time (number of shots) by 31% and 25%, respectively. The **GREEDY** algorithm does not work that well in this 2D problem, because the blanking area varies in both horizontal and vertical directions and the native first-bin-best-fit packing very easily get stuck in local optima.

Due to two-dimensional optimization, the runtime of our approach is much longer than 1D problem, comparatively. It takes a few hundred seconds, but is still satisfactory. The design of stencil is only a one-time process before projecting large volume of chips by EBL. Several minutes preprocessing time is relatively very tiny in the whole manufacturing procedure.

Table 3: Result Comparison for 2D problem

ckts	NO-OVERLAP			GREEDY			our approach		
	#shot	#char	CPU	#shot	#char	CPU	#shot	#char	CPU
2D-1	23319	676	1.3	26832	625	2.3	16877	803	466
2D-2	29368	576	1	25977	642	2.6	20141	750	447
2D-3	32399	526	0.9	30411	558	2.5	<23850	688	424
2D-4	35410	474	0.8	31930	531	2.7	25278	660	416
total	120496	2252	4	115150	2356	10.1	86146	2901	1755
ratio	1.40	1	0.002	1.33	1.05	0.006	1	1.30	1

6. CONCLUSION

In this paper, we have developed two algorithms for overlapping aware stencil design in electronic beam lithography. The experimental results show 51% reduction on the total projection time, compared to the conventional design when the characters are not overlapped.

7. ACKNOWLEDGEMENT

The authors would also like to thank Dr. Gi-Joon Nam at IBM Austin Research for helpful discussions on this problem.

8. REFERENCES

[1] Andrew B. Kahng, Chul-Hong Park, Xu Xu, and Hailong Yao. Layout decomposition for double patterning lithography. In *Proc. Int. Conf. on Computer Aided Design*, November 2008.

[2] Kun Yuan, Jae-Seok Yang, and David Z. Pan. Double patterning layout decomposition for simultaneous conflict and stitch minimization. In *Proc. Int. Symp. on Physical Design*, March 2009.

[3] Jae-Seok Yang, Katria Lu, Minsik Cho, Kun Yuan, and David Z. Pan. A New Graph-theoretic, Multi-objective Layout Decomposition Framework for Double Patterning Lithography. In *Proc. Asia and South Pacific Design Automation Conf.*, Janunary 2010.

[4] Aki Fujimur. Beyond Light: The Growing Importance of E-Beam. In *Proc. Int. Conf. on Computer Aided Design*, November 2009.

[5] Aki Fujimur. Design for E-Beam: Getting the Best Wafers Without the Exploding Mask Costs. In *Proc. Int. Symp. on Quality Electronic Design*, March 2010.

[6] Akira Fujimura, Takashi Mitsuhashi, Kenji Yoshida, Shohei Matsushita, Larry Lam Chau, Tam Dinh Thanh Nguyen, and Donald MacMillen. Stencil Design and Method for Improving Character Density for Cell Projection Charged Particle Beam Lithography. In *US Patent*, Jan. 2010.

[7] Hans Co Pfeiffer. New Prospects for Electron Beams as Tools for Semiconductor Lithography. In *Proc. of SPIE*, May 2009.

[8] Takeshi Fujino, Yoshihiko Kajiya, and Masaya Yoshikawa. Character-build standard-cell layout technique for high-throughput character-projection EB lithography . In *Proc. of SPIE*, July 2005.

[9] Makoto Sugihara, Taiga Takata, Kenta Nakamura, Yusuke Matsunaga, and Kazuaki Murakami. A CP mask development methodology for MCC systems. In *Proc. of SPIE*, May 2006.

[10] Makoto Sugihara, Kenta Nakamura, Yusuke Matsunaga, and Kazuaki Murakami. CP mask optimization for enhancing the throughput of MCC systems. In *Proc. of SPIE*, Octerber 2005.

[11] Yusuke Matsunaga Makoto Sugihara and Kazuaki Murakami. Technology mapping technique for enhancing throughput of multi-column-cell systems. In *Proc. of SPIE*, March 2007.

[12] Makoto Sugihara. Optimal character-size exploration for increasing throughput of MCC lithographic systems. In *Proc. of SPIE*, Feb. 2009.

[13] Lin Xie, Azadeh Davoodi, and Kewal K. Saluja. Post-silicon diagnosis of segments of failing speedpaths due to manufacturing variation. In *Proc. Design Automation Conf.*, June 2010.

[14] Sean Shi and David Pan. Wire Sizing and Shaping with Scattering Effect for Nanoscale Interconnection. In *Proc. Asia and South Pacific Design Automation Conf.*, Jan 2006.

[15] Yongchan Ban, Savithri Sundareswaran, and David Z. Pan. Total Sensitivity Based Standard Cell Layout Optimization. In *International Symposium on Physical Design (ISPD)*, 2010.

[16] Hiroshi Murata, Kunihiro Fujiyoshi, Shigetoshi Nakatake, and Yoji Kajitani. VLSI Module Placement Based on Rectangle-Packing by theSequence-Pair. In *IEEE Trans. on Computer-Aided Design of Integrated Circuits and Systems*, Decemember 1996.

[17] Xiaoping Tang, Ruiqi Tian, and Martin Wong. Fast Evaluation of Sequence Pair in Block placement by Longest Common Subsequence Computation. In *Proc. Design, Automation and Test in Eurpoe*, March 2000.

[18] Saurabh H. Adya and Igor L. Markov. Fixed-outline Floorplanning : Enabling Hierarchical Design. In *IEEE Trans. on Very Large Scale Integration (VLSI) Systems*, Decemember 2003.

[19] http://www.akira.ruc.dk/~keld/research/LKH.

More Realistic Power Grid Verification Based on Hierarchical Current and Power Constraints

[2]Chung-Kuan Cheng, [2]Peng Du, [2]Andrew B. Kahng, [1]Grantham K. H. Pang,
[§][1]Yuanzhe Wang, [1]Ngai Wong
[1] Dept of Electrical & Electronic Engineering, The University of Hong Kong, Hong Kong
{gpang,yzwang,nwong}@eee.hku.hk
[2] Dept of Computer Science & Engineering, University of California, San Diego, La Jolla, CA
{ckcheng,pedu,abk}@ucsd.edu *

ABSTRACT

Vectorless power grid verification algorithms, by solving linear programming (LP) problems under current constraints, enable worst-case voltage drop predictions at an early design stage. However, worst-case current patterns obtained by many existing vectorless algorithms are time-invariant (i.e., are constant throughout the simulation time), which may result in an overly pessimistic voltage drop prediction. In this paper, a more realistic power grid verification algorithm based on hierarchical current and power constraints is proposed. The proposed algorithm naturally handles general RCL power grid models. Currents at different time steps are treated as independent variables and additional power constraints are introduced; this results in more realistic time-varying worst-case current patterns and less pessimistic worst-case voltage drop predictions. Moreover, a sorting-deletion algorithm is proposed to speed up solving LP problems by utilizing the hierarchical constraint structure. Experimental results confirm that worst-case current patterns and voltage drops obtained by the proposed algorithm are more realistic, and that the sorting-deletion algorithm reduces runtime needed to solve LP problems by > 85%.

Categories and Subject Descriptors: B.7.2 [Design Aids]: Simulation

General Terms: Algorithms.

Keywords:

Power grid, worst-case voltage drop, hierarchical current and power constraints, sorting-deletion algorithm

1. INTRODUCTION

With decreasing feature size and increasing complexity of integrated circuits, IR and LdI/dt voltage drops on power grids are becoming increasingly significant, which may result in longer gate delays and logic errors. Thus, power grid verification is becoming an indispensable procedure to guarantee a functional and robust chip design. However, the extremely large size of power grid

*§ Author to whom correspondence should be addressed.

Permission to make digital or hard copies of all or part of this work for personal or classroom use is granted without fee provided that copies are not made or distributed for profit or commercial advantage and that copies bear this notice and the full citation on the first page. To copy otherwise, to republish, to post on servers or to redistribute to lists, requires prior specific permission and/or a fee.
ISPD'11, March 27–30, 2011, Santa Barbara, California, USA.
Copyright 2011 ACM 978-1-4503-0550-1/11/03 ...$10.00.

models (from tens of thousands to millions of nodes or circuit elements) renders traditional simulation tools such as SPICE inefficient. Much work has been done to find efficient methods for power grid simulation and optimization [1,4,8,9,12–14,17].

Most existing power grid verification algorithms fall into the category of time-domain simulation. These algorithms model power grids as RC(L) circuits and model currents drawn by transistors and logic gates as ideal time-varying current sources. Nodal voltages can be solved given the waveforms of current sources [1,8,9,12,17]. Yet, such methods are not always feasible for two reasons. First, there may exist too many current sources, with each current source having various patterns. Hence it is expensive to determine which patterns result in the worst-case voltage drop. Second, one may wish to perform an early-stage power grid verification before the design of specific functional blocks, in which case current waveform information is unknown.

To facilitate early-stage power grid verification, a class of *vectorless* algorithms has been proposed [2,3,5–7,11,15,16]. These algorithms determine the worst-case voltage drop by solving linear programming (LP) problems under a set of current constraints. The vectorless method is based on DC analysis in [7] and is extended to transient analysis in [3, 6]. An approximate matrix inversion method and a convex dual algorithm are proposed in [5] and [15], respectively, to speed up the LP solution. An impulse response-based approach considering the transition time of current sources is presented in [2]. Some vectorless algorithms rely on the assumption that the system matrix is an M-matrix [10] and hence cannot handle power grid models with inductors. Additionally, some methods assume constant current patterns; since such patterns may keep their peak values throughout the simulation period, voltage drop prediction may be overly pessimistic.

In this paper, we propose a novel algorithm, based on hierarchical current and power constraints, which generates more realistic time-varying current patterns and provides less pessimistic voltage drop predictions. Our main contributions are as follows.

1. The proposed algorithm is based on modified nodal analysis (MNA) and thereby naturally handles general RCL power grid models.

2. Currents at different time steps are treated as independent variables in LP problems, and additional power constraints are introduced which restrict the energy consumption of a current source. The current patterns solved are time-varying and do not stay at their peak values all the time. Consequently the proposed algorithm generates more realistic current patterns and provides less pessimistic voltage drop predictions.

159

3. A sorting-deletion algorithm is proposed which exploits hierarchical constraint structure for greater efficiency than standard LP methods. The time needed to solve the LP problems is reduced by $> 85\%$.

The paper is organized as follows. Background is introduced in Section 2. Hierarchical current and power constraints and LP optimization problem formulations are proposed in Section 3. Problem reductions and a sorting-deletion algorithm are proposed in Section 4. Experimental results are given in Section 5, and Section 6 draws conclusions.

2. BACKGROUND

2.1 RCL Power Grid Model

A typical 3D power grid consists of several metal layers, with each layer containing either horizontal or vertical conductors. Conductors in different layers are connected to each other by vias at their intersection points. External power supplies are connected to conductors of the top layer. Power drains, such as logic gates, transistors and memory units, are connected to conductors of the bottom layer. Such power grid structures are usually modeled as RCL circuits. Each conductor segment is modeled as a resistor in series with an inductor and each grid node is connected to the ground through a capacitor. External power supplies are modeled as ideal constant voltage sources and current drains are modeled as ideal time-varying current sources. The power grid model may or may not be regular.

Assume there exist a total of N grid nodes that are not terminals of ideal voltage sources. If only the voltage drops at these N nodes are of interest, a *revised* circuit model can be generated by short-circuiting all ideal voltage sources and reversing the directions of all ideal current sources [7]. Nodal voltages of the revised circuit are voltage drops of the original circuit. The MNA equation of the revised circuit can be written as

$$C\dot{x}(t) + Gx(t) = Hu(t). \qquad (1)$$

Here, $x(t) \in \mathbb{R}^n$ is the state vector of nodal voltages and inductor currents; n is the total number of nodes, voltage sources and inductors; $u(t) \in \mathbb{R}^m$ is the vector of current sources; $C \in M_{n,n}(\mathbb{R})$ is a diagonal matrix with its diagonal elements being capacitances and inductances; $G \in M_{n,n}(\mathbb{R})$ is a matrix of conductances and "± 1"; $H \in M_{n,m}(\mathbb{R})$ is the 0-1 current distribution matrix. Note that each row of H may contain more than one "1", which means that more than one current source can be attached to a single node. On the other hand, each column of H contains exactly one "1", which corresponds to the position at which a current source is attached.

2.2 Previous Vectorless Power Grid Verification Methods

Vectorless power grid verification is first proposed in [7], where only DC analysis is considered. The worst-case voltage drop is solved from LP problems under current constraints. In [3, 6], the algorithm is extended to transient analysis. [3] uses geometry-based methods to solve the LP problems. This algorithm achieves lower computational complexity with some sacrifice of accuracy. The current constraints in [3] are time-independent, and current patterns obtained there are constant throughout the simulation time span. [6] takes inductors into consideration and is applicable to general RCL power grid models. In [2] an impulse response-based algorithm is proposed which considers the transition time of current sources. With the transition time constraints, the current patterns

generated are more realistic, the method is inefficient for large instances. In [5] an approximate matrix inversion method is proposed to more quickly formulate the LP problems faster. The small entries in the approximate matrix are set to zero, so this method introduces added inaccuracy in predicting the voltage drops. In [15], a dual formulation is proposed in which the dual problems are convex with fewer variables. Solving these "reduced" convex optimization problems is expected to be more efficient than solving the original LP problems. The problem formulations of [3, 5, 15] are based on M-matrix assumptions (i.e., the system matrix of the model is an M-matrix).

3. HIERARCHICAL CONSTRAINTS AND LINEAR PROGRAMS

Given a power grid model and current and power constraints, our objective is to predict the worst-case voltage drops on the power grid by solving LP problems. In this section, we focus on how to formulate the LP problems with hierarchical current and power constraints. The next section will focus on how to efficiently solve the LP problems.

3.1 Transient Analysis

In this subsection, backward Euler-based transient analysis is performed to derive the relationship between voltage drops and currents. Similar derivations appear in [5]. Using backward Euler method, (1) can be discretized as

$$\left(G + \frac{C}{\Delta t}\right) x(t + \Delta t) = \frac{C}{\Delta t} x(t) + Hu(t + \Delta t). \qquad (2)$$

Under the stability assumption (all the poles of (1) are distributed on the left half of complex plane), the system matrix $G + \frac{C}{\Delta t}$ is invertible. Define

$$\mathcal{M} = \left(G + \frac{C}{\Delta t}\right)^{-1} \frac{C}{\Delta t}, \qquad (3a)$$

$$\mathcal{N} = \left(G + \frac{C}{\Delta t}\right) H. \qquad (3b)$$

We have

$$x(t + \Delta t) = \mathcal{M}x(t) + \mathcal{N}u(t + \Delta t). \qquad (4)$$

By dividing the simulation time span into k_t time steps, and assuming a zero initial state (i.e., $x(0) = 0$), the voltage drops at the last time step can be represented as

$$x(k_t \Delta t) = \sum_{k=1}^{k_t} \mathcal{M}^{k_t - k} \mathcal{N}u(k\Delta t). \qquad (5)$$

3.2 Hierarchical Constraints

In practice the peak value of a current source is usually bounded, i.e., $u_i(t) \leq I_{L,i}$. Local current constraints can be formulated by combining all these inequalities as

$$0 \leq u(t) \leq I_L \quad \text{or} \quad 0 \leq u(k\Delta t) \leq I_L, \qquad (6)$$

where $I_L \in \mathbb{R}^m$ is a vector with its ith element being the upper bound of the ith current source. Here, and in what follows, "$\leq$" is taken to be element-wise. On the other hand, global current constraints are formulated as

$$Uu(t) \leq I_G \quad \text{or} \quad Uu(k\Delta t) \leq I_G, \qquad (7)$$

where $U \in M_{p,m}(\mathbb{R})$ is a 0-1 matrix. Each inequality in (7) corresponds to a certain functional block, i.e., the total current of a

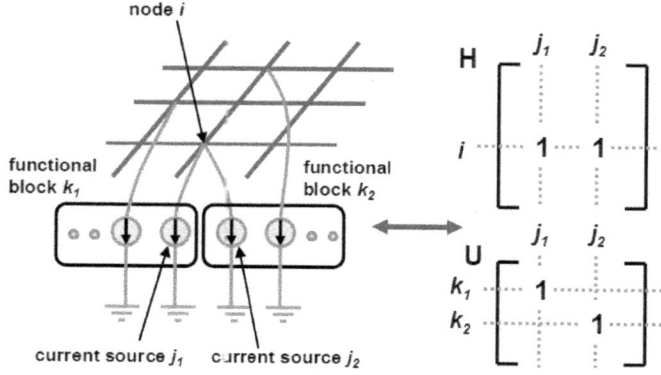

Figure 1: Block-level current constraints based on total currents of functional blocks.

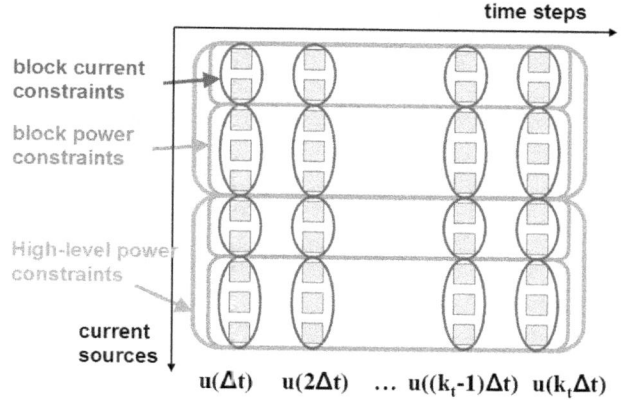

Figure 2: Illustration of hierarchical current and power constraints. Each column represents a current source vector at a single time step. Unlike current constraints, power constraints define an upper bound of the sum of currents at different time steps.

functional block is bounded. The global current constraints here are different from [5,7] in the sense that each column of U contains at most one nonzero entry. This implies that one current source belongs to only one specific functional block (i.e., appears in only one inequality). Hence if two different functional blocks draw currents from one grid node, the current drawn from this node should be modeled as two independent current sources. This agrees with the fact that more than one "1" may exist in one row of H, as shown in Fig. 1.

Besides current constraints, novel power constraints are introduced which restrict the average power consumption of a functional block:

$$U\left(\sum_{k=1}^{k_t} u(k\Delta t)\right) \leq \frac{k_t}{V_{dd}} P_B. \qquad (8)$$

Here V_{dd} is the voltage value of the external power supply. $P_B \in \mathbb{R}^p$ is a vector with its ith element being the power limit of the ith functional block. The power constraint is reasonable as power consumption of a functional block is usually bounded in the design requirements.

If by design some functional blocks have interactions with each other, high-level power constraints can be introduced which restrict the total power consumption of certain groups of functional blocks:

$$[1^{st} \text{ level}]: U_1 U\left(\sum_{k=1}^{k_t} u(k\Delta t)\right) \leq \frac{k_t}{V_{dd}} P_{T1};$$

$$\cdots \qquad (9)$$

$$[r^{th} \text{ level}]: U_r U_{r-1} \cdots U_1 U\left(\sum_{k=1}^{k_t} u(k\Delta t)\right) \leq \frac{k_t}{V_{dd}} P_{Tr}.$$

Here $U_1 \in M_{p,p}(\mathbb{R})$, $U_2 \in M_{p_2,p_1}(\mathbb{R})$, ..., $U_r \in M_{p_r,p_{r-1}}(\mathbb{R})$ are 0-1 matrices with each column containing at most one "1".

With the property that $U, U_1, ..., U_r$ containing at most one nonzero entry in every column, (6)-(9) constitute a group of *hierarchical* constraints, as depicted in Fig. 2. Based on this special hierarchical structure, solving LP problems can be significantly simplified by applying a sorting-deletion algorithm, which will be detailed in Section 4.3.

3.3 LP Formulation

Linear programs to solve the worst-case voltage drop can be formulated as

$$\max_{\substack{i=1,...,N \\ k'=1,...,k_t}} x_i(k'\Delta t) = \sum_{k=1}^{k'} c_{i,k',k} u(k\Delta t)$$

$$\text{s.t.} \begin{cases} 0 \leq u(k\Delta t) \leq I_L, \quad Uu(k\Delta t) \leq I_G, \\ U\left(\sum_{k=1}^{k_t} u(k\Delta t)\right) \leq \frac{k_t}{V_{dd}} P_B, \\ U_{r'}U_{r'-1}\cdots U\left(\sum_{k=1}^{k_t} u(k\Delta t)\right) \leq \frac{k_t}{V_{dd}} P_{Tr'} \ (r'=1...r). \end{cases}$$

$$(10)$$

Here $c_{i,k',k}$ is the ith row of $\mathcal{M}^{k'-k}\mathcal{N}$.

Equation (10) in fact contains $N \times k_t$ LP problems. We denote the solution of one single LP problem as $\Phi(i,k)$. The worst-case voltage drop is recognized as $\max\{\Phi(i,k)\}$ $(i=1,...,N, k'=1,...,k_t)$. However, solving all these LP problems is prohibitively expensive. Therefore, the problem size is first reduced in both time and space domains. Then, an efficient parallel algorithm is proposed to calculate $c_{i,k',k}$'s. Moreover, a sorting-deletion algorithm is proposed which is more efficient than standard LP algorithms. The details will be elaborated in the next section.

4. PROBLEM REDUCTION AND SORTING-DELETION ALGORITHM

4.1 Problem Reduction in Both Time and Space Domains

LEMMA 4.1. *For any integers* i, k_1, k_2 *with* $1 \leq i \leq N$ *and* $1 \leq k_1 < k_2 \leq k_t$, *we have* $\Phi(i,k_1) \leq \Phi(i,k_2)$.

PROOF. Assume that the maximum value $\Phi(i,k_1)$ is reached when currents are $u^{(1)}(k\Delta t)$ for $k=1,...,k_t$, where $u^{(1)}(k\Delta t)$ satisfy the constraints in (10). Then, for the LP problem (i,k_2), let $u^{(2)}(k\Delta t) = 0$ for $k = 1,...,k_2 - k_1$ and $u^{(2)}(k\Delta t) = u^{(1)}((k-k_2+k_1)\Delta t)$ for $k = k_2 - k_1 + 1,...,k_t$. Notice that since $\max\{u_i^{(2)}(k\Delta t)|k = 1,...,k_t\} \leq \max\{u_i^{(1)}(k\Delta t)|k =$

$1, \ldots, k_t\} \leq I_{L,i}$, we have $u^{(2)}(k\Delta t)$ satisfies the current constraints. Besides, since $\sum_{k=1}^{k_t} u^{(2)}(k\Delta t) = \sum_{k=1}^{k_t - k_2 + k_1} u^{(1)}(k\Delta t) \leq \sum_{k=1}^{k_t} u^{(1)}(k\Delta t)$, $u^{(2)}(k\Delta t)$ also satisfies the power constraints. With the observation that the value of $c_{i,k',k}$ is determined by $k' - k$ only (i.e., $c_{i,k'_1,k_1} = c_{i,k'_2,k_2}$ if $k'_1 - k_1 = k'_2 - k_2$), we have

$$
\begin{aligned}
x_i(k_2\Delta t) &= \sum_{k=1}^{k_2} c_{i,k_2,k} u^{(2)}(k\Delta t) \\
&= 0 + \sum_{k=k_2-k_1+1}^{k_2} c_{i,k_2,k} u^{(2)}(k\Delta t) \\
(j \overset{\Delta}{=} k - k_2 + k_1) &= \sum_{j=1}^{k_1} c_{i,k_2,j+k_2-k_1} u^{(2)}((j+k_2-k_1)\Delta t) \\
&= \sum_{j=1}^{k_1} c_{i,k_1,j} u^{(1)}(j\Delta t) = \Phi(i, k_1)
\end{aligned}
$$

Therefore $\Phi(i, k_2) \geq \Phi(i, k_1)$. $\square$

The main idea of this proof is that the longer the time span is, the worse the voltage drop can be. If the worst-case voltage drop at t_0 is obtained under a specific current pattern, the same voltage drop can also be obtained after t_0 by translating the same current pattern. Lemma 4.1 indicates that to obtain the worst-case voltage drop, we only have to solve LP problems (10) when $k = k_t$, which significantly reduces the computation load in the time domain.

Furthermore, we need not solve (10) for all $i = 1, \ldots, N$. In practice the most significant voltage drops often appear at nodes having longest distances to voltage sources. For example, the largest voltage drops of a mesh-like power grid with voltage sources at corners are most likely to occur near the center of the power grid. Hence we can choose a set of nodes (the indices of which form a set Ω) farthest from voltage sources and solve LP problems only for nodes belonging to Ω. Or we can first perform a DC analysis-based vectorless verification as in [7] and choose Ω to be the set of nodes with the largest voltage drops. This works in practice as the solutions of the DC analysis-based algorithm, although potentially inaccurate, are able to provide a rough picture of voltage drops and identify nodes where the worst-case is most likely to occur. It is also possible to determine Ω based on previous experience, or choose Ω to be the set of nodes which are critical to circuit performance. In any event, we need only to solve (10) for $i \in \Omega$, and the cardinality of Ω can be made much smaller than N, i.e., $|\Omega| \ll N$.

As a result, the LP problems (10) can be reduced to

$$
\max_{i \in \Omega} x_i(k_t\Delta t) = \sum_{k=1}^{k_t} c_{i,k} u(k\Delta t)
$$
$$
\text{s.t.} \begin{cases} 0 \leq u(k\Delta t) \leq I_L, \quad U u(k\Delta t) \leq I_G, \\ U\left(\sum_{k=1}^{k_t} u(k\Delta t)\right) \leq \frac{k_t}{V_{dd}} P_B, \\ U_{r'} U_{r'-1} \cdots U\left(\sum_{k=1}^{k_t} u(k\Delta t)\right) \leq \frac{k_t}{V_{dd}} P_T \ (r' = 1 \ldots r), \end{cases}
$$
(11)

where $c_{i,k}$ is the ith row of $\mathcal{M}^{k_t-k} \mathcal{N}$. If all nodes must be considered (i.e. $\Omega = \{1, \ldots, N\}$), we can solve $c_{i,k}$ in parallel as proposed in the next subsection to reduce the runtime.

4.2 Efficient Calculation of Coefficients

By definition

$$
c_{i,k} = e_i^T \left[\left(G + \frac{C}{\Delta t}\right)^{-1} \frac{C}{\Delta t} \right]^{k_t-k} \left(G + \frac{C}{\Delta t}\right)^{-1} H, \quad (12)
$$

where $e_i \in \mathbb{R}^n$ is the ith elementary unit vector. Directly solving $c_{i,k}$ by computing the matrix inverse is prohibitively expensive. Now we propose an efficient method which costs only one sparse-LU decomposition and k_t forward/backward substitutions.

Performing transposition on both sides of (12), we have

$$
c_{i,k}^T = H^T \left(G^T + \frac{C^T}{\Delta t}\right)^{-1} \left[\frac{C^T}{\Delta t} \left(G^T + \frac{C^T}{\Delta t}\right)^{-1}\right]^{k_t-k} e_i.
$$
(13)

Assuming that $G^T + \frac{C^T}{\Delta t} = L_d U_d$ (sparse-LU decomposition), (13) can be written as

$$
c_{i,k}^T = H^T U_d^{-1} L_d^{-1} \underbrace{\left(\frac{C^T}{\Delta t}\right) U_d^{-1} L_d^{-1} \cdots \left(\frac{C^T}{\Delta t}\right) U_d^{-1} L_d^{-1}}_{k_t - k \text{ times}} e_i.
$$
(14)

Note that L_d^{-1} (U_d^{-1}) multiplied to a vector is in fact a forward (backward) substitution. Hence computing (14) for all $k = 1, \ldots, k_t$ only involves one sparse-LU decomposition, k_t matrix-vector multiplications and k_t forward (backward) substitutions. The matrix-vector multiplication is extremely efficient as $\frac{C}{\Delta t}$ is a diagonal matrix. Solving all the $c_{i,k}$'s ($i \in \Omega$) requires one sparse-LU and $k_t|\Omega|$ forward (backward) substitutions and matrix-vector multiplications. Note that $c_{i,k}$ can be solved in parallel as calculations of $c_{i,k}$'s for different i are independent of each other. The calculation of (14) is equivalent to the numerical computation of transient analysis and thus is numerically robust.

4.3 Sorting-Deletion Algorithm

Consider one LP problem in (11) (a specific $i \in \Omega$), with all the $c_{i,k}$ ($k = 1, \ldots, k_t$) vectors computed following Section 4.2. It can be seen that the objective function is a linear combination of all the currents at all time steps. The coefficient of the jth current source at time step k is the jth entry of the vector $c_{i,k}$. To simplify notations, we reorder the subscripts of coefficients and variables as

$$
\begin{aligned}
\tilde{c}_1 &= e_1^T c_{i,1}; & \cdots & & \tilde{c}_m &= e_m^T c_{i,1}; \\
\vdots & & \ddots & & \vdots & \\
\tilde{c}_{(k_t-1)m+1} &= e_1^T c_{i,k_t}; & \cdots & & \tilde{c}_{k_t m} &= e_m^T c_{i,k_t};
\end{aligned}
$$
(15a)

$$
\begin{aligned}
\tilde{u}_1 &= u_1(\Delta t); & \cdots & & \tilde{u}_m &= u_m(\Delta t); \\
\vdots & & \ddots & & \vdots & \\
\tilde{u}_{(k_t-1)m+1} &= u_1(k_t\Delta t); & \cdots & & \tilde{u}_{k_t m} &= u_m(k_t\Delta t).
\end{aligned}
$$
(15b)

Consequently, any constraint in (11) is in the form of $\sum_{i \in \mathcal{L}} \tilde{u}_i \leq \ell$, where $\mathcal{L}$ is a subset of the indices $\{1, \ldots, k_t m\}$. Assume the total number of constraints is κ_t. In the case of (11) $\kappa_t = mk_t + pk_t + p + \sum_{r'=1}^{r} p_{r'}$. (11) can be rewritten as

$$
\max \ x_{i_{node}} = \left(\sum_{i=1}^{mk_t} \tilde{c}_i \tilde{u}_i\right) \quad \text{s.t.} \quad \sum_{i \in \mathcal{L}_\kappa} \tilde{u}_i \leq \ell_\kappa (\kappa = 1, \ldots, \kappa_t),
$$
(16)

where $i_{node} \in \Omega$ is the index of a node.

LEMMA 4.2. *The maximum of (16) is hit when all the $\tilde{u}_i$'s associated with negative $\tilde{c}_i$'s are set to zero.*

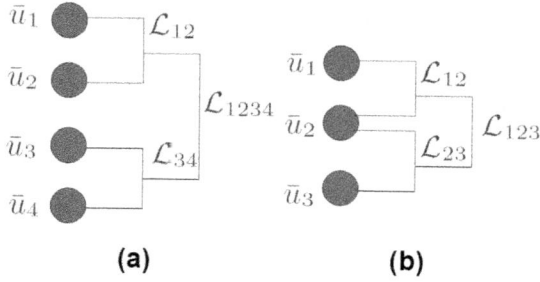

Figure 3: Different current and power constraints: (a) has a hierarchical structure as $\mathcal{L}_{12} \cap \mathcal{L}_{34} = \emptyset$, $\mathcal{L}_{12} \subset \mathcal{L}_{1234}$ **and** $\mathcal{L}_{34} \subset \mathcal{L}_{1234}$; **(b) is not hierarchical as** $\mathcal{L}_{12} \not\subset \mathcal{L}_{23}$ **and** $\mathcal{L}_{12} \not\supset \mathcal{L}_{23}$ **and** $\mathcal{L}_{12} \cap \mathcal{L}_{23} \neq \emptyset$.

PROOF. Assume the maximum is hit at a feasible point $\tilde{u}$ with at least one $\tilde{u}_j$ associated with a negative $\tilde{c}_j$ is set to a positive number. It is readily verified that the new point $\tilde{u}'$ obtained by setting $\tilde{u}_j$ to zero is still feasible (as the sum of any subset of $\tilde{u}'_i$'s is equal to or smaller than that of $\tilde{u}_i$'s). On the other hand, $x|_{\tilde{u}} < x|_{\tilde{u}'}$, which conflicts with the assumption that the maximum is hit at $\tilde{u}$. $\square$

There exist negative $\tilde{c}_i$'s due to the existence of inductors. Sometimes the voltage drops on inductors caused by the "change" of currents are more significant than static voltage drops on resistors, so the worst-case voltage drop has a negative correlation with some current points. With Lemma 4.2, we can set all the $\tilde{u}_i$'s associated with negative $\tilde{c}_i$'s to zero and then delete these $\tilde{u}_i$ from the constraints of (16). Constraint κ is also deleted if $\mathcal{L}_\kappa$ becomes empty after deleting $\tilde{u}_i$'s. Suppose after the deletions there exist $\bar{k}$ variables and $\bar{\kappa}$ constraints. Reorder remaining $\tilde{c}_i$'s such that $\bar{c}_1 \geq \cdots \geq \bar{c}_{\bar{k}} > 0$ and reorder $\bar{u}_i$'s accordingly. (16) can be reformulated as

$$\max x_{i_{node}} = \sum_{i=1}^{\bar{k}} \bar{c}_i \bar{u}_i \quad \text{s.t.} \quad \sum_{i \in \mathcal{L}_\kappa} \bar{u}_i \leq \ell_\kappa \quad (\kappa = 1, \ldots, \bar{\kappa}) \tag{17}$$

Now we digress to take a look at the constraints of (17), which have a hierarchical structure as depicted in Fig. 2. The hierarchical constraints structure implies that for any two sets $\mathcal{L}_{\kappa_1}$ and $\mathcal{L}_{\kappa_2}$, one of the following equations must hold: (i) $\mathcal{L}_{\kappa_1} \cap \mathcal{L}_{\kappa_2} = \emptyset$; (ii) $\mathcal{L}_{\kappa_1} \subset \mathcal{L}_{\kappa_2}$; (iii) $\mathcal{L}_{\kappa_1} \supset \mathcal{L}_{\kappa_2}$. See Fig. 3.

DEFINITION 4.3 (BOUNDARY POINT). *Suppose* $\bar{u} = [\bar{u}_1, \ldots, \bar{u}_{\bar{k}}]^T$ *is a feasible point under the constraints of (17).* $\bar{u}$ *is called a boundary point if for any* $1 \leq i \leq \bar{k}$ *and any* $\epsilon > 0$, $[\bar{u}_1, \ldots, \bar{u}_i + \epsilon, \ldots, \bar{u}_{\bar{k}}]^T$ *is not feasible.*

We look at an illustrative example as shown in Fig.3(a). The constraints in Fig.3(a) can be written explicitly as:

$$0 < \bar{u}_1 < l_1, \quad 0 < \bar{u}_2 < l_2, \quad 0 < \bar{u}_3 < l_3, \quad 0 < \bar{u}_4 < l_4;$$
$$\bar{u}_1 + \bar{u}_2 < l_{12}, \quad \bar{u}_3 + \bar{u}_4 < l_{34}; \tag{18}$$
$$\bar{u}_1 + \bar{u}_2 + \bar{u}_3 + \bar{u}_4 < l_{1234}$$

The upper bounds of $\{\bar{u}_1, \bar{u}_2, \bar{u}_3, \bar{u}_4\}$ and its subsets are:

$$\mathcal{B}(\bar{u}_1) = \min\{l_1, l_{12}, l_{1234}\}, \quad \mathcal{B}(\bar{u}_2) = \min\{l_2, l_{12}, l_{1234}\},$$
$$\mathcal{B}(\bar{u}_3) = \min\{l_3, l_{34}, l_{1234}\}, \quad \mathcal{B}(\bar{u}_4) = \min\{l_4, l_{34}, l_{1234}\};$$
$$\mathcal{B}(\bar{u}_1 + \bar{u}_2) = \min\{\mathcal{B}(\bar{u}_1) + \mathcal{B}(\bar{u}_2), l_{12}, l_{1234}\};$$
$$\mathcal{B}(\bar{u}_3 + \bar{u}_4) = \min\{\mathcal{B}(\bar{u}_3) + \mathcal{B}(\bar{u}_4), l_{34}, l_{1234}\};$$
$$\mathcal{B}(\bar{u}_1 + \bar{u}_2 + \bar{u}_3 + \bar{u}_4) = \min\{\mathcal{B}(\bar{u}_1 + \bar{u}_2) + \mathcal{B}(\bar{u}_3 + \bar{u}_4), l_{1234}\}; \tag{19}$$

LEMMA 4.4. *The sum of elements (coordinates) of any boundary point is the same, i.e.,* $\sum_{i=1}^{\bar{k}} \bar{u}_i$ *is a constant as long as* $[\bar{u}_1, \ldots, \bar{u}_{\bar{k}}]^T$ *is a boundary point.*

PROOF. Consider a general (arbitrary) hierarchical constraint structure, which can be represented by a "tree". Let the depth of the "tree" be d and each node of the tree have at most w children. Denote the jth node in the i level as $\mathcal{L}_{i,j}$. Assuming there exists a boundary point $(\bar{u}_1, \ldots, \bar{u}_{\bar{k}})$ with $\sum_{i=1}^{\bar{k}} \bar{u}_i < \mathcal{B}(\sum_{i=1}^{\bar{k}} \bar{u}_i)$, let $\epsilon = \mathcal{B}(\sum_{i=1}^{\bar{k}} \bar{u}_i) - \sum_{i=1}^{\bar{k}} \bar{u}_i > 0$. So there exists at least one child (w.l.o.g., we assume this is the first child) with $\sum_{i \in \mathcal{L}_{2,1}} \bar{u}_i \leq \mathcal{B}(\sum_{i \in \mathcal{L}_{2,1}} \bar{u}_i) - \epsilon/w$. Otherwise, $\sum_{i=1}^{\bar{k}} \bar{u}_i > \mathcal{B}(\sum_{i=1}^{\bar{k}} \bar{u}_i) - \epsilon$. Perform this deduction to the bottom of the tree, we conclude that there exists at least one child (w.l.o.g.. we assume it is the first child of its parent) with $\bar{u}_1 \leq \mathcal{B}(\bar{u}_1) - w^{-d}\epsilon$. Thus we conclude that $(\bar{u}_1 + w^{-d}\epsilon, \ldots, \bar{u}_{\bar{k}})$ is feasible. This contradicts the fact that $(\bar{u}_1, \ldots, \bar{u}_{\bar{k}})$ is a boundary point. $\square$

The main idea behind this lemma is that all the boundary points belong to the plane $\bar{u}_1 + \cdots + \bar{u}_{\bar{k}} = \mathcal{B}(\bar{u}_1 + \cdots + \bar{u}_{\bar{k}})$. Now we return to the LP problem (17), and show that the solution computed by the sorting-deletion algorithm (Algorithm 1) is optimal. We begin with the following two lemmas.

Algorithm 1 : Sorting-deletion algorithm

for $i = 1, \ldots, \bar{k}$ **do**
    (1) Select all the sets $\mathcal{L}_\kappa$ that satisfy $i \in \mathcal{L}_\kappa$. The subscripts of these $\mathcal{L}_\kappa$ form a set $\mathcal{K}_i$;
    (2) Set $\bar{u}_i$ to be $\min\{\ell_\kappa | \kappa \in \mathcal{K}_i\}$;
    (3) $\ell_\kappa = \ell_\kappa - \bar{u}_i$ for all $\kappa \in \mathcal{K}_i$;

LEMMA 4.5. *The solution computed by the sorting-deletion algorithm is a boundary point.*

PROOF. Assume that the solution $[\bar{u}_1, \ldots, \bar{u}_{\bar{k}}]^T$ computed by the sorting-deletion algorithm is not a boundary point. Then there exist an integer $i \in [1, \bar{k}]$ and $\epsilon > 0$ such that $[\bar{u}_1, \ldots, \bar{u}_i + \epsilon, \ldots, \bar{u}_{\bar{k}}]^T$ is feasible. However, from Algorithm 1 we know that $\bar{u}_i = \min\{\ell_\kappa | \kappa \in \mathcal{K}_i\}$. If the ith variable is set to be $\bar{u}_i + \epsilon$, at least one constraint is violated. Therefore, $[\bar{u}_1, \ldots, \bar{u}_i + \epsilon, \ldots, \bar{u}_{\bar{k}}]^T$ is not feasible which leads to a contradiction. $\square$

LEMMA 4.6. *Any optimal solution of the LP problem (17) is a boundary point.*

PROOF. Let $\bar{u} = [\bar{u}_1, \ldots, \bar{u}_{\bar{k}}]^T$ be any optimal solution to LP problem (17). Assume $\bar{u}$ is not a boundary point for the sake of contradiction. Then there exists an integer $i \in [1, \tilde{k}]$ and $\epsilon > 0$ such that $[\bar{u}_1, \ldots, \bar{u}_i + \epsilon, \ldots, \bar{u}_{\tilde{k}}]^T$ is feasible. Since $\bar{c}_i > 0$ for all $1 \leq i \leq \tilde{k}$, we have $\bar{c}_1 \bar{u}_1 + \ldots + \bar{c}_i (\bar{u}_i + \epsilon) + \ldots + \bar{c}_{\tilde{k}} > \bar{c}_1 \bar{u}_1 + \ldots + \bar{c}_i \bar{u}_i + \ldots + \bar{c}_{\tilde{k}}$. Hence, $[\bar{u}_1, \ldots, \bar{u}_i + \epsilon, \ldots, \bar{u}_{\tilde{k}}]^T$ is a better solution for (17) than $\bar{u}$, a contradiction. $\square$

THEOREM 4.7. *The solution computed by the sorting-deletion algorithm is an optimal solution of the LP problem (17).*

PROOF. Suppose the solution computed by the sorting-deletion algorithm is $\bar{u}^{(1)} = [\bar{u}_1^{(1)}, \ldots, \bar{u}_{\tilde{k}}^{(1)}]^T$ and an optimal solution of the LP problem is $\bar{u}^{(2)} = [\bar{u}_1^{(2)}, \ldots, \bar{u}_{\tilde{k}}^{(2)}]^T$. Assume the two points are different. From Lemma 4.4, Lemma 4.5 and Lemma 4.6 we know that $\sum_{i=1}^{\tilde{k}} \bar{u}_i^{(1)} = \sum_{i=1}^{\tilde{k}} \bar{u}_i^{(2)}$.

Suppose j is the lowest index at which $\bar{u}^{(1)}$ and $\bar{u}^{(2)}$ are different. We have $\bar{u}_j^{(2)} < \bar{u}_j^{(1)}$. Otherwise, $\bar{u}_j^{(2)} > \bar{u}_j^{(1)}$, and at least one constraint $\mathcal{L}_\kappa$ ($\kappa \in \mathcal{K}_j$) is not satisfied. As $\sum_{i=1}^{\tilde{k}} \bar{u}_i^{(1)} = \sum_{i=1}^{\tilde{k}} \bar{u}_i^{(2)}$, there exists at least one $j' > j$ such that $\bar{u}_{j'}^{(2)} > \bar{u}_{j'}^{(1)} \geq 0$. As for $\forall \kappa_1, \kappa_2 \in \mathcal{K}_j$, $\mathcal{L}_{\kappa_1} \bigcap \mathcal{L}_{\kappa_2} \neq \emptyset$ (both contain j), we have $\mathcal{L}_{\kappa_1} \subset \mathcal{L}_{\kappa_2}$ or $\mathcal{L}_{\kappa_1} \supset \mathcal{L}_{\kappa_2}$. Assume w.l.o.g. that $\mathcal{L}_1 \subset \ldots \subset \mathcal{L}_\mu$ (μ is the cardinality of $\mathcal{K}_j$). Suppose $\mathcal{L}_{\kappa_1}$ is the first (i.e. smallest) set among $\mathcal{L}_1, \ldots, \mathcal{L}_\mu$ that contains $\bar{u}_{j'}^{(2)} > 0$ ($j' > j$), we adapt the optimal point $\bar{u}^{(2)}$ by setting $\bar{u}_j^{(2)}$ to $\bar{u}_j^{(2)} + \delta$ and $\bar{u}_{j'}^{(2)}$ to $\bar{u}_{j'}^{(2)} - \delta$ with $\delta \triangleq \min\{\bar{u}_j^{(1)} - \bar{u}_j^{(2)}, \bar{u}_{j'}^{(2)}\}$. If no set among $\mathcal{L}_1, \ldots, \mathcal{L}_\mu$ contains such $\bar{u}_{j'}^{(2)} > 0$ ($j' > j$), choose the first $\bar{u}_{j'}^{(2)} > 0$ ($j' > j$) and perform the similar adaptation. It is readily verified that the adapted point still satisfies all the constraints and thus is feasible. On the other hand, because $\bar{c}_j \geq \bar{c}_{j'}$, we have $\bar{c}_1 \bar{u}_1^{(2)} + \cdots + \bar{c}_j (\bar{u}_j^{(2)} + \delta) + \cdots + \bar{c}_{j'} (\bar{u}_{j'}^{(2)} - \delta) + \cdots + \bar{c}_{\tilde{k}} \bar{u}_{\tilde{k}} \geq \bar{c}_1 \bar{u}_1^{(2)} + \cdots + \bar{c}_j \bar{u}_j^{(2)} + \cdots + \bar{c}_{j'} \bar{u}_{j'}^{(2)} + \cdots + \bar{c}_{\tilde{k}} \bar{u}_{\tilde{k}}$. Therefore the adapted point is also an optimal point. After repeatedly performing such adaptation, $\bar{u}_j^{(1)} = \bar{u}_j^{(2)}$. Then the first difference appears at a position after j. Perform all the steps so on and so forth we have $\bar{u}^{(1)} = \bar{u}^{(2)}$. As $\bar{u}^{(2)}$ is an optimal point, $\bar{u}^{(1)}$ is also an optimal solution. $\square$

The intuition behind this theorem is that the optimal solution is obtained by giving the variable associated with the largest coefficient the largest possible value.

4.4 Algorithm Flow and Complexity

The algorithm flow for the worst-case voltage drop prediction is summarized as Algorithm 2. Its computational complexity is analyzed as follows.

1. Computing $c_{i_{node}, k}$'s. Computing $c_{i_{node}, k}$'s for each $i_{node} \in \Omega$ requires one sparse-LU decomposition and k_t forward or backward substitutions. As the system matrix is an n by n sparse matrix with $O(n)$ nonzero entries mainly distributing near the diagonal line, this procedure has a complexity of $O(n^\alpha k_t)$, with $1 < \alpha < 2$.

2. Sorting $\bar{c}_i$'s. For each LP problem there exist mk_t variables and coefficients. Employing the most efficient sorting algorithm, this procedure has a complexity of $O(mk_t \log(mk_t))$.

3. Sorting-deletion algorithm. The sorting-deletion algorithm determines $\bar{u}_i$ one at a time and then subtract the value from the constraints. In practice each $\bar{u}_i$ involves only several constraints, i.e., $|\mathcal{K}_i| < 10$. This procedure has a complexity of $O(\bar{k})$ with $\bar{k} \leq mk_t$.

In summary, the complexity of Algorithm 2 is dominated by the computation of $c_{i_{node}, k}$'s. If executed in sequence, the overall

---

Algorithm 2 : Worst-case voltage drop prediction

1: Set up hierarchical current and power constraints based on previous experience and/or design requirements;
2: **for** $\forall i_{node} \in \Omega$, execute 3 : 12 (in *parallel*)
3:     Calculate $c_{i_{node}, k}$'s following Section 4.2;
4:     Set up the (reduced) LP problem as (11);
5:     Set all $\tilde{u}_i$'s associated with negative $\bar{c}_i$'s to zero and delete them from constraint sets $\mathcal{L}_i$'s;
6:     Sort the coefficients $\bar{c}_i$'s in the descending order and reformulate the LP problem as (17);
7:     **for** $i = 1 : \bar{k}$ **do**
8:         Select all the sets $\mathcal{L}_\kappa$ that satisfy $i \in \mathcal{L}_\kappa$. The subscripts of these $\mathcal{L}_\kappa$ form a set $\mathcal{K}_i$;
9:         Set $\bar{u}_i$ to be $\min\{\ell_\kappa | \kappa \in \mathcal{K}_i\}$;
10:         $\ell_\kappa = \ell_\kappa - \bar{u}_i$ for all $\kappa \in \mathcal{K}_i$;
11:     **end**
12:     Compute $x_{i_{node}} = \sum_{i=1}^{\bar{k}} \bar{c}_i \bar{u}_i$;
13: **end**
14: Worst-case voltage drop is $\max\{x_{i_{node}} | i_{node} \in \Omega\}$.

---

complexity is $O(n^\alpha k_t |\Omega|)$. If executed in parallel, the overall complexity is $O(n^\alpha k_t)$.

5. EXPERIMENTAL RESULTS

We generate two 3D power grids as benchmarks. Each of the power grids has four metal layers and is modeled as an equivalent RCL circuit. Basic parameters of the power grids and corresponding RCL circuits are recorded in Table 1. The simulation time is $0 - 1ns$ and is divided into 100 intervals ($k_t = 100$) with each interval being $10ps$. LP problems are set up based on local current constraints and global (including block-level current, block-level power and high-level power) constraints. Sizes of the resulting LP problems are also recorded in Table 1.

The LP problems are solved for every node $i_{node} \in \Omega$ ($|\Omega| = 100$), first by standard LP methods and then by the proposed sorting-deletion algorithm. The program is executed on a Linux workstation with 3.0GHz 8-core Intel Xeon CPU and 16G memory. CPU times are reported in Table 2. The voltage drops are omitted in this table as they are the same for both methods. From the table we conclude that solving the LP problems is speeded up by approximately $> 7\times$ using the proposed algorithm when there exist no power constraints. If there exist power constraints, the standard LP method does not work because the iteration number exceeds an acceptable number. An intuitive explanation for this phenomenon is that in the LP problem without power constraints, the coefficient matrix of the inequality contains one entry in each column (like a "diagonal" matrix). Thus using standard methods to solve LP problems without power constraints is much faster than to solve LP problems with power constraints. The speedups for both single node case and multiple node case are roughly the same because each node is solved independently.

To show that omitting power constraints may result in an overly pessimistic voltage drop prediction, we solve the LP problems (using sorting-deletion algorithm) for nodes belonging to Ω both with and without the power constraints (pc's). The results are shown in Table 3. The worst-case current patterns of some specific current sources are plotted in Fig. 4(a) & 4(b). It can be seen from Table 3 that omitting power constraints may result in an 30% overestimation of the worst-case voltage drop. Fig. 4(a) & 4(b) show that current sources keep their peak current values in a much longer time period if power constraints are omitted, which are not realistic.

Table 1: Parameters of the power grids used in the experiment

	Power grid models						LP problems	
	Nodes (N)	Sources (m)	Matrix size (n)	No. of R's	No. of C's	No. of L's	Variables	$|\Omega|$
Power grid 1	75,762	37,881	113,499	54,350	37,684	37,684	3.7M	100
Power grid 2	980,313	490,157	146,9755	608,792	394,444	394,444	690M	100

The worst-case current patterns with power constraints are more realistic. To provide intuitive pictures of how voltage drop at the node where worst-case voltage drop occurs changes, we perform transient simulation using the worst-case current waveforms. It can be seen from Fig. 5(a) & 5(b) that omitting power constraints does result in overly pessimistic voltage drop predictions.

To show the impact of the constraint structure on voltage drops and CPU times, we do experiments on power grid 1 (one single node) using different constraint structures. The results are shown in Table 4. Both standard method and sorting-deletion algorithm are used. Sorting-deletion algorithm does not apply for non-hierarchical constraint structure ("×" in Table 4). Standard methods do not work for non-hierarchical constraints and constraints with pc's because the iteration number exceeds an acceptable value (CPU time too large). Table 4 indicates that more power constraint levels result in lower voltage drop prediction. In practice the number of constraint levels should be determined based on design requirements or experience. Table 4 also indicates that CPU time of sorting-deletion algorithm does not increase significantly with hierarchical levels.

Table 3: Voltage drop predictions with and without power constraints

		Without pc's	With pc's	Over es-timation	Percen-tage
Power grid 1	Average for Ω	62.3 mV	46.9 mV	15.4 mV	33%
	Worst-case	63.4 mV	48.1 mV	15.3 mV	32%
Power grid 2	Average for Ω	80.2 mV	61.1 mV	19.1 mV	31%
	Worst-case	81.3 mV	63.2 mV	18.1 mV	29%

Table 4: Voltage drop and CPU times for different constraint structures

		non-hier	w/o pc's	L-1 pc's	L-2 pc's	L-3 pc's
Standard method	Voltage drop (mV)	—	61.5	—	—	—
	CPU time (s)	—	7.14	—	—	—
Sorting deletion	Voltage drop (mV)	×	61.5	45.7	37.4	33.2
	CPU time (s)	×	0.74	0.83	0.90	0.96

[2]Here "non-hier" represents non-hierarchical constraints. L-1 (L-2, L-3) pc's represent hierarchical constraints with $r = 1$ ($r = 2$, $r = 3$) level(s) of power constraints.

6. CONCLUSIONS

A more realistic early-stage power grid verification algorithm based on hierarchical current and power constraints has been proposed in this paper. The proposed algorithm does not rely on the M-matrix assumption and thus naturally handles general RCL power grid models. Besides, currents at different time steps are treated as independent variables in LP problems and additional power constraints are introduced which restrict the energy consumed by certain current sources. As a result, worst-case current patterns solved by the proposed algorithm are more realistic in the sense that they are time-varying and cannot keep peak values all the time. Consequently, the worst-case voltage drop prediction is less pessimistic. Moreover, a sorting-deletion algorithm is proposed which significantly speeds up the solutions of LP problems. Experimental results have verified that the proposed algorithm generates more realistic worst-case current patterns and voltage drops. Utilizing the proposed sorting-deletion algorithm, the CPU time needed to solve LP problems is reduced by >85%.

7. REFERENCES

[1] T. Chen and C. Chen. Efficient large-scale power grid analysis based on preconditioned krylov-subspace iterative methods. In *DAC*, pages 559–562, 2001.

[2] P. Du, X. Hu, S. Weng, A. Shayan, X. Chen, E. Engin, and C. Cheng. Worst-case noise prediction with non-zero current transition times for early power distribution system verification. In *ISQED*, pages 624–631. IEEE, 2010.

[3] I. Ferzli, F. Najm, and L. Kruse. A geometric approach for early power grid verification using current constraints. In *ICCAD*, pages 40–47, 2007.

[4] J. Fu, Z. Luo, X. Hong, Y. Cai, S. Tan, and Z. Pan. A fast decoupling capacitor budgeting algorithm for robust on-chip power delivery. In *ASPDAC*, pages 505–510, 2004.

[5] A. Ghani and F. Najm. Fast vectorless power grid verification using an approximate inverse technique. In *DAC*, pages 184–189, 2009.

[6] N. Ghani and F. Najm. Handling inductance in early power grid verification. In *ICCAD*, page 134, 2006.

[7] D. Kouroussis and F. Najm. A static pattern-independent technique for power grid voltage integrity verification. In *DAC*, pages 99–104, 2003.

[8] S. Nassif and J. Kozhaya. Fast power grid simulation. In *DAC*, pages 156–161, 2000.

[9] S. Pant, D. Blaauw, V. Zolotov, S. Sundareswaran, and R. Panda. A stochastic approach to power grid analysis. In *DAC*, pages 171–176, 2004.

[10] R. Plemmons. M-matrix characterizations. I–nonsingular M-matrices. *Linear Algebra and its Applications*, 18(2):175–188, 1977.

[11] H. Qian, S. Nassif, and S. Sapatnekar. Early-stage power grid analysis for uncertain working modes. *IEEE Trans. Comput.-Aided Design Integr. Circuits Syst.*, 24(5):676–682, 2005.

[12] H. Qian, S. Nassif, and S. Sapatnekar. Power grid analysis using random walks. *IEEE Trans. Comput.-Aided Design Integr. Circuits Syst.*, 24(8):1204–1224, 2005.

[13] Y. Wang, C. U. Lei, G. K. H. Pang, and N. Wong. MFTI: Matrix-Format Tangential Interpolation for Modeling Multi-Port Systems. In *DAC*, pages 683–686, 2010.

[14] Y. Wang, Z. Zhang, C. K. Koh, G. K. H. Pang, and N. Wong. PEDS: Passivity Enforcement for Descriptor Systems via Hamiltonian-Symplectic Matrix Pencil Perturbation. In *ICCAD*, pages 800–807, 2010.

[15] X. Xiong and J. Wang. An efficient dual algorithm for vectorless power grid verification under linear current constraints. In *DAC*, pages 837–842, 2010.

[16] W. Zhang, W. Yu, X. Hu, L. Zhang, R. Shi, H. Peng, Z. Zhu, L. Chua-Eoan, R. Murgai, T. Shibuya, et al. Efficient power network analysis considering multidomain clock gating. *IEEE Trans. Comput.-Aided Design Integr. Circuits Syst.*, 28(9):1348–1358, 2009.

[17] M. Zhao, R. Panda, S. Sapatnekar, and D. Blaauw. Hierarchical analysis of power distribution networks. *IEEE Trans. Comput.-Aided Design Integr. Circuits Syst.*, 21(2):159–168, 2002.

Table 2: Runtime comparison of standard LP algorithms and the proposed sorting-deletion algorithm

			Without pc's			With pc's		
			Standard method	Proposed algorithm	Speed-up	Standard method	Proposed algorithm	Speed-up
Power grid 1	Single node	Setup	9.86 sec	9.86 sec	—	9.86 sec	9.86 sec	—
		Solving	6.08 sec	0.71 sec	8.56×	—[1]	0.77 sec	—
	$|\Omega|$ nodes	Setup	901 sec	901 sec	—	901 sec	901 sec	—
		Solving	577 sec	70.2 sec	8.22×	—[1]	76.5 sec	—
Power grid 2	Single node	Setup	278 sec	278 sec	—	278 sec	278 sec	—
		Solving	74.4 sec	9.91 sec	7.51×	—[1]	10.87 sec	—
	$|\Omega|$ nodes	Setup	417 min	417 min	—	417 min	417 min	—
		Solving	120 min	15.4 min	7.83×	—[1]	17.1 min	—

[1] Here the standard LP solver does not work because the iteration number is too large and exceeds "MaxIter".

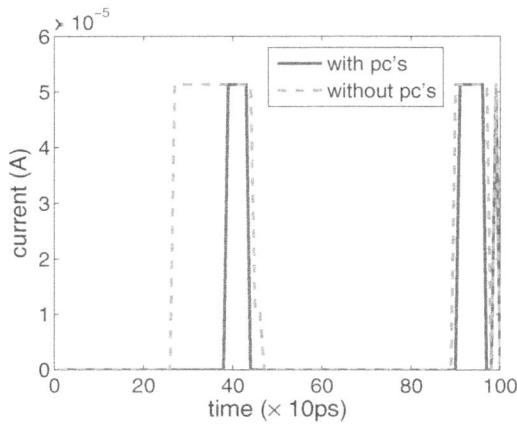
(a) Current source 18,941 of power grid 1

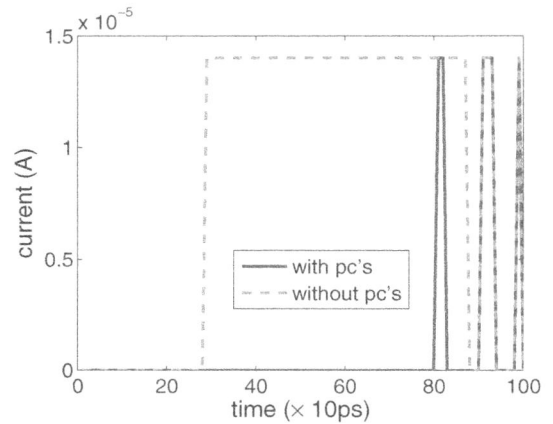
(b) Current source 113,990 of power grid 2

Figure 4: Worst-case current patterns with and without power constraints at some nodes of power grid 1 and power grid 2.

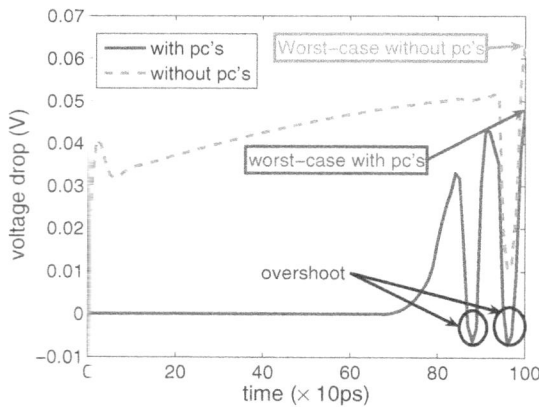
(a) Power grid 1

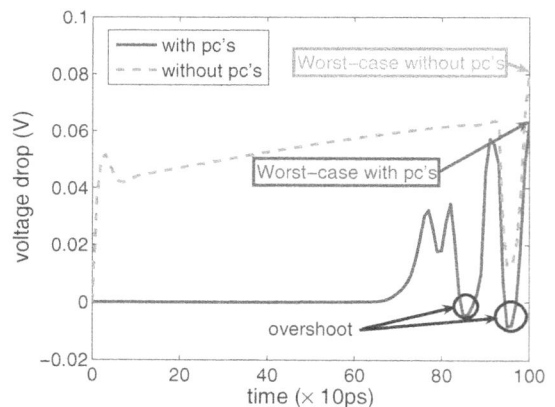
(b) Power grid 2

Figure 5: Voltage drop pattern at the node where worst-case voltage drop occurs for both power grid 1 and power grid 2. The values of the red and blue curves at $t = 1ns$ are the worst-case voltage drops with and without power constraints.

Lagrangian Relaxation for Gate Implementation Selection

Yi-Le Huang
Department of ECE
Texas A&M University
College Station, TX 77843
kiwe@tamu.edu

Jiang Hu
Department of ECE
Texas A&M University
College Station, TX 77843
jianghu@ece.tamu.edu

Weiping Shi
Department of ECE
Texas A&M University
College Station, TX 77843
wshi@ece.tamu.edu

ABSTRACT

In a typical circuit optimization flow, one essential decision is to select the implementation for each gate according to a cell library. An implementation implies specific gate size, threshold voltage, etc. The selection normally needs to handle multiple and often conflicting objectives. An effective approach for multi-objective optimization is Lagrangian relaxation (LR), which has been adopted in continuous gate sizing. When LR is applied to the gate implementation selection, the Lagrangian dual problem is no longer convex like in continuous gate sizing, and conventional sub-gradient method becomes inefficient. In this paper, we propose a projection-based descent method and a new technique of Lagrangian multiplier distribution for solving the Lagrangian dual problem in discrete space. Experimental results demonstrate that our approach leads to significantly better solution quality and faster convergence compared to the sub-gradient method.

Categories and Subject Descriptors

B.7 [**Integrated Circuits**]: Design Aids

General Terms

Algorithms, Design

Keywords

Gate sizing, Lagrangian relaxation, Optimization, Low power, Threshold voltage

1. INTRODUCTION

In deep sub-micron technologies, minimization of leakage power becomes a main concern as we try to combat the increase in the overall circuit power consumption. In addition, power consumption directly relates to battery life, reliability, packaging and heat removal cost. Therefore, how to efficiently handle trade-off between circuit performance and power consumption becomes a big issue in current design flow. Gate sizing [1] has been one of the most popular methods for circuit optimization, such as area/timing and area/power/timing, for long time. Besides, in recent years people start to pay attention to leakage power and try

Permission to make digital or hard copies of all or part of this work for personal or classroom use is granted without fee provided that copies are not made or distributed for profit or commercial advantage and that copies bear this notice and the full citation on the first page. To copy otherwise, or republish, to post on servers or to redistribute to lists, requires prior specific permission and/or a fee.
ISPD'11, March 27–30, 2011, Santa Barbara, California, USA.
Copyright 2011 ACM 978-1-4503-0550-1/11/03...$10.00.

to manage leakage power by using different threshold voltage (Vt) levels [2-4]. Gates with higher Vt level are used on non-critical paths to reduce leakage power and those on critical paths work under lower Vt level for retaining desired performance. There are many similarities between discrete gate sizing and Vt assignment, and hence they can be easily combined with each other for combinational circuit optimization [5-10].

Most of simultaneous gate sizing and Vt assignment methods are either sensitivity-based heuristics [2,8] or mathematical programming methods [3,9,11]. In sensitivity-based heuristics, people use sensitivity functions to choose gate size and Vt level. Usually, the sensitivity function only considers local information, like the efficiency of trading power for performance for single gate. However, the greedy nature makes sensitivity-based methods easily to fall into local optima.

In [1], continuous gate/transistor sizing is formulated as geometric programming. In [11,12], Lagrangian relaxation (LR) is used to solving the sizing problem and proved to converge and guarantee optimality. Nowadays, circuits are implemented by gates in standard cell library provided by foundries or library companies. Sizes and Vt options of logic gates are limited and discretely specified in the standard cell library. In [13, 14], continuous solutions are rounded to the nearest discrete options in a library but the rounding may result in large rounding errors if options in standard cell library are highly discrete [15]. In [16], simultaneous discrete gate sizing and Vt assignment is solved by Lagrangian relaxation along with a dynamic programming (DP)-like method without rounding.

In general, LR is effective on handling multiple conflicting objectives or complex constraints. The effectiveness is obtained by transforming the original optimization into a Lagrangian sub-problem and a Lagrangian dual problem. The efficiency of LR is successfully demonstrated in continuous gate sizing [11, 12, 17]. The work of [16] employs LR for solving simultaneous discrete gate sizing and Vt assignment. It mainly focuses on the algorithm of solving the Lagrangian sub-problem, and solves the dual problem using sub-gradient method like in continuous gate sizing [11,12]. For discrete cases, the Lagrangian dual problem is no longer convex like in many continuous cases. Therefore, the conventional sub-gradient method becomes inefficient.

In this work, we attempt to solve the problem of gate implementation selection. A gate implementation means specific gate size, Vt level, etc. according to a cell library. We also adopt the LR approach to handle power-performance tradeoff. Our focus is on new techniques for solving the Lagrangian dual problem. We propose a projection-based descent method and a new technique of Lagrangian multiplier distribution. Compared to the conventional sub-gradient method, our algorithm leads to not only considerably better solution quality but also faster convergence. Our algorithm can satisfy tight timing constraints for which sub-gradient method fails. When timing constraints are loose, our algorithm results in about 33% power reduction than sub-gradient method.

The rest of this paper is organized as follows. In Section II, we introduce some notations and terminology used in this paper. The detailed problem formulation is also given. In Section III, we briefly present how Lagrangian relaxation solves constrained optimization problem. In Section IV, we show how to improve Lagrangian relaxation with our algorithm for dual problem under discrete gate sizes and Vt levels and our algorithm for sub-problem is in Section V. In Section VI, we show experimental results compared with sub-gradient method under two different timing requirements. The convergence of our algorithm is demonstrated by results of an ISCAS85 benchmark.

2. PRELIMINARIES

A combinational logic circuit is described by a directed acyclic graph (DAG) $G(V, E)$, where V is a set of nodes representing circuit components, including logic gates X, input drivers S and output loads T, $V = X \cup S \cup T$, and E is a set of edges indicating the wire connection between components. Each edge $e_{ij} \in E$ indicates the connection between v_i and $v_j \in V$ and logic signal propagates from v_i to v_j. Each logic gate $v_i \in V$ can have a gate size from set W_i and a Vt level from set U_i. Therefore, v_i has total $|W_i| \times |U_i|$ possible options. The gate implementation selection problem is to assign $w_i \in W_i$ and $u_i \in U_i$ for all $v_i \in X$ such that the total power consumption is minimized subject to timing constraints.

Here we only consider two major types of power consumption, dynamic and leakage power. Dynamic power P_{dyn} consumption is proportional to switching factor β, clock frequency f_{clk}, load capacitance C_{load} and square of supply voltage level V_{dd}. Detailed equation of dynamic power is $P_{dyn} = \frac{1}{2}\beta V_{dd}^2 f_{clk} C_{load}$. On the other hand, leakage power $P_{leakage}$ consumption is related to supply voltage and off current I_{off} of gates, where I_{off} is given by cell library depending on size and Vt level. Detailed equation of leakage power consumption is $P_{leak} = V_{dd} \times I_{off}$. There are still some other types of power consumption, like short circuit power, and they are relatively small so we ignore them in this work.

In this work, we take Elmore delay model as our delay model and model circuit components as resistance-capacitance (RC) circuits [18]. A logic gate $v_i \in X$ is modeled as input capacitance c_i and output resistance r_i plus intrinsic delay D_{int_i}, as shown in Figure 1. A wire segment is modeled as a π-type RC circuit. The wire model for a wire $e_{ij} \in E$ is shown in Figure 2 where $l_{e_{ij}}$ is length of e_{ij} and r_{unit} and c_{unit} are the unit length wire resistance and capacitance, respectively. The delay associated with a resistor is calculated by its resistance times its downstream capacitance. The delay for a path is the sum of the delay on resistors which it passes through. The Elmore delay model is relatively simple and proven to be applicable to distributed network of resistors and capacitors. However, our work can also deal with more complex delay model easily. Here we assume that the arrival time a at primary inputs and required arrival time q at primary outputs are given. The timing information, a and q, for all gates can be obtained easily

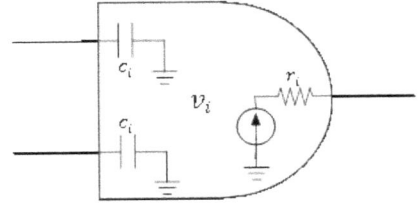

Figure 1. RC gate model.

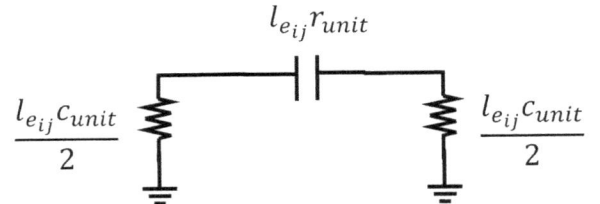

Figure 2. RC wire model.

by static timing analysis [19]. Then, slack $q - a$ indicates the timing criticality of components. The overall circuit timing is characterized by the minimum slack among components.

The size of a gate $v_i \in X$ affects intrinsic delay, input capacitance, output resistance, and dynamic and leakage power consumption. The Vt level of a gate also changes input capacitance and leakage power. We use $r_i^{w_i u_i}$ to represent output resistance of $v_i \in X$ under $w_i \in W_i$ and $u_i \in U_i$ and other variables are also symbolized by the same rule. All the information of logic gates is defined in a cell library. A gate with bigger size usually results in larger input capacitance, smaller output resistance, bigger intrinsic delay and more power consumption. A gate of higher Vt level will be converted to a gate model with bigger output resistance and less power consumption.

Given a combinational circuit, we want to solve the problem of minimizing the total circuit power consumption by selecting gate size and Vt level for each gate subject to the timing constraints that no negative slack in the circuit. We formulate the problem as a constrained optimization problem. These constraints are formulated with arrival time a and only applied on components rather than paths to reduce the number of constraints. We call the constrained optimization problem primal problem pp.

$$pp : Minimize \sum_{v_i \in X} \alpha_i p_i$$

$$a_i \leq A_i \qquad v_i \in T$$
$$a_j + D_i \leq a_i \qquad \forall v_j \in fanin(v_i), \qquad v_i \in X$$
$$D_i \leq a_i \qquad v_i \in S$$
$$w_i \in W_i, u_i \in U_i$$

A_i and α_i are given required arrival time and weighting factor of v_i, respectively. p_i is power consumption of v_i derived according to gate size and Vt level of v_i.

$$p_i = \frac{1}{2}\beta V_{dd}^2 f_{clk} c_i^{w_i} + V_{dd} \times I_{off}(w_i, u_i) \qquad (1)$$

$I_{off}(w_i, u_i)$ is a function representing off current of gate based on gate size and Vt level.

D_i is gate delay of v_i derived based on gate size, Vt level and downstream capacitance of v_i.

$$D_i = elmore\_delay(w_i, u_i, C_{ds}) \qquad (2)$$

C_{ds} is downstream capacitance of v_i and $elmore\_delay(w_i, u_i, C_{ds})$ is a function calculating Elmore delay of v_i based on its gate size, Vt and downstream capacitance.

3. LAGRANGIAN RELAXATION

Lagrangian relaxation method is a well-known approach for solving optimization problems with complex constraints. In [12], a timing constrained area optimization problem is solved by Lagrangian relaxation for continuous gate sizing. In Lagrangian relaxation method, constraints are relaxed and incorporated into the objective function by multiplying with Lagrangian multiplier vector $\vec{\lambda}$. A Lagrangian multiplier λ must be non-negative for each related constraint. We will associate Lagrangian multiplier λ_{ji} with the arrival time constraint of the wire connection from v_j to v_i, where v_j is a fan-in gate of v_i. Furthermore, we will associate Lagrangian multiplier u_i with gate v_i. Then the new objective function becomes

$$L_\lambda = \sum_{v_i \in X} \alpha_i p_i + \sum_{\substack{v_j \in fanin(v_i) \\ v_i \in T}} \lambda_{ji} (a_i - A_i)$$

$$+ \sum_{\substack{v_j \in fanin(v_i) \\ v_i \in X}} \lambda_{ji} (a_j + D_i - a_i) + \sum_{v_i \in S} \lambda_{0i} (D_i - a_i) \qquad (3)$$

For a given vector $\vec{\lambda}$, a new optimization problem only has size and Vt constraints, which is called Lagrangian sub-problem LRS/λ.

$$LRS/\lambda : \; Minimize \; L_\lambda$$

$$subject \; to \; w_i \in W_i, u_i \in U_i$$

Let $Q(\vec{\lambda})$ denote the function of the optimal solution of LRS/λ with respect to $\vec{\lambda}$, the Lagrangian dual problem LDP is defined as

$$LDP : \; Maximize \; Q(\vec{\lambda})$$

$$subject \; to \; \vec{\lambda} \geq 0$$

To reduce the complexity of LRS/λ, Kuhn-Tucker (KKT) conditions are applied, requiring $\frac{\partial L_\lambda}{\partial a_i} = 0$ at the optimal solution for $v_i \in X$. Applying $\frac{\partial L_\lambda}{\partial a_i} = 0$, the optimal conditions for $\vec{\lambda}$

$$\sum_{v_k \in fanout(v_i)} \lambda_{ik} = \sum_{v_j \in fanin(v_i)} \lambda_{ji} \; , \forall v_i \in X \qquad (4)$$

The optimal conditions say that for all logic gates and input resistors, the sum of Lagrangian multipliers on the wire connected from its fan-in gates must be equal to the sum of Lagrangian multipliers on the wire connected to its fan-out gates. Using the result of KKT conditions, the equation L_λ is further reduced to

$$L_\lambda = \sum_{v_i \in X} \alpha_i p_i + \sum_{v_i \in X} (\sum_{v_j \in fanin(v_i)} \lambda_{ji}) D_i + \sum_{v_i \in S} \lambda_{0i} D_i$$

$$+ \sum_{v_i \in T} (\sum_{v_j \in fanin(v_i)} \lambda_{ji}) A_i \qquad (5)$$

Replacing $\sum_{v_j \in fanin(v_i)} \lambda_{ji}$ by μ_i for $\forall v_i \in X$ and λ_{0i} by μ_i for $\forall v_i \in S$, the problem can be written as

$$L_\mu = \sum_{v_i \in X} \alpha_i p_i + \sum_{v_i \in V} \mu_i D_i \qquad (6)$$

The term $\sum_{v_i \in T}(\sum_{v_j \in fanin(v_i)} \lambda_{ji}) A_i$ is fixed when solving LRS/λ, so it can be ignored in L_μ. It is obvious that there is only power and gate delay left in L_μ without arrival time. The part of L_μ affected by a gate is independent from that affected by others. Therefore, the problem LRS/λ can be solved much easier than the original objective function with arrival time. In [12], for a given $\vec{\lambda_e}$, the problem LRS/λ is solved optimally by a greedy algorithm when the gate sizes are continuous.

As shown in [12], the dual problem LDP can be solved by sub-gradient method. The sub-gradient direction is used for replacing gradient direction in steepest descent method if the problem is not differentiable. Given that the step size satisfies the following conditions: $\lim_{n \to \infty} \rho_n = 0$ and $\sum_{n=1}^{\infty} \rho_n = \infty$, the sub-gradient method will converge to the optimum in continuous solution space. The equations for updating Lagrangian multipliers associated are shown as followed

$$\lambda_{ji} = \begin{cases} \lambda_{ji} + \rho_n(a_j - A_i), & v_i \in T \\ \lambda_{ji} + \rho_n(a_j + D_i - a_i), & v_i \in X, \\ \lambda_{ji} + \rho_n(D_i - a_i), & v_i \in S \end{cases}$$

$$where \; v_j \in fanin(v_i), \qquad j = 0 \; if \; v_i \in S$$

After updating $\vec{\lambda}$ by sub-gradient method, $\vec{\lambda}$ must reside in the feasible region where all the element of $\vec{\lambda}$ must be non-negative. Usually, updated $\vec{\lambda}$ outside the feasible region will be projected to the nearest feasible vector.

4. IMPROVED ALGORITHM FOR LDP

In [17], it is shown that Lagrangian relaxation methods are easily weakened by the practical implementation details. The sub-gradient direction does not always guide Lagrangian relaxation method to the optimal solution under discrete cases. Here we want to find a better way to solve LDP to make Lagrangian relaxation method to converge faster with better solution quality under discrete solution space.

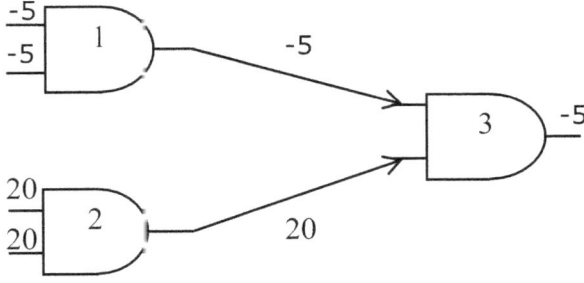

Figure 3. A circuit.

4.1 Drawbacks of Sub-gradient Method

In the sub-gradient method, $\overrightarrow{\Delta\lambda}$ is decided by slack. Under discrete solution space, circuit timing on some sensitive paths may change drastically and oscillate between huge positive and negative slack. Sub-gradients on those paths are relatively large and make Lagrangian relaxation jumping over the optimal solution. On the other hand, if small slack happen on less sensitive paths, sub-gradients on those paths are small. The small steps for those paths cause many iterations to approach optimal solutions. Hence, in our algorithm we find $\overrightarrow{\Delta\lambda}$ considering slack and sensitivities of paths thus avoiding the drawbacks of sub-gradient method.

Furthermore, in some cases the sum of sub-gradients on fan-in edges of a gate is positive while that on fan-out edges is negative, or vice versa. In that case, it is hard to satisfy the KKT condition. Here we take gate 3 in Figure 3 as an example. We assume that slack on e_{13}, e_{23} and fan-out edge is -5, 20 and -5 respectively, and hence sub-gradients on those edges are 5, -20 and 5. Therefore, the sum of sub-gradients on fan-in edges is -15 and that on fan-out edges is 5. It is not well defined that how to make KKT conditions satisfied on gate 3 and what value $\Delta\mu_3$ should be. By our algorithm, we will assign multipliers for e_{13} and e_{23} such that the slack on them are expected to be the same. We can handle those situations well.

4.2 Rationale of Our Algorithm

Here two new key concepts used in our algorithm are proposed. By following them, our algorithm can solve Lagrangian dual problem better than sub-gradient method.

1. Lagrangian multipliers are assigned according to the sensitivity $\frac{\partial\mu}{\partial a}$ where μ is the Lagrangian multiplier and a is the timing.

In sub-gradient method, sensitivities of different paths are assumed to be the same. However, in reality due to different structures, sensitivities of different paths should be different. In our algorithm, the sensitivities for different paths are different. Furthermore, the sensitivity allows us to predict what is the resulting timing when we applying a specific value of Lagrangian multiplier on a gate.

2. To reduce power, slacks on non-critical paths are reduced by assigning multipliers according to sensitivities and timing.

In static timing analysis, arrival time of a gate is the maximal arrival time among its fan-in edges. It is no help for timing of the gate to reduce the arrival time for those non-timing critical edges. By making arrival times among fan-in edges of a gate equal, we will not waste power for those non- critical paths.

ALGORITHM LRGIS

1. sort V in a reverse topological order;
2. $T_i = \emptyset$ for all $v_i \in V$;
3. repeat
4.   static_timing_analysis;
5.   for each $v_i \in V$
6.   record (μ_i, a_i) in T_i for estimating sensitivity;
7.   prune (μ_j, a_j) in T_i if $\exists (\mu_k, a_k)$ in T_i such that
* $\mu_j \geq \mu_k$ and $a_j \geq a_k$;*
8.   if $(v_i \in PO)$
9.   calculate $\Delta\mu_i$ from slack and sensitivity
*   of v_i;*
10.   else
11. $\Delta\mu_i$ equals the sum of $\Delta\lambda$ on fan-out edges
12.   distribute $\Delta\mu_i$ to $\Delta\lambda$s of fan-in edges such that
*   predicted arrival times of fan-in edges are all*
*   equal;*
13.   update $\vec{\lambda}$ and $\vec{\mu}$;
14.   solve sub-problem for circuit solution based on $\vec{\mu}$;
15. until $\Delta L_u <$ a bound ε or reach a number of iterations

Figure 4. The highlight of our algorithm.

In addition, the optimal constraints of KKT conditions for $\vec{\lambda}$ of gates are very similar to the flow constraints for nodes in flow network, in flow must being equal to out flow. In order to keep KKT conditions hold when solving LDP, we updating Lagrangian multipliers by using the same idea of distributing flows in flow network, as showed in [11]. At first, we calculate flows, $\Delta\mu$'s, at source nodes and then distribute flows toward sinks in circuit, $G(V, E)$, without edge capacity limits. With that, we can guarantee that the KKT conditions are always met.

4.3 Overview of Our Algorithm

In this section, our algorithm is highlighted in Figure 4. In line 1 and 2, gates are sorted in reserve topological order and table T for all gates are initialized at first. The main part of our algorithm is between line 3 and 15, which repeat until ΔL_μ less than a bound or reaching the iteration limit. For each iteration, static timing analysis is performed to get timing information of circuit. Then, we distribute $\Delta\mu$ in reverse topological order. In line 6 and 7, T for estimating sensitivity is updated and pruning is performed inside it. From line 8 to 11, $\Delta\mu$ of a gate is calculated and then in line 12 distributed to fan-in edges such that the expected arrival times of fan-in edges are all the same. By this step, KKT conditions are guaranteed to be satisfied and power dissipation on non-critical paths can be reduced. When the distribution is done, multipliers are updated. Then, sub-problem is solved based on updated multipliers to find the circuit solution.

4.4 Timing vs. Lagrangian Multiplier

We will discuss the behavior of timing and its dependence on Lagrangian multipliers, which is the main basis for our algorithm for solving the Lagrangian dual problem. In this discussion, the timing is characterized by arrival times but the conclusion can be applied with slack as well. When solving LDP, if $\Delta\mu$ is distributed to a gate, it will be distributed in the sub-circuit rooted by the gate. Therefore, the effect of distributing $\Delta\mu$ to the gate is equal to that

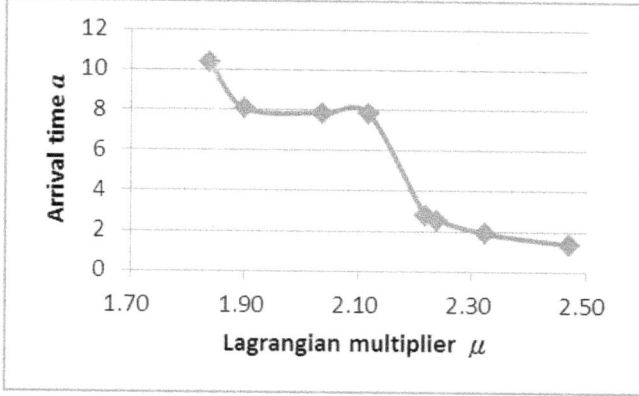

Figure 5. Lagrangian multiplier and arrival time curve.

Figure 6. Pseudo code of descent function η'.

of distributing $\Delta\mu$ to the sub-circuit. Then, due to the distribution of $\Delta\mu$ in the sub-circuit, timing inside the sub-circuit may change and arrival time of the gate may also change. Thus, the arrival time of a gate will change along with its Lagrangian multiplier and there must be a relationship between Lagrangian multiplier and arrival time of a gate. Since for each Lagrangian multiplier, the arrival time a is uniquely defined, the relationship can be modeled as a function $a = \eta(\mu)$ for each gate. Each gate has its own η function. Then the first order differentiation of η function means the ratio of trading Lagrangian multiplier for arrival time, $\Delta a = \eta'(\mu)\Delta\mu$. If function η' is known, it is more reasonable to calculate $\Delta\mu = (q - a)/\eta'(\mu)$ for a gate. This $\Delta\mu$ is what the gate needs now to make its constraint satisfied. This approach is more accurate than sub-gradient method.

Unfortunately, discrete gate sizes and Vt levels result in non-smooth Lagrangian multiplier and arrival time curve. The curve shown in Figure 5 is extracted from simulation results of a gate in C432 benchmark. It shows that the curve of Lagrangian multiplier and arrival time is not only non-smooth but complex so that we cannot use either a linear function, or even quadratic function, to model it. In the following sections, we will solve this problem by our projection-based descent method. The function η' for each gate is predicted by a history-based method.

4.5 Projection-Based Descent Method

We propose a projection-based descent method for solving the Lagrangian dual problem. It predicts the timing vs. μ behavior based on the history of previous iterations and then speculates the direction and step size of the descent move. Roughly speaking, we take the direction of a weighted average of historic gradient, with more emphasis on recent history. This direction is equivalent to the gradient of a smoothed $a = \eta(\mu)$ function. For the convenience of presentation, we denote this direction as η', and call it descent function. The gradients are estimated through finite difference. In order to perform the history-based projection, we maintain a table T storing (μ, a) pairs of previous iterations for each gate. Those pairs stored in the tables are sorted by μ value. It is reasonable to assume that the descent direction η' is more correlated with the gradients of recent history, or gradients at the value of μ close to that of current iteration. Here, we use weight variables γ to present the correlations and the sum of all the weight variables is one. Then, the return value of descent function is the sum of product of each slope between two consecutive pairs in the table T and its corresponding weight variable. The descent function $\eta'(T, \mu_{cur})$ can be written as

$$\eta'(T, \mu_{cur}) = \sum_{(\mu_{j-1}, a_{j-1}),(\mu_j, a_j) \in T} \left(\gamma_j \frac{a_{j-1} - a_j}{\mu_{j-1} - \mu_j} \right) \qquad (7)$$

where $\gamma_j \propto \left| \frac{1}{\frac{\mu_{j-1}+\mu_j}{2} - \mu_{cur}} \right|$. The pseudo code of our history based method for calculating descent function is shown in Figure 6.

Figure 7 is an example to show the calculation of descent function. There are three pairs, (5,0), (3,1) and (2.8,2), in a table T and μ_{cur} is 2. Then the return value of $\eta'(T, 2)$ is calculated by

$$\left(\frac{5-3}{0-1} \left| \frac{1}{\frac{0+1}{2}-2} \right| + \frac{3-2.8}{1-2} \left| \frac{1}{\frac{1+2}{2}-2} \right| \right) / \left(\left| \frac{1}{\frac{0+1}{2}-2} \right| + \left| \frac{1}{\frac{1+2}{2}-2} \right| \right) = -1.5 \qquad (8)$$

The solid curve is formed by the data in T and the dashed line is the prediction based on the return value of $\eta'(T, 2)$. The slope of the dashed line is -1.5.

In order to optimize μ and a, we perform pruning in each table. Generally, a greater value of Lagrangian multiplier allows more power for timing, resulting in less arrival time for a gate. Therefore, if a greater Lagrangian multiplier results in greater arrival time, it means that the Lagrangian multipliers in the fan-in cone of the gate are not well distributed. Some power is spent to speed up some unnecessary paths and those timing improvements have no benefit for the timing of the fan-in cone. We define that

$$(\mu_i, a_i) \text{ is inferior to } (\mu_j, a_j) \text{ if } \mu_i \geq \mu_j \text{ and } a_i \geq a_j \qquad (9)$$

and those interior pairs will be pruned.

On the other hand, the sizes of tables keep growing with the increase of simulation iterations. To prevent from memory explosion, we need to remove some data which is less useful to reduce table size. We know that finally the vector of Lagrangian

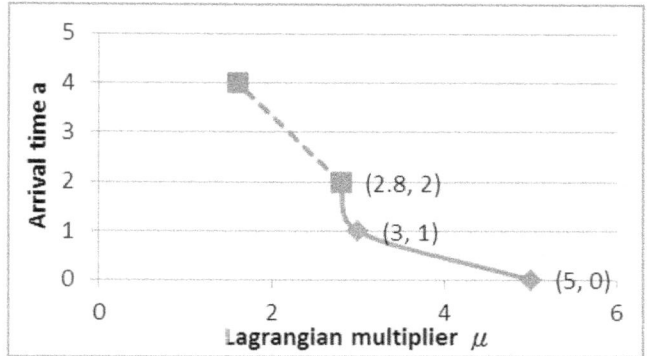

Figure 7. An example of descent function calculation.

```
FUNCTION local optimal sizing
Input : combinational circuit G, cell library and μ⃗
Output : size and Vt level for all gates in G

for all the gates vᵢ ∈ X in G
    for j ∈ Wᵢ and k ∈ Uᵢ
        L = αᵢpᵢʲᵏ + μᵢrᵢʲᵏCᵢ;
        for vᵧ ∈ fanin(vᵢ)
            L+= μᵧrᵧcᵢʲᵏ;
        if (L < L_best)
            L_best = L;
            wᵢ = j;
            uᵢ = k;
for all the gates vᵢ ∈ X in G
    implement vᵢ by wᵢ and uᵢ;
```

Figure 8. Pseudo code of our local optimal sizing

multipliers will converge so that the Lagrangian multipliers only change in a small region. The information which is far away from current operating region has little influence when calculating descent function. Therefore, the pair (μ_i, a_i) with greatest value of Lagrangian multiplier difference, $|\mu_i - \mu_{cur}|$, is the least useful data so it will be removed when the size of the table exceeds a user-defined limit

4.6 Distribution of Lagrangian Multipliers

The value of Lagrangian multipliers will affect solutions for gates. The initial vector of Lagrangian multiplier is a zero vector. We build up the circuit from the initial solution with least power. By distributing Lagrangian multipliers, we can speed up the circuit to satisfy the timing constraints without wasting power.

Each time the multipliers at the primary outputs are updated, the updated multipliers are distributed to other parts of the circuit to satisfy flow conservation. The distribution proceeds in a reverse-topological order. Alternatively, one can update the multipliers at primary inputs based on required arrival time and then distribute them to the remaining circuit in a topological order.

In each iteration, we calculate the change of Lagrangian multipliers $\Delta\mu$ at primary output by

$$\Delta\mu_i = \frac{q_i - a_i}{\eta'(T_i, \mu_i)} \qquad (10)$$

The symbol $a_i, q_i, T_i \, \mu_i$ is arrival time, required arrival time, data table and Lagrangian multiplier of v_i, respectively. Then those $\Delta\mu$'s are propagated toward primary inputs in a reverse-topological order. For each logic gate, it receives $\Delta\mu$s from its fan-out gates and then propagates the sum of $\Delta\mu$ it receives to its fan-in gates. The difference between $\Delta\mu$ propagation and common flow propagation in flow network is that $\Delta\mu$ can be negative. In general, $\Delta\mu$ is negative if the constraint on a gate is met. Although $\Delta\mu$ can be negative, μ must be non-negative during distribution.

When distributing $\Delta\mu$'s, it is not always possible to assign as much $\Delta\mu$ as a gate needs to satisfy its constraint without violating KKT conditions. In static timing analysis, the arrival time of a gate is the maximum arrival time among its fan-in gates. It is not useful to reduce arrival time on those non-critical fan-in gates. Thus for a gate, when distributing its $\Delta\mu$ for fan-in gates, the main goal is to make their arrival times equal. Therefore, we will not

waste power on the non-critical sub-circuits by assigning too large number of Lagrangian multiplier to them. At first, we calculate the expected arrival time a_{exp} for all the fan-in gates such that the sum of $\Delta\mu$'s distributing to them is equal to what the gate received from its fan-in gates. The equation for calculating a_{exp} for $v_i \in V$ is

$$\sum_{v_j \in fanin(v_i)} \frac{a_{exp} - a_j}{\eta'(T_j, \mu_j)} = \sum_{v_k \in fanout(v_i)} \Delta\lambda_{ki} = \Delta\mu_i \qquad (11)$$

With a_{exp}, we can calculate $\Delta\mu$ for each fan-in gate and also guarantee KKT satisfied. Then the $\Delta\mu$ for each fan-in gate is passed by the wire. Therefore, the value of $\Delta\lambda_{ji}$ means the $\Delta\mu$ distributed from v_i to v_j. The equation for calculating $\Delta\lambda$ on the fan-in wire of a gate is

$$\Delta\lambda_{ji} = \frac{a_{exp} - a_j}{\eta'(T_j, \mu_j)}, \forall v_j \in fanin(v_i) \qquad (12)$$

When distributing $\Delta\mu$ of a gate, we regard $\Delta\mu$ as flows passing among the gate and its fan-in gates. We make sure that flow conservation is satisfied in the distribution.

5. HEURISTIC FOR LRS/λ

For Lagrangian sub-problem, we apply the local optimal sizing for each gate individually like [12]. We find size and Vt level for a gate while keeping all other gates fixed. For a gate $v_i \in X$, objective function L_λ can be written as

$$L_\mu = \alpha_i p_i^{w_i u_i} + \left(\sum_{v_j \in fanin(v_i)} \mu_j r_j \right) c_i^{w_i} + \mu_i r_i^{w_i u_i} C_i$$

$$+ terms \; independent \; of \; v_i \qquad (13)$$

Due to discrete sizes and Vt levels, we cannot solve L_μ by any mathematical method. Here we use a table look-up method to find the solution for each gate. For a gate $v_i \in X$, we have known $r_i^{w_i u_i}$ and $c_i^{w_i u_i}$ for different combination of size $w_i \in W_i$ and Vt level $u_i \in U_i$. Then we evaluate L_μ with all the combinations of size and Vt level to find the size and Vt level combination resulting in minimum L_λ. We show our local optimal sizing for solving Lagrangian sub-problem in Figure 8. Due to the nature of greedy method, we need to iteratively solve the Lagrangian sub-problem until converge.

However, sometimes descent function may be not accurate due to some suddenly timing change when size or Vt level of gates change. Therefore, we need to do some adjustment for $\Delta\mu$s when iteratively solving Lagrangian sub-problem. Here no new $\Delta\mu$s comes into $G(V, E)$ at sources, only redistributing $\Delta\mu$ by descent function with updated table data of each gate. Our result shows that the minor change for Lagrangian multipliers will not delay convergence too much.

6. EXPERIMENTAL RESULTS

In this work, ISCAS85 benchmarks are synthesized by SIS [20] and placed by mPL [21] for comparison. The cell library is based on $70nm$ technology. For each logic gate, there are eight size options, 1x, 2x, 4x, 8x, 12x, 16x, 24x and 32x, and three Vt levels. Therefore, there are total 24 different implementations for each logic gate. The V_{dd} is set to 0.9V. Elmore delay model and an analytical leakage power model [5] are used for delay and power calculation. At first, we set all the gates to minimum power consumption implementation, i.e., smallest size and highest Vt

172

Table 1. Experimental results of our algorithm and sub-gradient method under loose and tight timing constraints.

testcase	# of gates	Loose timing constraints								Tight timing constraints							
		Initial setup		Sub-gradient method			Our algorithm			Initial setup		Sub-gradient method			Our algorithm		
		power	slack	power	slack	run time	power	slack	run time	power	slack	power	slack	run time	power	slack	run time
chain	11	9.3	-215.5	27.2	5.0	0.0	27.2	5.0	0.1	9.3	-295.6	60.3	-13.9	0.0	104.4	0.1	0.4
c432	289	221.7	-8033.3	249.8	17.0	0.8	238.7	18.0	1.1	221.7	-10379.8	832.4	-33.1	0.8	803.7	0.6	1.2
c499	539	418.8	-4198.2	874.5	614.0	1.5	498.8	7.0	2.2	418.8	-5389.7	1545.4	-11.5	1.5	1522.8	1.7	2.2
c880	340	259.4	-3219.2	327.8	231.0	0.9	279.8	1.0	1.3	259.4	-4239.1	515.5	-31.7	0.9	549.4	15.6	1.7
c1355	579	426.6	-4084.3	736.4	38.0	1.7	522.1	3.0	2.5	426.6	-5353.7	1470.0	-5.3	1.7	1403.9	7.6	2.5
c1908	722	582.8	-5716.4	878.0	22.0	2.2	666.7	70.0	3.1	582.8	-7286.4	1452.7	-12.8	2.2	1402.7	5.9	3.2
c2670	1082	725.1	-12969.7	760.0	711.0	2.8	734.2	115.0	4.0	725.1	-16177.1	1465.9	-32.8	2.8	1312.6	9.1	4.1
c3540	1208	994.5	-5873.4	2012.5	718.0	3.7	1147.6	7.0	5.5	994.5	-7369.0	2650.5	-116.6	3.7	3016.6	20.0	5.5
c5315	2440	1941.7	-8156.4	3165.6	1033.0	7.4	2171.9	17.0	10.8	1941.7	-9956.3	3627.4	-199.0	7.4	4088.7	7.8	10.9
c6288	2342	1819.7	-7786.7	3951.3	310.0	7.5	2518.8	13.0	11.5	1819.7	-10476.1	6305.5	-29.4	7.6	5382.4	3.4	11.4
c7552	3115	2390.0	-16899.7	3897.8	1386.0	9.7	2445.5	107.0	14.3	2390.0	-21197.9	6875.7	-97.2	9.7	5433.9	20.6	14.8
Sum		9790		16881		38.25	11251		56.34	9790				38.38			57.82
# of violation			11		0			0			11		11			0	

level, and λ for all constraints is zero. We compare our algorithm with a Lagrangian relaxation based method using sub-gradient for dual problem. The initial vector of Lagrangian multiplier and the method for Lagrangian sub-problem are the same. Our experiments focus on the comparison between our Lagrangian dual problem method using descent function and sub-gradient method in [12].

Table 1 shows the experimental results of our algorithm and sub-gradient method under both loose and tight timing constraints. We set 90% of original circuit delay as loose constrains. Then, we choose timing constraints which cause violations when using sub-gradient method as tight constraints. The experimental results show that our algorithm reduces 33% power consumption on average with only 32% run time overhead compared with sub-gradient method under loose timing constraints. Then we run experiments with tight timing constraints. The results show that our algorithm can find the solution with positive slack for all the benchmarks but sub-gradient method cannot. Therefore, the experimental results demonstrate the effectiveness of our algorithm for either power reduction or circuit performance.

In addition, we show the detailed power and slack information of C432 benchmark iteratively. In Figure 9 and 10, the results show that the power consumption and slack of our algorithm only change in small region and converge faster than sub-gradient method. Due to less reference data, the inexact estimation of our descent function causes oscillation only for a few iterations at beginning. Therefore, our improved algorithm for solving Lagrangian dual problem make Lagrangian relaxation stable and converge faster to better result.

7. CONCLUSION

In this work, we propose an improved Lagrangian relaxation method for simultaneous discrete gate sizing and Vt assignment. The main idea of this work is that we distribute Lagrangian multipliers based on not only slack but sensitivity of timing and Lagrangian multipliers by our descent function. The Lagrangian multiplier distributed on each gate is more accurate so that timing

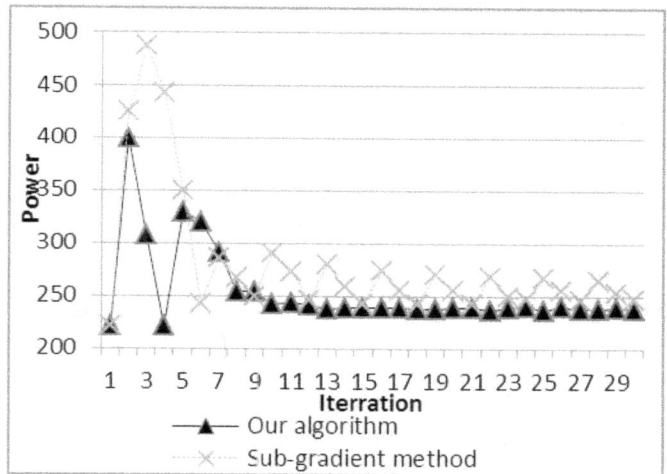

Figure 9. Detailed power information.

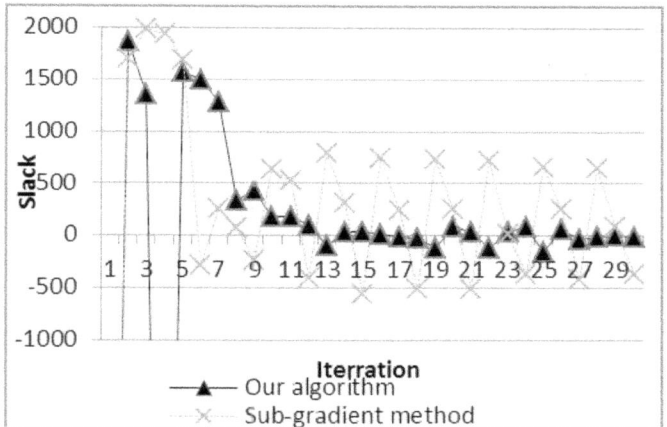

Figure 10. Detailed slack information

constraints may be just satisfied and hence no extra power will be wasted. The experimental results show that our algorithm can improve 33% in power consumption under the same timing constraints than a Lagrangian relaxation method using sub-gradient method. In addition, our algorithm can also find feasible solution but sub-gradient cannot when tight timing constraints are given. As a result, our improved Lagrangian relaxation method is powerful enough to handle discrete sizes and Vt levels with good solution quality and tolerable run time cost.

8. REFERENCES

[1] J. Fishburn and A. Dunlop. "TILOS: A posynomial programming approach to transistor sizing," *In Proc. of ICCAD*, pp. 326–328, 1985.

[2] L. Wei, Z. Chen, K. Roy, and V. De. "Design and optimization of dual threshold circuits for low voltage low power application," *IEEE Trans. VLSI*, vol. 7, no. 1, pp. 16–24, Mar. 1999.

[3] V. Sundararajan and K.K. Parhi. "Low power synthesis of dual threshold voltage CMOS circuits," *In Proc. of ISLPED*, pp. 139–144, 1999.

[4] V. Khandelwal, A. Davoodi, and A. Srivastava. "Simultaneous Vt selection and assignment for leakage optimization," *IEEE Trans. VLSI*, vol. 13, no. 6, pp. 762–765, 2005.

[5] M. Ketkar and S. S. Sapatnekar. "Standby power optimization via transistor sizing and dual threshold voltage assignment," *In Proc. of ICCAD*, pp. 375–378, 2002.

[6] D. Nguyen, A. Davare, M. Orshansky, D. Chinnery, B. Thompson, and K. Keutzer. "Minimizion of dynamic and static power through joint assignment of threshold voltages and sizing optimization," *In Proc. of ISLPED*, pp. 158–162, 2003.

[7] S. Shah, A. Srivastava, D. Sharma, D. Sylvester, D. Balaauw, and V. Zolotov. "Discrete Vt assignment and gate sizing using a self-snapping continuous formulation," *In Proc. of ICCAD*, pp. 704–710, 2005.

[8] S. Sirichotiyakul, T. Edwards, C. Oh, J. Zuo, A. Dharchoudhury, R. Panda, and D. Blaauw. "Stand-by power minimization through simultaneous threshold voltage selection and circuit sizing," *In Proc. of DAC*, pp. 436–441, 1999.

[9] H. Chou, Y. Wang, and C. Chen. "Fast and effective gate-sizing with multiple-Vt assignment using generalized Lagrangian relaxation," *In Proc. of ASPDAC*, pp. 381–386, 2005.

[10] T.-H. Wu, L. Xie, and A. Davoodi. "A parallel and randomized algorithm for large-scale dual-Vt assignment and continuous gate sizing," *In Proc. of ISLPED*, pp. 45–50, 2008.

[11] J. Wang, D. Das, and H. Zhou. "Gate sizing by Lagrangian relaxation revisited," *In Proc. of ICCAD*, pp. 111-118, 2007.

[12] C. Chen, C. C. N. Chu, and D. F. Wong. "Fast and exact simultaneous gate and wire sizing by Lagrangian relaxation," *IEEE Trans. CAD*, vol. 18, no. 7, pp. 1014–1025, 1999.

[13] K. Kasamsetty, M. Ketkar, and S.S. Sapatnekar. "A new class of convex functions for delay modeling and its application to the transistor sizing problem [CMOS gates]," *IEEE Trans. CAD*, vol. 13, no. 6, pp. 779–788, 2000.

[14] S. Roy, W. Chen, and C.C. Chen. "ConvexFit: An optimal minimum-error convex fitting and smoothing algorithm with application to gate-sizing," *In Proc. of ICCAD*, pp. 196–204, 2005.

[15] S. Hu, M. Ketkar, and J. Hu. "Gate sizing for cell library based designs," *In Proc. of DAC*, pp. 847–852, 2007.

[16] Y. Liu and J. Hu, "A new algorithm for simultaneous gate sizing and threshold voltage assignment," *In Proc. of ISPD*, pp. 27-34, 2009.

[17] H. Tennakoon and C. Sechen. "Gate sizing using Lagrangian relaxation combined with a fast gradient-based pre-processing step," *In Proc. of ICCAD*, pp. 395-402, 2001.

[18] J. Shyu, J. P. Fishburn, A. E. Dunlop, and A. L. Sangiovanni-Vincentelli. "Optimization-based transistor sizing," *IEEE J. Solid-State Circuits*, vol. 23, no. 2, pp. 400 - 409, April, 1988.

[19] R. Chadha and J. Bhasker. *Static Timing Analysis for Nanometer Designs*, Springer, 2009.

[20] E.M. Sentovich, K.J. Singh, L. Lavagno, C. Moon, R. Murgai, A. Saldanha, H. Savoj, P.R. Stephan, R.K. Brayton, and A.L. Snagiovanni. "SIS: a system for sequential circuit synthesis," Electron. Res. Lab., Univ. California, Berkeley, CA, Mem. UCB/ERL M92/41, May 1992.

[21] CPMO-constrained placement by multilevel optimization. http://ballade.cs.ucla.edu/cpmo. Computer Science Department, UCLA.

Stochastic Analog Circuit Behavior Modeling by Point Estimation Method

Fang Gong
University of California, Los Angeles
Electrical Engineering Department
Los Angeles, CA 90095, US
gongfang@ucla.edu

Hao Yu
Nanyang Technological University
Electrical and Electronic Engineering
haoyu@ntu.edu.sg

Lei He
University of California, Los Angeles
Electrical Engineering Department
Los Angeles, CA 90095, US
lhe@ee.ucla.edu

ABSTRACT

Stochastic device parameter variations have dramatically increased beyond the scale of 65nm and can significantly lead to large mismatch for analog circuits. To estimate unknown analog circuit behavior in performance space under the given stochastic variations in parameter space, many state-of-art approaches have been developed recently. However, either Gaussian distribution or response surface model (RSM) with analytical formulae has to be assumed when connecting performance space and parameter space. A novel point-estimation based approach has been proposed in this paper to capture arbitrary stochastic distributions for analog circuit behaviors in performance space. First, to evaluate high-order moments of circuit behavior in an accurate fashion, the point-estimation method has been applied with only a few number of simulations. Then, probability density function (PDF) of circuit behavior can be efficiently extracted by the obtained high-order moments. This method is further extended for multiple parameters under linear complexity. Extensive numerical experiments on a number of different circuits have demonstrated that the proposed point-estimation method can provide up to 181X runtime speedup with the same accuracy, when compared with Monte Carlo method. Moreover, it can further achieve up to 15X speedup over the RSM-based method such as APEX with the similar accuracy.

ACM Classification Keywords: B.7.2: - Integrated Circuits-Design Aids
General Terms: Algorithms, Performance
Authors Keywords: Behavior Modeling, Point Estimation, Circuit simulation.

1. INTRODUCTION

As semiconductor industry enters into nano-technology node, large process variations become inevitable and hence pose a serious threat to both analog circuit design and man-ufacturing [1, 2, 3, 4]. Device variables in parameter space, such as the effective channel length and threshold voltage of transistors, can deviate significantly from nominal values due to large uncertainties from chemical mechanical polishing (CMP), etching, lithography and etc. Under such circumstance, circuit behaviors in performance space can differ from the nominal case by a large margin, which may further lead to high loss of yield. With the process variation of device variables in parameter space, it is desirable to extract the unknown distribution of variable circuit behaviors in performance space. A robust circuit design and yield enhancement are especially important for analog circuits. One critical but missing link here is how to find an efficient yet accurate mapping between parameter space and performance space.

Note that the local random or stochastic variation is the most difficult one to be calculated, which is also called *mismatch* for the behavior modeling of analog circuits. In the past decade, many stochastic techniques had been proposed, such as Monte Carlo simulation, linear regression [1], stochastic orthogonal polynomials (SoPs) expansion [5, 3], response surface modeling based approaches [2, 4] and etc. The most general approach is to apply the Monte Carlo (MC) simulation, which samples all variable parameters and then calculate stochastic analog circuit behaviors by a large number of repeated simulations. As such, MC is too time-consuming to be afforded for the post-layout verifications beyond 65nm. To relieve the computational complexity, linear regression method [1] has been deployed to approximate the circuit behavior in performance space by a linear function of a number of normally distributed process variables in parameter space, or Gaussian distribution. This approach is efficient because of using analytical formula to obtain the circuit behavior. However, this approach cannot approximate non-normal (non-Gaussian) distributions and might lead to the loss of accuracy. With the use of different stochastic orthogonal polynomials (SoPs), SoP based methods can model process variations with non-Normally distributed random variables. The unknown distribution of circuit behaviors in performance space can be estimated by solving SoP expansion coefficients [5, 3]. However, the SoP-based methods require knowing the type of the stochastic distribution of the circuit behavior. In practice, one known parameter distribution in parameter space usually becomes unknown in performance space after the mapping.

In order to capture unknown random distribution after

Permission to make digital or hard copies of all or part of this work for personal or classroom use is granted without fee provided that copies are not made or distributed for profit or commercial advantage and that copies bear this notice and the full citation on the first page. To copy otherwise, to republish, to post on servers or to redistribute to lists, requires prior specific permission and/or a fee.
ISPD'11, March 27–30, 2011, Santa Barbara, California, USA.
Copyright 2011 ACM 978-1-4503-0550-1/11/03 ...$10.00.

the mapping, response-surface-method (RSM) based methods [6, 4, 2] have been developed. One most important work developed recently is asymptotic probability extraction (APEX) [2] with the use of asymptotic waveform evaluation [7]. This approach assumes a polynomial function of all process parameters and further applies moment matching to extract the random distribution of circuit behavior (e.g. delay, gain, etc.). Nevertheless, the limitations of RSM-based approaches can be summarized in two-fold. First, the circuit behavior in performance space has become one strongly nonlinear function for random device variables. As such, the extraction by RSM has become computationally expensive. Second, it is prohibitive to evaluate high-order moments $E(f^k)$ with analytical formula of f, especially when the number of random variables and the moment order k increase. As such, the approaches based on RSM still cannot mitigate the super-linearly complexity while remaining accuracy for large-scale problems.

In this paper, a new mapping algorithm is developed to obtain the arbitrary circuit behavior in performance space from the arbitrary device variable in parameter space. High-order moments of circuit behavior f are first estimated by Point Estimation (PE) method to efficiently characterize the high-order moments $E(f^k)$ by weighted-sum of a few sampled simulations. Therefore, one can significantly improve the efficiency and accuracy of APEX [2] without the need to assume RSM inputs. As a result, the distribution of circuit behavior f in performance space can be efficiently obtained by its moments $E(f^k)$, calculated from the PE method in the parameter space. Moreover, a normalized PDF function is introduced so that to enhance the accuracy by eliminating the potential round-off error. In addition, this approach has been extended to consider the case with multiple parameters.

Extensive experiments on a number of different circuits are performed to demonstrate the validity and efficiency of our proposed algorithm. The contributions of this paper are further clarified as follows. First, although point estimation method has widely been applied for reliability analysis [8, 9], it can only estimate at most four moments (e.g. the mean, the variance, the skewness, the kurtosis) with empirical analytical formulae, and hence remains unclear how to estimate moments with higher order. In this paper, a modified point estimation method is developed to approximate higher order moments in a systematic manner. Moreover, unlike the observed super-linearly complexity in RSM based methods, our proposed method can be extended to deal with multiple parameters with linear complexity, which is significant for large-scale analog circuits.

The rest of this paper is organized as follows. In Section 2, we first review the mathematical formulation of the PDF estimation and the moments used in response-surface-model (RSM) methods. In Section 3, we introduce the point estimation (PE) method and further propose a new high-order moments evaluation via PE. We also discuss one normalized PDF technique in Section 4 to reduce error and further present experimental results in Section 5. This paper concludes in Section 6.

2. BACKGROUND

2.1 Mathematical Formulation

We consider circuit behavior f with multiple random vari-

ables of process variations $(x_1, x_2, \cdots, x_n)$, which can be expressed as $f(x_1, x_2, \cdots, x_n)$. As such, parameter space can be defined as the space $\mathbb{R}^n$ bounded by the min and max of all random variables, and performance space $\mathbb{R}$ consists of all possible behavior merits.

As a result of uncertainties in process technology, random variables can deviate from their nominal values and lead to variational circuit behavior. Our purpose is to extract unknown distribution (e.g. PDF/CDF functions) of circuit behavior by mapping the variable parameter distributions in parameter space into performance space.

To this end, the probabilistic moments in both spaces should be defined according to probability theory [10, 11]:

$$
\begin{aligned}
m_f^p = E(f^p) &= \int_{-\infty}^{+\infty} (f^p \cdot pdf(f)) df \\
m_x^p = E(x^p) &= \int_{-\infty}^{+\infty} (x^p \cdot pdf(x)) dx
\end{aligned}
\tag{1}
$$

where m_f^p is the p-th moment of circuit behavior f in performance space, and m_x^p is the p-th moment of random variable x in parameter space.

> LEMMA 1. *Suppose $pdf(f)$ is continuous in performance space. Then $pdf(f)$ can be determined uniquely by high order moments $E(f^k)$ $(k = 1, 2, \cdots, m)$.*

PROOF. Let $\Phi(\omega)$ is the Fourier transform [12] of $pdf(f)$ and can be written as:

$$
\begin{aligned}
\Phi(\omega) &= \int_{-\infty}^{+\infty} \left(pdf(f) \cdot e^{-j\omega f} \right) df \\
&= \int_{-\infty}^{+\infty} \left(pdf(f) \cdot \sum_{p=0}^{+\infty} \frac{(-j\omega f)^p}{p!} \right) df \\
&= \sum_{p=0}^{+\infty} \frac{(-j\omega)^p}{p!} \cdot \int_{-\infty}^{+\infty} (f^p \cdot pdf(f)) df. \\
&= \sum_{p=0}^{+\infty} \frac{(-j\omega)^p}{p!} \cdot m_f^p.
\end{aligned}
\tag{2}
$$

As such, $\Phi(\omega)$ can be expanded with high order moments m_f^p, and $pdf(f)$ can be extracted from Inversion Fourier transform of $\Phi(\omega)$. In other words, there is a one-to-one correspondence between high order moments m_f^p and $pdf(f)$. □

Notice that m_x^p in (1) can be computed accurately in parameter space with known $pdf(x)$. Therefore, it is the key problem to find an efficient mapping between m_x^p and m_f^p in order to extract $pdf(f)$ in performance space.

2.2 Preliminary of PDF Calculation

The techniques to extract $pdf(f)$ with moments $E(f^k)$ have been proposed in [2, 7], which will be reviewed in what follows. First, time moments for f can be defined as:

$$
\widehat{m}_f^k = \frac{(-1)^k}{k!} \cdot \int_{-\infty}^{+\infty} f^k \cdot pdf(f) df.
\tag{3}
$$

It is clear that $\widehat{m}_f^k$ is defined in performance space but different from m_f^k in (1) due to a scaling factor $(-1)^k/k!$.

On the other hand, consider a linear time-invariant (LTI) system H, and its time moments can also be defined as[7]:

$$\widehat{m}_t^k = \frac{(-1)^k}{k!} \cdot \int_{-\infty}^{+\infty} t^k \cdot h(t)dt. \qquad (4)$$

where t is the time variable and $h(t)$ is impulse response of LTI system H. So, impulse response $h(t)$ can be an optimal approximation to $pdf(f)$ if we treat t as circuit behavior f and make $\widehat{m}_t^k$ equal to $\widehat{m}_f^k$. Furthermore, time moments in (4) can be expressed as [7]:

$$\widehat{m}_t^k = -\sum_{r=1}^{M} \frac{a_r}{b_r^{k+1}}. \qquad (5)$$

Where a_r and b_r $(r = 1, \cdots, M)$ are the residues and poles of this LTI system, respectively. As such, the impulse response of the LTI system can be simplified as:

$$h(t) = \begin{cases} \sum_{r=1}^{M} a_r \cdot e^{b_r^{k+1} \cdot t} & (t \geq 0) \\ 0 & (t < 0) \end{cases} \qquad (6)$$

In general, there are three steps to calculate $h(t)$ as an approximation to $pdf(f)$:

- Mapping m_x^k in parameter space into performance space as $\widehat{m}_f^k$ in (3).

- Make $\widehat{m}_f^k$ equal to $\widehat{m}_t^k$ and solve nonlinear equation system in (5) for residues a_r and poles b_r.

- Compute impulse response $h(t)$ in (6) with residues a_r and poles b_r.

Clearly, one needs to find an efficient mapping between parameter space and performance space to obtain the stochastic circuit behavior. Within this mapping, the most challenging step is the evaluation of high-order moments. Although the RSM based methods, such as APEX, assume that one nonlinear function can be found to approximate the mapping, they might become unaffordable to calculate the high-order moments for large-scale stochastic problems. To this end, we have developed a modified point estimation (PE) method to perform the mapping for the calculation of high-order moments.

3. HIGH ORDER MOMENTS ESTIMATION

In this section, we discuss how to evaluate high order moments of circuit behavior m_f^k by mapping parameter moments m_x^k from parameter space into performance space via Point Estimation (PE) method.

3.1 Moments via Point Estimation

For illustration purpose, we consider circuit behavior $f(x)$ with single variable parameter x. Usually it is impractical to compute m_f^k as (1) because $pdf(f)$ is unknown. The other straightforward way is to use Taylor expansion, which involves high order derivatives. But there is no way to guarantee the existence of high order derivatives of $f(x)$.

As such, we propose to leverage the Point Estimation method to compute high order moments[8, 9], which approximates m_f^k with a weighted sum of sampling values of

$f(x)$. Assume $\tilde{x}_j$ $(j = 1, \cdots, p)$ are estimating points of random variable, and P_j are corresponding weights. In this way, the k-th order moment of $f(x)$ can be approximated as:

$$m_f^k = \int_{-\infty}^{+\infty} f^k \cdot pdf(f)df \approx \sum_{j=1}^{p} P_j \cdot f(\tilde{x}_j)^k. \qquad (7)$$

However, [8, 9] only provide empirical analytical formulae of $\tilde{x}_j$ and P_j for first four moments. Therefore, it is significant but remains unknown how to determine $\tilde{x}_j$ and P_j for higher order moments systematically.

3.2 Estimating Points and Weights

To this end, we start with (7) in performance space, but it is impossible to compute $\tilde{x}_j$ and P_j since both sides are unknown. Thus, we need to reformulate the problem in parameter space, where random variable x and its distribution (e.g. PDF function $pdf(x)$) are known beforehand.

According to classic probability theory[10, 11], we have following theorem:

> THEOREM 1. *Let x and $f(x)$ are both continuous random variables, and their PDFs are $pdf(x)$ and $pdf(f)$, respectively. Suppose $\int f^k(x) \cdot pdf(x)dx$ exists. Then*
>
> $$E(f^k(x)) = \int f^k(x) \cdot pdf(f)df = \int f^k(x) \cdot pdf(x)dx$$

As such, the moments of circuit behavior $f(x)$ in performance space can be calculated in parameter space. For example, the k-th order moment of $f(x)$ in equation (7) becomes:

$$m_f^k = \int f(x)^k \cdot pdf(x)dx \approx \sum_{j=1}^{m} P_j \cdot f(\tilde{x}_j)^k. \qquad (8)$$

On the other hand, the k-th order moments of random variable x can be written as:

$$m_x^k = \int x^k \cdot pdf(x)dx \approx \sum_{j=1}^{m} P_j' \cdot (\tilde{x}_j')^k. \qquad (9)$$

It is obvious that estimating points $\tilde{x}_j$ and corresponding weights P_j in (8) are the *same* as $\tilde{x}_j'$ and P_j' in (9) because they are all defined in parameter space. Therefore, we can calculate $\tilde{x}_j'$ and P_j' from equation (9) with m_x^k obtained from (1), and then estimate m_f^k using (8).

Now, the problem is how to solve for $\tilde{x}_j'$ and P_j' systematically. Since there are total $2m$ unknowns in (9), we need to build $2m$ equations using first $2m$ moments of random variable, which can be rewritten as:

$$\begin{aligned} \sum_{j=1}^{m} P_j' &= 1 = m_x^0 \\ \sum_{j=1}^{m} P_j' \cdot \tilde{x}_j' &= E(x) = m_x^1 \\ \sum_{j=1}^{m} P_j' \cdot (\tilde{x}_j')^2 &= E(x^2) = m_x^2 \\ &\cdots \\ \sum_{j=1}^{m} P_j' \cdot (\tilde{x}_j')^{2m-1} &= E(x^{2m-1}) = m_x^{2m-1} \end{aligned} \qquad (10)$$

Note that the right-hand-side of above nonlinear system are first $2m$ moments of x in the behavior domain and can be calculated exactly with known $pdf(x)$ and definition in (1).

This nonlinear system (10) can be solved using algorithm proposed in [7]. In what follows, we briefly describe this algorithm.

Assume residues $a_j = P_j'$ and poles $b_j = 1/\tilde{x}_j'$, the equations (10) can be reformulated as:

$$
\begin{bmatrix}
a_1 + a_2 + \cdots a_m \\
\frac{a_1}{b_1} + \frac{a_2}{b_2} + \cdots \frac{a_m}{b_m} \\
\frac{a_1}{b_1^2} + \frac{a_2}{b_2^2} + \cdots \frac{a_m}{b_m^2} \\
\vdots \\
\frac{a_1}{b_1^{2m-1}} + \frac{a_2}{b_2^{2m-1}} + \cdots \frac{a_m}{b_m^{2m-1}}
\end{bmatrix}
=
\begin{bmatrix}
m_x^0 \\
m_x^1 \\
m_x^2 \\
\vdots \\
m_x^{2m-1}
\end{bmatrix}
\quad (11)
$$

The system matrix of (11) is the well-known Vandermonde matrix and can be divided into two parts:

$$M \cdot v = rhs_{low}; \quad M \cdot \Lambda^{-q} \cdot v = rhs_{upper}$$

where rhs_{low} consists of the low order moments ($k = 0, 1, \cdots, m-1$), and rhs_{upper} contains the high order moments ($k = m, m+1, \cdots, 2m-1$). Λ^{-1} is a diagonal matrix of $\{1/b_j\}$ ($j = 1, \cdots, m$). And M matrix ($m \times m$) can be expressed as:

$$
M = \begin{bmatrix}
1 & 1 & \cdots & 1 \\
b_1^{-1} & b_2^{-1} & \cdots & b_{m-1}^{-1} \\
b_1^{-2} & b_2^{-2} & \cdots & b_{m-1}^{-2} \\
\vdots & \vdots & \vdots & \vdots \\
b_1^{-(m-1)} & b_2^{-(m-1)} & \cdots & b_{m-1}^{-(m-1)}
\end{bmatrix}
$$

Therefore, the linear system $M \cdot v = rhs_{low}$ can be solved as $v = M^{-1} \cdot rhs_{low}$, and $rhs_{upper} = M \cdot \Lambda^{-q} \cdot M^{-1} \cdot rhs_{low}$. Since M is also a Vandermonde matrix that is the modal matrix for a system matrix in companion form, rhs_{upper} can be simplified as $rhs_{upper} = \hat{M}^{-q} \cdot rhs_{low}$, where $\hat{M}^{-1}$ is

$$
\hat{M}^{-1} = \begin{bmatrix}
0 & 1 & 0 & \cdots & 0 \\
0 & 0 & 1 & \cdots & 0 \\
\vdots & \vdots & \vdots & \vdots & \vdots \\
s_0 & s_1 & s_2 & \cdots & s_{m-1}
\end{bmatrix}. \quad (12)
$$

In this way, the eigenvalues of $\hat{M}^{-1}$ are $\{1/b_j\}$ in (11). To calculate $\hat{M}^{-1}$, we need to compute $\{s_t\}$ ($t = 0, \cdots, m-1$) with following equation system:

$$
\begin{bmatrix}
m_x^0 & m_x^1 & \cdots & m_x^{m-1} \\
m_x^1 & m_x^2 & \cdots & m_x^m \\
\vdots & \vdots & \vdots & \vdots \\
m_x^{m-1} & m_x^m & \cdots & m_x^{2m-2}
\end{bmatrix}
\begin{bmatrix}
s_0 \\
s_1 \\
\vdots \\
s_{m-1}
\end{bmatrix}
=
\begin{bmatrix}
m_x^m \\
m_x^{m+1} \\
\vdots \\
m_x^{2m-1}
\end{bmatrix}.
$$
$$(13)$$

When $\{1/b_j\}$ are available, the $\{a_j\}$ can be calculated from Equation (11). Therefore, the weights $\{P_j'\}$ and estimating points $\{\tilde{x}_j'\}$ can be computed systematically and used to compute m_f^k in equation (8).

3.3 Extension to multiple parameters

It is usually necessary to handle multiple variable parameters simultaneously in real-world problems. Thus, we discuss how to extend aforementioned techniques to deal with multiple parameters.

Existing methods [8, 9] model $m_{f(x_1,x_2,\cdots,x_n)}^k$ as a linear combination of moments $m_{f(x_i)}^k$, where $f(x_i)$ is a function of single variable x_i with other variables set equal to mean values. However, [8, 9] can only estimate first four moments using explicit analytical formulae as linear combination.

Consider circuit behavior $f(x_1, x_2, \cdots, x_n)$ where x_1, x_2, $\cdots, x_n$ are independent random variables, it is desirable to estimate $m_{f(x_1,x_2,\cdots,x_n)}^k$ that is the k-th order moment of $f(x_1, x_2, \cdots, x_n)$.

To estimate higher order moments of $f(x_1, x_2, \cdots, x_n)$ systematically, we derive the equation for $m_{f(x_1,x_2,\cdots,x_n)}^k$ as:

$$m_{f(x_1,x_2,\cdots,x_n)}^k = \sum_{i=1}^{n} g_i m_{f(x_i)}^k. \quad (14)$$

Moreover, g_i can be calculated as follows and detailed derivation can be referred to technical report:

$$g_i = c \cdot \frac{\partial (f(x_i))}{\partial x_i}$$

For each parameter x_i, we have its estimating points $\tilde{x}_j$ and corresponding $f(\tilde{x}_j)$. Hence, it is possible to calculate $\partial (f(x_i))/\partial x_i$ with finite difference method numerically.

Besides, the constant c can be computed with:

$$m_{f(x_1,x_2,\cdots,x_n)}^0 = \sum_{i=1}^{n} g_i m_{f(x_i)}^0 = \sum_{i=1}^{n} g_i \quad (15)$$

$$= \sum_{i=1}^{n} c \cdot \frac{\partial (f(x_i))}{\partial x_i} = 1 \Rightarrow c = 1 \bigg/ \sum_{i=1}^{n} \frac{\partial (f(x_i))}{\partial x_i}.$$

As such, high order moments of multi-variable function $m_{f(x_1,x_2,\cdots,x_n)}^k$ can be evaluated with moments of univariate function $m_{f(x_i)}^k$ efficiently. Extensive experiments can demonstrate its validity and efficiency.

3.4 Error Estimation

Theoretical maximum approximation error of point estimation method is analyzed in [13]. For the univariate case, the maximum approximation error to exact integral value in equation (1) can be governed by:

$$\left| \sum_{j=1}^{m} P_j \cdot f^k(\tilde{x}_j) - \int_{-\infty}^{+\infty} f^k(x) \cdot pdf(f)df \right| \leq \alpha \cdot k^{1/m}. \quad (16)$$

where α is a constant, and k is the order of moments. m is the number of estimating points $\tilde{x}_j$ for order k. As such, it implies that more estimating points for each variable should be used to reduce the estimation error of higher order moments.

4. PDF CALCULATION WITH MOMENTS

4.1 PDF/CDF Estimation with Moments

With high order moments m_f^k available, the next step is to compute residues $\{a_r\}$ as well as poles $\{b_r\}$ in (5) and impulse response $h(t)$ [2, 7] in (6). To do so, we calculate $\widehat{m}_f^k$ in (3) and make them equal to $\widehat{m}_t^k$ in (5). The nonlinear equation system becomes:

$$-\begin{bmatrix} \frac{a_1}{b_1} + \frac{a_2}{b_2} + \cdots \frac{a_M}{b_M} \\ \frac{a_1}{b_1^2} + \frac{a_2}{b_2^2} + \cdots \frac{a_M}{b_M^2} \\ \frac{a_1}{b_1^3} + \frac{a_2}{b_2^3} + \cdots \frac{a_M}{b_M^3} \\ \vdots \\ \frac{a_1}{b_1^{2M}} + \frac{a_2}{b_2^{2M}} + \cdots \frac{a_M}{b_M^{2M}} \end{bmatrix} = \begin{bmatrix} \widehat{m}_f^0 \\ \widehat{m}_f^1 \\ \widehat{m}_f^2 \\ \vdots \\ \widehat{m}_f^M \end{bmatrix}. \quad (17)$$

which has a Vandermonde matrix similar to (11) and can be solved with the same technique in [7]. With poles $\{b_r\}$ and residues $\{a_r\}$ available, PDF function can be approximated with equation (6).

Notice that impulse response $h(t)$ is zero for $t < 0$, but the PDF in real-life problems can be nonzero for $f \le 0$. In this case, PDF function can be shifted as [2] which can be demonstrated with our experiments.

4.2 Normalized PDF for Error Prevention

From (17), it is obvious that the accuracy of PDF approximation mainly depends on the accuracy of residues a_r, poles b_r and moments estimation. However, there are roundoff error within moment estimation and the PDF approximation, which can lead to instability issue. In order to prevent potential error, we propose to normalize PDF calculated from equation (6) to cancel out the potential roundoff error.

For illustration purpose, we take the roundoff error in moments as an example, and other roundoff error can be eliminated with the same way. Assume $\widehat{m}_f^k$ is the exact value of k-th time moment in equation (3), and $\tilde{m}_f^k$ is the estimated value of k-th time moment. Also, we assume $\tilde{m}_f^k = const \cdot \widehat{m}_f^k$ due to roundoff error, where $const$ is a scaling constant.

As such, when we use the direct solution in section 2, the scaling constant in both system matrix and right-hand-side vector of (15) can be canceled out. Hence, s_t ($t = 0, \cdots, m - 1$) and thus eigenvalues of equation (12) (that is poles $\{1/b_j\}$) are both exact values.

Next, the nonlinear equation system (17) becomes:

$$-\begin{bmatrix} \frac{a_1}{b_1} + \frac{a_2}{b_2} + \cdots \frac{a_M}{b_M} \\ \frac{a_1}{b_1^2} + \frac{a_2}{b_2^2} + \cdots \frac{a_M}{b_M^2} \\ \frac{a_1}{b_1^3} + \frac{a_2}{b_2^3} + \cdots \frac{a_M}{b_M^3} \\ \vdots \\ \frac{a_1}{b_1^{2M}} + \frac{a_2}{b_2^{2M}} + \cdots \frac{a_M}{b_M^{2M}} \end{bmatrix} = const \cdot \begin{bmatrix} \widehat{m}_f^0 \\ \widehat{m}_f^1 \\ \widehat{m}_f^2 \\ \vdots \\ \widehat{m}_f^M \end{bmatrix} \quad (18)$$

Which leads to $\tilde{a}_j = const \cdot a_j$, where a_j are exact values of residues. Therefore, PDF of $f(x)$ is approximated with:

$$pdf(f) = \sum_{r=1}^{M} \tilde{a}_r \cdot e^{\widehat{b}_r^{k+1} \cdot f} = \sum_{r=1}^{M} const \cdot a_r \cdot e^{\widehat{b}_r^{k+1} \cdot f} \quad (19)$$

In order to eliminate the scaling constant, we propose to normalize PDF of $f(x)$ as follows: First, we discretize $f(x)$ into discrete points $\{f_p(x)\}$ ($p = 1, \cdots, N$). As such, PDF on p-th discrete point $\{f_p(x)\}$ can be expressed as:

$$pdf(f_p) = \sum_{r=1}^{M} const \cdot a_r \cdot e^{\widehat{b}_r^{k+1} \cdot f_p} \quad (20)$$

To normalize it, we can divide it with the total value of

PDF on all discrete points as:

$$\begin{aligned} pdf_{norm}(f_p) &= \frac{\sum_{r=1}^{M} const \cdot a_r \cdot e^{\widehat{b}_r^{k+1} \cdot f_p}}{\sum_{p=1}^{N} \sum_{r=1}^{M} const \cdot a_r \cdot e^{\widehat{b}_r^{k+1} \cdot f_P}} \quad (21) \\ &= \frac{\sum_{r=1}^{M} a_r \cdot e^{\widehat{b}_r^{k+1} \cdot f_p}}{\sum_{p=1}^{N} \sum_{r=1}^{M} a_r \cdot e^{\widehat{b}_r^{k+1} \cdot f_p}} \end{aligned}$$

In this way, the scaling constant can be eliminated from the approximation of PDF and thus normalization improves the numerical stability of proposed algorithm.

4.3 Error Estimation

Since the Fourier transform in equation (3) is unique, it is equivalent to evaluate the error of $\Phi(\omega)$ in order to investigate the accuracy of PDF approximation with qth order moments. It is ideally to compare the difference between Fourier transform of estimated PDF and that of exact PDF. However, the exact PDF is usually not available. Instead, we use approximation with $n+1$ order moments as the exact value, and proceed to estimate the error.

$$\begin{aligned} Error &= \left| \frac{\Phi^{q+1}(\omega) - \Phi^q(\omega)}{\Phi^{q+1}(\omega)} \right| \quad (22) \\ &= \left| \frac{\sum_{p=0}^{q+1} \frac{(-j\omega)^p}{p!} \cdot m_f^p - \sum_{p=0}^{q} \frac{(-j\omega)^p}{p!} \cdot m_f^p}{\sum_{p=0}^{q+1} \frac{(-j\omega)^p}{p!} \cdot m_f^p} \right| \\ &= \left| \frac{(-j\omega)^{q+1}}{(q+1)!} \cdot \left(\sum_{p=0}^{q+1} \frac{(-j\omega)^p}{p!} \cdot \frac{m_f^p}{m_f^{q+1}} \right)^{-1} \right|. \end{aligned}$$

When $|m_f^p| \ge |m_f^{q+1}|$ ($p \le q+1$), above error estimation can become:

$$Error \le \left| \frac{(-j\omega)^{q+1}}{(q+1)!} \cdot \left(\sum_{p=0}^{q+1} \frac{(-j\omega)^p}{p!} \right)^{-1} \right|. \quad (23)$$

The same error estimation can be obtained for $|m_f^p| \le |m_f^{q+1}|$ by taking reciprocal of circuit behavior to shift the behavioral distribution.

As such, the error estimation can be used to measure the accuracy of the approximation with first q-th order moments. When the approximation order increases, the error estimation should move to higher order as required.

4.4 Complexity Analysis

The Monte Carlo simulation requires to generate massive samples to cover the entire parameter space evenly, so number of simulations is p^n where p is the number of samplings for every single variable and n is the total number of random variables.

As for RSM based methods, we take APEX as an example where RSM formulation is the most time-consuming part. For example, when RSM uses n random variables and k order polynomial function, it has total $C_n^1 + C_n^2 + \cdots + C_n^k$ terms and thus APEX requires $C_n^1 + C_n^2 + \cdots + C_n^k$ simulation samples. In other words, the complexity of APEX is $O(n^k)$.

When proposed algorithm handles total n variables and needs m estimating points for each variable, the total complexity is $(m-1)*n+1$ (usually $m \ll n$) or $O(n)$. $m-1$ denotes that estimating points of each variable includes the nominal point and nominal circuit simulation can be shared by all variables. Therefore, it has linear complexity.

5. EXPERIMENTAL RESULTS

We have implemented proposed algorithm in the MAT-LAB environment, and all experiments are carried out on a Linux server with a 2.4GHz Xeon processor and 4GB memory. We use a six-transistor SRAM cell and a two-stage operational amplifier to compare the accuracy and efficiency of proposed algorithm with APEX [2] and Monte Carlo simulation. As an illustration, we consider the threshold voltages of MOSFETs as independent random variables subject to process variations, but our algorithm can also handle other variation sources.

5.1 SRAM Cell

We first consider a typical design of 6T SRAM cell in Fig.(1) and investigate the access time failure of the SRAM cell during reading operation, which is determined by the voltage difference between BL_B and BL.

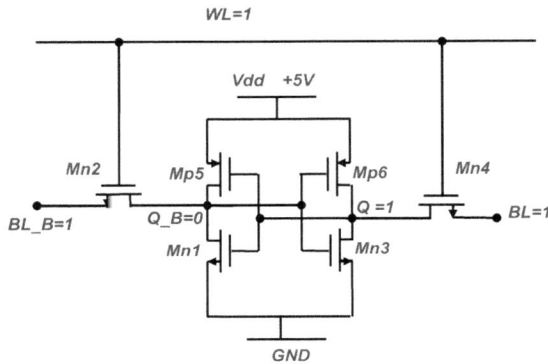

Figure 1: Schematic of SRAM 6-T Cell

Initially, both BL_B and BL are pre-charged to Vdd, while Q_B stores zero and Q stores one. When reading the SRAM cell, BL_B starts to discharge from Vdd and produces a voltage difference ΔV between it and BL. The time it takes BL_B to produce a large enough voltage difference is called access time. Since process variations are inevitable, the access time of manufactured SRAM cells can deviate from nominal value. When access time is larger than acceptable maximum value T_{max}, this leads to an access time failure.

In our experiment, we consider threshold voltages of all MOSFETs as independent variables which are normally distributed with 30% perturbation from nominal values. As such, there are perturbations to the nominal discharge trajectory on BL_B as shown in Fig.(2). Therefore, we can investigate the access time failure by capturing the random distribution of voltage on BL_B at T_{max} time-step: when voltage of BL_B at T_{max} is larger than its nominal value, the access time failure happens.

Note that both PEM and APEX can not capture high precision in the tail region of CDF/PDF, which is required

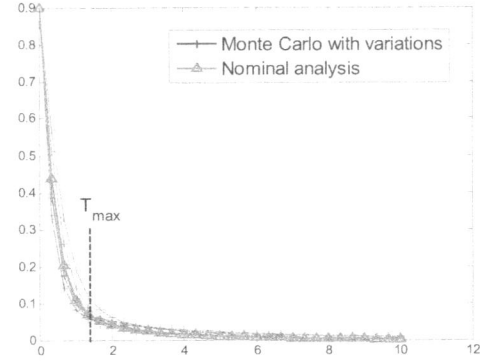

Figure 2: BL_B **discharge behavior with** V_{th} **variations**

to deal with rare event in an SRAM. Therefore, we focus on reducing average error of the performance distribution in the entire range. We start from univariate case to validate proposed algorithm and then extend it to multi-variable case. We have implemented three other methods for comparison:

- **Monte Carlo simulation (MC):** This is direct Monte Carlo simulation.

- **APEX:** Implementation of asymptotic probability extraction algorithm proposed in [2].

- **Point Estimation Method (PEM):** Proposed algorithm that leverages the point estimation method.

5.1.1 Univariate Case

First, we consider one random variable (e.g. threshold voltage variation on Mn_2) to compare the accuracy of APEX and PEM against MC. The random distributions (PDF function) from these methods are plotted in Fig.(3). Note that the histogram from Monte Carlo simulation has been normalized to eliminate the effect of total number of samplings.

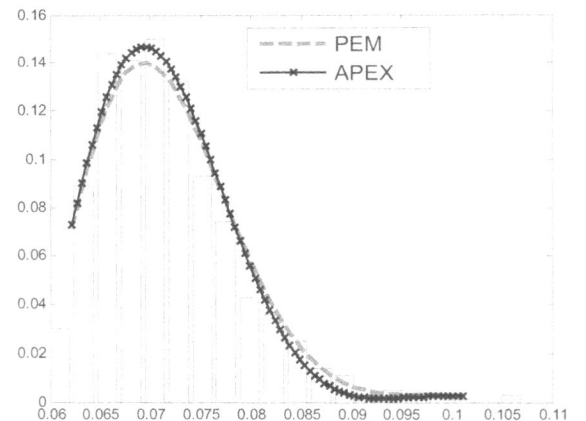

Figure 3: random distributions of BL_B **voltage at** T_{max}(**Monte Carlo result has been normalized**)

When compared with Monte Carlo results, PEM can provide better accuracy than APEX, especially in the peak and right tail regions. However, APEX has better efficiency even if the same order of moments are used: APEX needs only 8.59 second with quadratic function in response surface model, while PEM requires 17.71 seconds with seven estimating points for one variable.

It is because APEX use analytical formula which is suitable for low dimensions. It should be noticed that the number of required simulation samples in APEX will increases exponentially when more variables or a strongly nonlinear RSM required.

5.1.2 Multiple Variable Parameters

Next, we consider all threshold voltages of six transistors are independent random variables, which are normally distributed with 30% perturbation from nominal values. Similarly, we attempt to compare all three methods, but it is prohibitive to implement APEX as response surface model becomes very complicated, especially when high order response surface model is required.

Instead of original APEX, we calculate high order moments numerically using results from Monte Carlo simulations and extract PDF with technique in APEX, which is denoted as MMC+APEX. The random distributions from all methods are compared in Fig.(4).

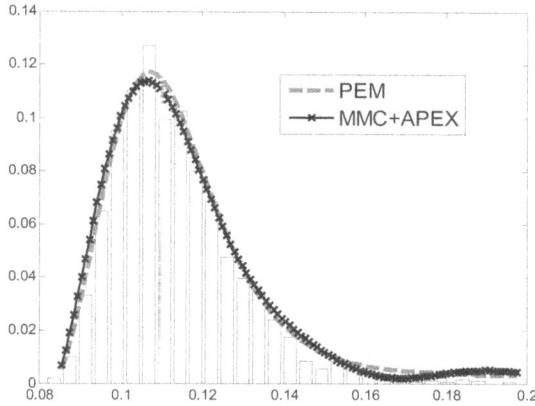

Figure 4: random distributions of BL_B voltage at T_{max} (Monte Carlo result has been normalized)

It is obvious that PEM can capture the exact distribution of BL_B voltage at T_{max}, which fits well with Monte Carlo simulation and MMC+APEX method. This can shows that PEM can achieve very high accuracy in the multi-variable problems. Also, we compare only the runtime of MC and PEM in Table 1, since runtime of MMC+APEX is almost the same as MC. In this table, PEM uses 5 estimating points for each variable parameter, and can achieve the same accuracy with $119.2X$ speedup over Monte Carlo method.

Table 1: Runtime Comparison of three methods

Method	Time (second)	Speedup
Monte Carlo (3×10^3)	7644	1x
PEM (5 point)	64.12	119.2x

To compare the efficiency with APEX, we can consider a SRAM cell under commercial 65nm CMOS process where

10 independent variables are used to model random variation for each transistor [14]. As such, RSM using quadratic function has 1830 coefficients and thus APEX requires 1830 simulation samples as discussed in Section 4.4. However, PEM only needs 121 simulation samples when 3 estimating points are used for each independent variable, and achieves $15X$ speedup over APEX.

5.2 Operational Amplifier

We further consider a two-stage operational amplifier in Fig. (5) and a negative feedback circuit in Fig.(6). We use this example to show that proposed method can estimate random distributions where circuit behavior is negative.

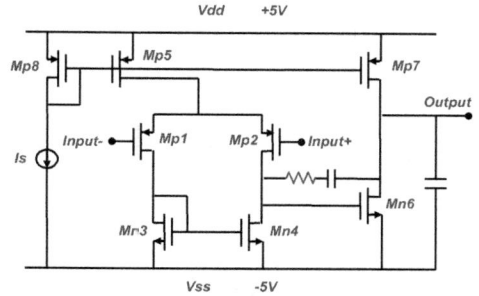

Figure 5: Schematic of Operational Amplifier

Similar to SRAM cell example, we consider threshold voltages of all MOSFETs as independent variables which are normally distributed with 30% perturbation from nominal values. Note that the threshold voltages of input transistor pair (Mp_1, Mp_2) should be kept the same to ensure the convergence of nonlinear system solver in circuit simulators.

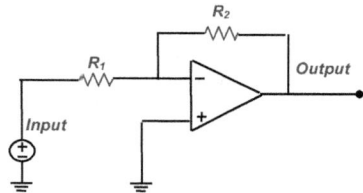

Figure 6: Schematic of a unity gain feedback circuit

There are a number of op am specifications in time-domain and frequency-domain, such as slew rate, settling time, phase margin, input offset voltage and etc. In our experiment, we investigate the input offset voltage variation due to threshold voltage variations of MOSFETs.

In this experiment, we implement following methods for comparison purpose:

- **Monte Carlo simulation (MC):** This is direct Monte Carlo simulation.

- **Moments from Monte Carlo (MMC+APEX):** Estimate high order moments of input offset voltage from Monte Carlo simulation, and extract the PDF using techniques in [2, 7].

- **Point Estimation Method (PEM):** Proposed algorithm that leverages the point estimation method.

First, we validate the accuracy of PEM by comparing the random distributions of input offset voltage from different methods in Fig.(7). PEM employs 5 estimating points for each variable and achieve the correct PDF function by shifting distribution of circuit behavior into positive region and moving it back.

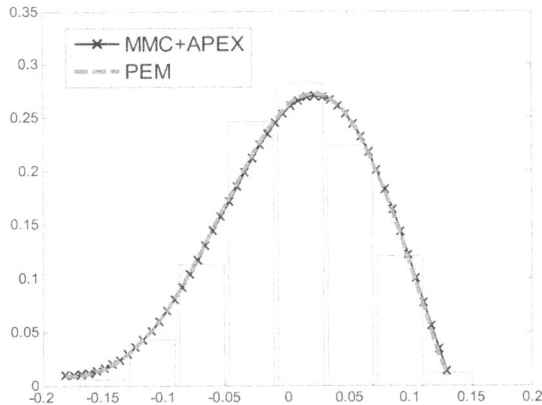

Figure 7: random distributions of input offset voltage (Monte Carlo result has been normalized.)

Since MMC+APEX method has the exact moment values from Monte Carlo simulations, it can provide very high accuracy. Also, PEM can offer the same accuracy with MMC+APEX method, which implies that PEM can achieve high accuracy of high order moments. On the other hand, we further validate accuracy of PDF function from PEM against the histogram from Monte Carlo simulation in Fig.(7), which fit with each other very well.

Moreover, we compare the runtime of Monte Carlo method and PEM in Table (2), and MMC+APEX is omitted because it has almost the same runtime as Monte Carlo method. Besides, we list PEM method with different number of estimating points for each variable to demonstrate it has linear scalability with the estimating points at the same time.

Table 2: Runtime Comparison between Monte Carlo simulation and PEM

Method	Time (second)	Speedup
Monte Carlo (3×10^3)	27765 (7.71 hours)	1x
PEM (3 point)	153.1 (0.04 hours)	181.5x
PEM (9 point)	525.84 (0.15 hours)	52.8x

From Table (2), PEM can provide up to hundreds times speedup over MC method. Moreover, the computational cost increases linearly with the number of estimating points for each variable.

6. CONCLUSION

In this paper, we have proposed one efficient point estimation (PE) based algorithm to extract the stochastic circuit behavior in performance space from parameter space. Our approach can perform an efficient evaluation of high-order moments of circuit behavior, and thus circumvent the use of response surface model (RSM) methods. This can dramat-

ically reduce the computational cost seen in APEX. Moreover, the proposed method can be extended to deal with multiple parameters under linear complexity. Experiments on a few different circuits have shown that the proposed method can provide up to 181X more runtime speedup with the same accuracy when compared with the Monte Carlo method. Also, it can achieve up to 15X speedup over the RSM based method such as APEX with the similar accuracy.

7. REFERENCES

[1] S. Nassif, "Modeling and analysis of manufacturing variations," *Proc. IEEE Custom Integrated Circuits Conf.*, pp. 223–228, 2001.

[2] X. Li, J. Le, P. Gopalakrishnan, and L. T. Pileggi, "Asymptotic probability extraction for non-normal distributions of circuit performance," in *Proc. IEEE/ACM Int. Conf. Computer-aided-design (ICCAD)*, pp. 2–9, 2004.

[3] S. Vrudhula, J. M. Wang, and P. Ghanta, "Hermite polynomial based interconnect analysis in the presence of process variations," *IEEE Tran. on Computer-aided-design (TCAD)*, pp. 2001–2011, 2006.

[4] X. Li, Y. Zhan, and L. Pileggi, "Quadratic statistical MAX approximation for parametric yield estimation of analog/RF integrated circuits," *IEEE Tran. on Computer-aided-design (TCAD)*, vol. 27, pp. 831–843, 2008.

[5] D. Xiu and G. E. Karniadakis, "The wiener-askey polynomial chaos for stochastic differential equations," *SIAM J. Sci. Comput.*, vol. 24, pp. 619–644, 2002.

[6] G. E. P. Box and N. R. Draper, "Empirical model building and response surfaces," *Wiley series In Probability and Mathematical Statistics, New York: John Wiley and Sons*, 1987.

[7] L. T. Pillage and R. A. Rohrer, "Asymptotic waveform evaluation for timing analysis," *IEEE Tran. on Computer-aided-design (TCAD)*, vol. 9, no. 4, pp. 352–366, 1990.

[8] E. Rosenblueth, "Point estimation for probability moments," *Proc. Nat. Acad. Sci. U.S.A.*, vol. 72, no. 10, pp. 3812–3814, 1975.

[9] Y.-G. Zhao and T. Ono, "New point estimation for probability moments," *Journal of Engineering Mechanics*, vol. 126, no. 4, pp. 433–436, 2000.

[10] A. Papoulis and S. Pillai, "Probability, random variables and stochastic processes," *McGraw-Hill*, 2001.

[11] M. H. DeGroot and M. J. Schervish, "Probability and statistics," *Addison Wesley*, 2011.

[12] A. V. Oppenheim, A. S. Willsky, and S. Hamid, "Signals and systems," *Prentice Hall*, 1996.

[13] S. Haber, "Numerical evaluation of multiple integrals," *SIAM Review*, vol. 12, pp. 481–525, 1970.

[14] X. Li and H. Liu, "Statistical regression for efficient high-dimensional modeling of analog and mixed-signal performance variations," in *Proc. ACM/IEEE Design Automation Conf. (DAC)*, pp. 38–43, 2008.

Author Index

www.ingramcontent.com/pod-product-compliance
Lightning Source LLC
Chambersburg PA
CBHW081528220326

41598CB00036B/6362